NATURAL RESOURCES MANAGEMENT

NATURAL RESOURCES MANAGEMENT

Editor

Mareike Kruckel

scitus academics

Natural Resources Management

Edited by **Mareike Kruckel**

This Book is Printed on Acid-Free Paper. ♾

ISBN: 9781681179148
Library of Congress Control Number: 2017948942

SCITUS Academics LLC,
616, Corporate Way, Suite 2, 4766,
Valley Cottage, NY 10989

www.scitusacademics.com

Notice

Printed in United States of America.

Preface

The recent concerns about the environment and natural resources have arisen because of the increasing awareness of the finite nature. What appeared as to be plentiful earlier, clean water and air, fuel energy, land for food, forests and trees are becoming increasingly scarce. Not just roads and buildings, but all the things we use or consume – food, clothes, books, toys, furniture, tools and vehicles – are obtained from resources on this earth. The only thing we get from outside is energy which we receive from the Sun. Even this energy is processed by living organisms and various physical and chemical processes on the earth before we make use of it. Energy is required for the evolution of life forms on earth. We obtain energy from different sources. The natural resources that we use today are the prime source of energy for day-to-day activities of humans, without which we would still be lurking around in the Stone Age. This work will emphasizes why there is a need to manage and conserve our natural resources. Why do we need to use our resources carefully? Because these are not unlimited and with the human population increasing at a tremendous rate due to improvement in health-care, the demand for all resources is increasing at an exponential rate. The management of natural resources requires a long-term perspective so that these will last for the generations to come and will not merely be exploited to the hilt for short term gains. This management should also ensure equitable distribution of resources so that all, and not just a handful of rich and powerful people, benefit from the development of these resources. Population increase, resource use conflicts, unsustainable use and harvesting of resources, technological development, climate change, and political despair have all put more pressure on natural resources leading to land degradation and poverty. To achieve a flourishing situation, we need to transform our mindset advocating incorporated and holistic strategies in managing our natural resources.

This volume 'Natural Resources Management' covers studies of natural resources exploration, evaluation and exploitation, including environmental and risk-related aspects on a wide variety of resources including minerals, coal, hydrocarbon, geothermal, water, and vegetation. It presents a variety of sustainable strategies and/or approaches including use of GIS and Remote Sensing technologies, decision support system models, involvement of stakeholders in major decisions regarding use of natural resources, community level initiatives, and use of surveillance and monitoring mechanisms.

Table of Contents

CHAPTER 1

Environmental Risk, Resilience and Migration: Implications for Natural Resource Management and Agriculture

Priya Deshingkar

School of Global Studies, University of Sussex, UK

ABSTRACT

This letter probes the causal links between migration, remittances and resilience to environmental change. Three case studies have been chosen, Western Mexico, the Central Plateau of Burkina Faso and Eastern India, where satellite imagery shows recent regeneration of vegetative cover and where there is evidence of high rates of migration. The findings are analysed through a framework that draws on concepts of ecological anthropology, new economics of labour theories and livelihood analyses of migration drivers and impacts.

1. INTRODUCTION

The links between climate change and migration are hotly debated with alarmist projections on the one hand which estimate that between 200 million (Myers 2005) and a billion people (Christian Aid 2007) will be displaced by 2050 resulting in mass migration into Europe. On the other hand are more balanced assessments which, based on past experience and current migration patterns, foresee more mobility within developing regions rather than movements between continents (Foresight 2011). Migration is in fact already an important adaptation strategy for the poor coping with gradual onset climate stresses and shorter, sudden shocks (Tacoli 2009).

The three regions being investigated have a long history of migration which is linked to environmental change as well as other economic, social and demographic factors. Migration from Mexico to the United States Studies has been driven by a combination of changes in labour demand and supply, US immigration policy changes, migrant networks, reform of communal land tenure and the North American Free Trade Agreement (NAFTA) (Martin 2003). In Burkina Faso, climate change appears to have a stronger link with short-term migration, which is an important diversification strategy rather than long-term migration, because that requires more resources which impoverished households may not have (Henry *et al* 2004). Finally, in the case of Eastern India, rural–urban and rural–rural migration to nearby towns and agricultural fields has given tribal subsistence farmers an alternative to risky rainfed farming. Severe droughts can reduce yields by 40% (Pandey *et al* 2007) and migration is critical for coping in such conditions. But other reasons for migration include higher wages, room for negotiation with employers on wage rates, number of days work, meal quality and timing (Rogaly *et al* 2001).

Against this backdrop, this letter attempts to answer the following questions:

(1) Under what circumstances do households with migrants invest (capital and labour) in agriculture and natural resource management?

(2) How do household factors (assets, education, kinship networks, individual attributes) and contextual factors (agro-climatic conditions, markets, policies etc) shape migration and natural resource management practices and outcomes?

The letter begins with a discussion of the core concepts used. This is followed by a discussion of the three regions in terms of the factors that drive migration, the characteristics of migrant households, and the impacts of migration and remittances on natural resource management at the household level. Observed differences in impacts between regions or communities are explained by examining the household and contextual factors that have mediated these impacts. The letter concludes with a discussion based on the findings to identify causal pathways.

1.1. Analytical framework

A core concept in this letter is that of resilience which can be both social and ecological. Social resilience is the ability of groups or communities to cope with external stresses and disturbances as a result of social, political and environmental change. Ecological resilience is the ability of ecosystems to maintain themselves in the face of disturbance. Both are in fact closely related and need to be addressed together to analyse the impacts of environmental variability (shocks and stresses) on communities that are dependent on natural resources (Adger 2000).

Exactly how environmental shocks and stresses impact on different households depends on their assets at the household level (physical, natural, human, social and financial) and the opportunities and constraints posed by the policy and institutional context within which livelihood activities are pursued (the sustainable livelihoods framework developed by Chambers and Conway 1992, Scoones 1998 and others). Viewed through such a lens, migration is a livelihood strategy (de Haan 1999, Ellis 2003) in the same way that farming is not an aberration as portrayed in pessimistic discourses of migration. Research on migration and resilience has shown that migration impacts on the social resilience of individuals and communities, as well as on the sustainability of the underlying resource base (Adger *et al* 2002). Relevant here are also economic theories related to the relationship between population and natural resources which posit that increases in population pressure on land with fixed technology lead to diminishing returns which can lead to agricultural intensification (Boserup 1965 and Tiffen *et al* 1994), and migration (Bilsborrow and Delargy 1990) among other survival strategies. Only in extreme cases will the entire family migrate away altogether (Bilsborrow and Delargy 1990, p 129). Which particular response or combination of responses occurs depends on agro-climatic conditions and the market and policy context.

Neither economic theories nor livelihoods discourses on migration explicitly recognize the role of remittances in the process. This was treated in detail by the New Economics Labour Migration analysts: Stark (1982) and Stark and Bloom (1985) argue that farming households participate in migration as a strategy to overcome constraints on production and investment resulting from missing or incomplete credit and insurance markets in rural areas. Market imperfections require rural households to self-finance production and self-insure against production and income risks. This hypothesis has been tested and confirmed by others including Lucas (1985), Taylor and Yunez-Naude (1999) and Taylor and Martin (2001).

But merely having a migrant in the family does not guarantee improvements in agriculture. The literature shows that both agricultural intensification and disintensification are possible outcomes of migration.

1.2. Intensification

Migrant remittances have been identified as one of the factors enabling agricultural intensification. Evidence from Morocco (de Haas 2007) and Ghana (Tsegai 2004) shows for example that initial negative effects of migration caused by the loss of labour (see below) may be offset over time as the inflow of remittances allows households to purchase agricultural inputs. This is more likely to occur in the case of international remittances as these tend to be larger (de Haas 2007, Adams 2005, Wouterse 2008) and only if a favourable institutional, environmental and market context supports agricultural development (De Haan 2005, Taylor *et al* 1996).

1.3. Disintensification

Foremost among disintensification discourses is the lost labour argument, according to which migration deprives rural households of critical labour resulting in falling productivity. The loss of labour could potentially lead to the adoption of labour-saving strategies or the abandonment of labour intensive crops, or a decrease in labour inputs to agriculture and fall in productivity (Lipton 1980, Rubenstein 1992). However if the departing member did not participate in agriculture before s/he left, the impacts would be different and could in fact reduce consumption requirements and labour availability related to those. Zimmerer (1993) found that migration led to the abandonment of traditional soil conservation techniques in Bolivia. Furthermore, the departure of male and female migrants would have distinct effects on the agricultural activities of origin households. The departure of men has been blamed for the feminization of agriculture.

However others have argued that labour losses will not be felt in situations where there is widespread under-employment in agriculture and surplus labour (Vargas-Lundius and Villareal 2008). Clearly, household level impacts will depend on the size of the household, dependency ratios and availability of labour. A few studies show that women left behind may be able to get help from extended family and social networks for farm work (see David 1995 on the Sahel and Lamichhane 2008 on Nepal). Finally, seasonal migration may enable people to seasonally allocate labour to agriculture and migration as migration typically occurs after the main rainfed growing season (Deshingkar and Farrington 2009).

In some circumstances migration can lead to the abandonment of land. This can occur with prolonged absence of the migrant and inability to meet the labour requirements of agriculture. Jokisch and Lair (2002) find increasing abandonment of land by migrants from Ecuador going to the US. Research in Central Mexico shows that land abandonment is especially likely if the land is marginal (López *et al* 2006). However there is a generalized trend of land abandonment in marginal areas in industrialized societies which is not unique to migrant households (López *et al* 2006).

What all of these examples show is that a variety of outcomes are possible depending on household and contextual factors. These are explored in greater detail through the three case studies under review.

2. WESTERN MEXICO

Historically Western Mexico (Jalisco, Michoacan, Guanajuato, Nayarit and Zacatecas) has been the most important sending area for migration to the United States (Massey and Espinosa 1997). Mexico has suffered from long-lasting economic crises despite its rapid industrialization after the Second World War. Globalized markets for labour and agricultural commodities such as maize have resulted in lucrative earning opportunities in the US and decreasing profitability of smallholder farming in Mexico (Radel and Schmook 2008). Poorer households have limited access to formal credit to underwrite productive investments in education, farming, cattle raising, manufacturing or retail sales. Migration to the United States is an attractive means of acquiring capital quickly, through social networks that reduce the costs and risks of undocumented entry (Singer *et al* 1995, Taylor 1986, 1987).

Migration has impacted on agriculture and natural resources in a variety of ways. Hostetler (2007) documents six kinds of 'remittance landscapes' including transformations from (a) forest to pasture, (b) forest to agriculture, (c) forest to urban, (d) agriculture to pasture, (e) agriculture to urban and (f) change of agricultural system. Here (a), (d) and (f) could be regarded as agricultural disintensification, and (b) and (d) could regarded as intensification. An example of (b) is avocado cultivation in Michoacán, which grew extremely rapidly in the past decade and is now worth 800 million dollars in trade. (Rhoda and Burton 2010). While investing in avocado farming is not unique to migrants, remittances have allowed subsistence farmers to enter the market (Romero 2006). A example of (f) is the switch to less labour intensive crops (Schmook and Radel 2008).

The Mexico Migration Project provides some insights into the circumstances in which migrants do/do not invest in intensification. The project studied socio-economically diverse communities of different sizes covering 5500 households selected through a random sampling technique. The project found that the share of migration earnings spent on improving agricultural production, was less than 50% on average and varied widely between communities. (Durand and Massey 1992, p 27). This variation was explained by several factors. First, access to farmland, whether or not ejido land, and land quality are important determinants of migration and investment in agriculture. Ejidos are communal lands granted to specific families in the 1930s for the perpetual use of their family. Ejidos provide rural families with land but no access to the means to make it productive; thus they create a need for regular infusions of cash to finance the purchase of seeds, fertilizers, irrigation, labour, and other inputs used (see Massey *et al* 1987, 1993),1 confirming the hypothesis of Stark and others. But other studies show ejido lands being rented out by migrants (Heath 1992, p 700). Clearly other household characteristics matter as we discuss below.

Second, the level of remittances also matters: in the Mexico Migration Project, remittances were found to have a positive impact on agriculture only if they are 'substantial' (Quinn 2009). Also, research in ejido areas has shown that it is the relatively better endowed households who are more likely to migrate in the first place (Davis *et al* 2000, p 118) and they are therefore better placed to invest in land. Third, individual attributes such as the ability to invest money profitably have also been identified as important prerequisites (Cederstrom 1990). Fourth, the duration is also important: migrants who have been abroad for a longer period of time direct remittances towards 'productive' investment—including agricultural intensification and expansion—after household consumption needs have been satisfied (Stuart and Kearney 1981, Basok 2003, Cohen and Rodriguez 2005).

While there is clear evidence of the impacts of migration on agricultural intensification through capital investment, there is little discussion of soil and water conservation, which suggests that either there is insufficient local knowledge and government support for soil and water conservation (SWC) techniques and/or there is insufficient labour at the household level to invest in SWC because of migration. These issues need further probing. While improved agricultural yields would reduce the vulnerability of poor farmers and improve their resilience to shocks, the impacts on the ecological functions of the systems where such intensification has occurred are not fully explored.

3. BURKINA FASO

Burkina Faso is a poor, densely populated country in the West African semi-arid tropics. Agriculture is the main source of subsistence for the population but is risky due to unreliable rainfall and poor quality soils. Migration has long been an important coping strategy in the Sahel (Adams and Mortimore 1997) allowing people to manage climate variability. Migration patterns differ by region and ethnic group: the Mossi from the Central Plateau region have historical links with Cote d'Ivoire which continue to the present (although migration to that country is much reduced after political upheaval). The Fulani, who live in the north, are mainly cattle and goat herders and Côte d'Ivoire has been an important market for them too. The Lobi in the south migrate from one rural area to another to find land to rent.

NOAA satellite imagery shows that the Sahel as a region suffered from prolonged drought starting with severe droughts in 1982–5. Drought conditions persisted until 1993 and this was followed by a relatively wet and green period in 1994–2003 (Anyamba and Tucker 2005).

However the greening is not uniform, suggesting that factors other than rainfall may have contributed to a differential greening response (Herrmann *et al* 2005). The spots where this is likely to have occurred have been identified as, inter alia, parts of Senegal, Mauritania, Mali, Niger, the Central Plateau of Burkina Faso and large portions of Chad. Herrmann *et al* identify improvements in SWC techniques as a possible cause.

According to Reij *et al* (2005) the Central Plateau was characterized by expanding declining cereal yields, a reduction in vegetation, falling ground-water levels and mass outmigration in the early 1980s. This, according to Reij and colleagues, led to the adoption of SWC techniques. Impact studies conducted a few years later found significant increases in crop yields, more tree cover and greater availability of forage for livestock, Improved soil fertility and rising ground-water tables (Reij *et al*2005).

One study makes the link between migration and the adoption of SWC measures. A survey of 115 migrant and 135 non-migrant households by Konsiega in 2004 found that migrant households had a significantly higher average adoption rate of stone bunds. Yields had increased by 40–100% depending on the region, the rainfall in the year of the study and the spacing of stone bunds. The choice of which field was protected by stone bunds was partly the decision of the field owner, partly dependent on receiving project benefits and partly related to the physical characteristics of the field (Konseiga 2004, p 7). However SWC is more labour intensive (rather than capital intensive) which suggests that these activities were done either by migrants in between trips, returning migrants or other members of the household. Reij *et al* (2005) notes that the huge success of SWC was made possible by substantial public investment in SWC between 1985 and 2000. It is possible that households with migrants were better educated and were better able to access support under government and NGO schemes. This issue needs further investigation.

The link between migrant remittances and intensification through capital investment has been made but the impacts differ between different types of migration. Research by Wouterse (2008) in the Central Plateau covering 223 households across four villages showed that cropping was the main economic activity, but migration within Africa and to Europe was common among young men from large households. The education level of migrants was extremely basic and did not differ by destination. More than 90% of the migrants stayed away for more than a year and seasonal migration was rare. Until recently, the primary destination for regional migrants was Côte d'Ivoire, but this has nearly vanished due to political instability. Many Burkinabé now migrate to the capital city Ouagadougou. Both international and regional migrants remit but the former remit roughly six times the amount that regional migrants remit (400 000 FCFA as opposed to 50 000 FCFA). The research indicates that intercontinental migration discourages labour intensive activities but facilitates livestock investment, which is capital intensive but not as labour intensive as cropping. Regional migrants remain engaged in low return, labour intensive activities such as subsistence farming. No mention is made of SWC adoption rates. According to Wouterse, this indicates that continental migration is a survival strategy stemming from a lack of wealth but positively related to household size.

4. JHARKHAND IN EASTERN INDIA

The Eastern Indian state of Jharkhand was historically populated by forest dwelling tribals2. Roughly 28% of its 7.9 million ha is under forest and 24% under cultivation. Although poverty has declined from 44% to 33% in the last decade, there is widespread poverty in rural areas, especially among tribals who constitute 23% of the population. Tribals are the most deprived segment of the Indian population with the lowest levels of human development indicators, resulting from a combination of poor governance, low levels of education and other multiple deprivations.

Migration was first documented in the late 19th century when tribals went to work in Indigo production in Bengal. Migration later shifted to the tea plantations in Assam, railway and road-building projects in other parts of the country and as far as Burma (De Haan 2002). It is likely that large-scale migration was triggered in part by the Indian Forest Act of 1878 which restricted the rights of local communities to forests. Deforestation has continued unabated through the 20th century driven by government licences granted to industrialists and contractors. The livelihoods of rural people were further affected by the nationalization of the trade in minor forest products in the 1970s (Jewitt 2008, p 73).

A majority of the cultivators in Jharkhand combine farming with seasonal migration. Most are small and marginal farmers growing upland varieties on poor soils. Yields are too low to sustain an average household through the year, so migration of some or all household members has become an important coping strategy.

A study of 12 villages across the state show that on average, one-third of the households have at least one member migrating (Dayal and Karan 2003). Around 70% of these are poor tribals with low levels of education and landholding. While 98% of the households in Dayal and Karan's sample reported an improvement in their lives because of migration, roughly 22%, mainly those with more than five acres of land, invested in agriculture. They were able to use skills that they had acquired in the green revolution areas of Punjab and Haryana where they migrated for farm work. This indicates that productive investment is likely to occur when the assets required for such investment and knowledge/skills are available. Other studies have also shown that migrants use earnings to buy fertilizers and draught animals but on a very small scale (Nitya, personal communication). Land is an important indicator of status in tribal society and own account cultivation continues to be a high priority for many of those who migrate (Rogaly 2003, p 624). In fact migration earnings are being invested in agriculture to reduce the need to migrate (Rogaly , personal communication). It has to be mentioned here that the quantum of remittances involved is much smaller compared to international migrants and the bulk of earnings are spent on consumption. Over time, and if current trends in urban development continue, migration could provide an important source of capital for subsistence farmers struggling to manage in unpredictable environmental conditions. Analysis of more than 10 years of rainfall and temperature data for the state has revealed an increase in occurrence of extreme weather events with a continuous rise in temperatures (Wadood and Kumari 2009). However, the implications of this for ecological resilience are less clear, as increased use of agrochemicals can lead to pollution of soil and water systems.

Recent research on indigenous SWC practices shows that a range of techniques have been practised traditionally, including stone-cum-earthen bunding and brushwood weirs, but many have now become less prevalent due to migration and the shortage of labour (Dey and Sarkar 2011). It is not clear whether there are government programmes that can effectively include households with migrant labour and this is an area for policy consideration.

Improvements in farm productivity could in theory reduce the area farmed and this could reduce pressure on forests. In fact, Jharkhand is one of the few states in India that has registered an increase in its forest area (Food Survey of India 2009). Many of the districts showing increased forest cover such as Palamu, Purbi Singbhum, Dumka, Hazaribagh and Gumla are also high migration areas where people have long combined agriculture, forest produce collection and migration to diversity income sources. But as discussed above, migration can lead to an intensification of agriculture through capital investment on larger farms, but also have adverse impacts on soil and water conservation.

5. DISCUSSION

The three regions covered show that there is diversity in migration patterns and impacts even within a region. Broad categories such as international migrants are not homogeneous but are in fact composed of individuals with diverse characteristics and households with diverse endowments, which have implications for the way in which migration impacts on agriculture and resilience to environmental shocks and stresses.

The evidence from Mexico indicates that households with certain characteristics and endowments such as ejido land, receiving substantial remittances, possessing business acumen and belonging to a household that has already met its consumption needs, all improve the chances of intensification of agriculture. In Burkina Faso, migration was positively associated with soil and water conservation in only one study. It is possible that this may have occurred due to their ability to access external support under SWC programmes. Other studies show a positive correlation between high return migration to Europe and capital investment in agriculture, whereas migration within the region did not generate earnings that were sufficient to enable rural households to move out of low return and labour intensive agriculture. Finally, in the case of Jharkhand, only those farmers with larger plots of land were able to invest remittances in agriculture. Soil and water conservation appear to have been negatively impacted by migration.

These case studies provide insights into factors at the household level, and even the level of the individual, as well as the broader context in shaping decisions to invest labour and capital in agriculture. They also indicate agricultural intensification through capital investments may improve the resilience of poor households to economic and environmental shocks but this does not necessarily mean an improvement in ecological resilience of the system.

Given these processes, what are the implications for a future world with increased climate variability and possibly higher levels of migration? What would increased migration mean for improving resilience through investment in agriculture?

A great deal of uncertainty remains in predictions of rainfall, storms and other processes based on GCM models, for example in the Sahel (Caminade *et al* 2009) and South Asia (Wang *et al* 2010). In the face of such uncertainty, migration can lead to a number of outcomes including intensification and disintensification. Experience from other parts of the world has shown that remittances are not invested in agriculture if policy conditions are not conducive, as is seen in the case of Albania (Miluka *et al* 2010). Early introduction of policies to introduce appropriate technologies and backup will improve the chances of remittances being invested in agriculture and natural resources. For poorer households where migration is a coping strategy, more stresses coming from slow onset climate impacts could mean a greater reliance on migration to cope with and adapt to change. Ultimately, policymakers need to heed the advice given by the recent Foresight report on migration and environmental change which is to view migration as a form of adaptation to environmental change, and an effective way to build long-term resilience (Foresight 2011).

FOOTNOTES

1 The Mexican government has introduced reforms aimed at improving the returns to ejidos in the 1990s but migration to the US has continued. A survey of ejido households in the late 1990s indicated that off-farm employment continued to be important (De Janvry and Sadoulet 2001).

2 Indian official term for indigenous populations.

REFERENCES

1. Adams R H 2005 Remittances, household expenditure and investment in Guatemala World Bank Policy Research Working Paper No. 3532
2. Adams W M and Mortimore M J 1997 Agricultural intensification and flexibility in the Nigerian Sahel Geograph. J. 163 150–60
3. Adger W N 2000 Ecological and environmental economics Environmental Science for Environmental Management 2nd edn ed T O'Riordan (Harlow: Longman) pp 93–118

4. Adger W N, Kelly P M, Winkels A, Huy L Q and Locke C 2002 Migration, remittances, livelihood trajectories and social resilience Ambio 31 358–66
5. Anyamba A and Tucker C J 2005 Analysis of Sahelian vegetation dynamics using NOAA-AVHRR NDVI data from 1981–2003 J. Arid Environ. 63 596–614
6. Basok T 2003 Mexican seasonal migration to Canada and development: a community based comparison Int. Migration 41 2
7. Bilsborrow R E and Delargy P F 1990 Land use, migration, and natural resource deterioration: the experience of Guatemala and the Sudan Population and Development Review (Supplement: Resources, Environment, and Population: Present Knowledge, Future Options vol 16) pp 125–47
8. Boserup E 1965 The Conditions of Agricultural Growth (London: George Allen and Urwin)
9. Caminade C, Morse A and Jones A 2009 European models reliability over West Africa: from seasonal forecasting to climate scenarios EGU General Assembly 2009 (Vienna, April 2009) p 6275 http://meetings.copernicus.org/egu2009
10. Cederstrom T 1990 Migrant remittances and agricultural development Cult. Agricult. 11 2–7
11. Chambers R and Conway G 1992 Sustainable rural livelihoods: practical concepts for the 21st century IDS Discussion Paper (Brighton: IDS) p 296
12. Christian Aid 2007 Human tide: the real migration crisis A Christian Aid Report (London: Christian Aid)
13. Cohen J H and Rodriguez L 2005 Remittance outcomes in rural Oaxaca, Mexico: challenges, options and opportunities for migrant households Populat. Space Place 11 1
14. David R 1995 Changing Places? Women, Resource Management and Migration in the Sahel Case Studies from Senegal, Burkina Faso, Mali and Sudan (London: SOS Sahel)
15. Davis B, de Janvry A, Sadoulet E and Diehl T 2000 Policy Reforms and Poverty in the Mexican Ejido Sector (available at www.farmfoundation.org/news/articlefiles/859-davis.pdf)
16. Dayal H and Karan A K 2003 Labour Migration From Jharkhand (New Delhi: Institute for Human Development)
17. De Haan A 1999 Livelihoods and poverty: the role of migration. A critical review of literature J. Dev. Stud. 36 1–47
18. De Haan A 2002 Migration and livelihoods in historical perspective: a case study of Bihar, India J. Dev. Stud. 38 115–42
19. De Haan A 2005 Migration in the development studies literature: has it come out of marginality? Paper presented at the UNUWIDER Jubilee Conf. Future of Development Economics
20. de Haas H 2007 The myth of invasion: irregular migration from West Africa to the Maghreb and the European Union IMI Research Report (Oxford: International Migration Institute, University of Oxford)
21. De Janvry A and Sadoulet E 2001 Income strategies among rural households in México: the role of off-farm activities World Develop. 29 467–80
22. Deshingkar P and Farrington J (ed) 2009 Circular Migration And Multilocational Livelihood Strategies In Rural India (New Delhi: OUP)
23. Dey P and Sarkar A K 2011 Revisiting indigenous farming knowledge of Jharkhand (India) for conservation of natural resources and combating climate change Ind. J. Tradition. Knowledge 10 71–9
24. Durand J and Massey D S 1992 Mexican migration to the United States: a critical review Latin Am. Res. Rev. 27 3–42
25. Ellis F 2003 A Livelihoods Approach to Migration and Poverty Reduction (Norwich: Department for International Development (DFID))
26. Foresight 2011 Migration and Global Environmental Change: Final Project Report (London: Government Office for Science)
27. FSI 2009 State of India's Forests 2009 (Dehradun: Forest Survey of India)

28. Heath J R 1992 Evaluating the impact of Mexico's land reform on agricultural productivity World Dev.20 695–711
29. Henry S, Bruno S and Cris B 2004 The impact of rainfall on the first out-migration: a multi-level event-history analysis in Burkina Faso Populat. Environ. 25 423–60
30. Herrmann S M, Assaf A and Tucker C J 2005 Recent trends in vegetation dynamics in the African Sahel and their relationship to climate Glob. Environ. Change 15 394–404
31. Hostetler S 2007 Land use changes and transnational migration: the impact of remittances in Western Mexico Thesis no. 3730 École Polytechnique Fédérale de Lausanne (available from http://biblion.epfl.ch/EPFL/theses /2007 /3730/EPFL_TH3730.pdf)
32. Jewitt S 2008 Political ecology of Jharkhand conflicts Asia Pacific Viewpoint 49 68–82
33. Jokisch B D and Lair B M 2002 One last stand? Forests and change on Ecuador's eastern Cordillera Geograph. Rev. 92 235–56
34. Konseiga A 2004 Adoption of agricultural innovations in the sahel: the role of migration in food security 2004 Food and Nutrition Security Policies Regional Workshop in Mali
35. Lamichhane A R 2008 Women's Coping Strategies in the Context of Male Out-Migration: A Case Study of Kaski District, Nepal (Bangkok: Asian Institute of Technology)
36. Lipton M 1980 Migration from rural areas of poor countries: the impact on rural productivity and income distribution World Develop. 8 1–24
37. López E, Bocco G, Mendoza M, Velázquez A and Auirre-Rivera R 2006 Peasant emigration and land-use change at the watershed level: a GIS-based approach in Central Mexico Agricult. Syst. 90 62–78
38. Lucas R 1985 The distribution of wages and employment in rural Botswana The Household Economy of Botswana: An African Case ed D Chevlchovsky (Washington, DC: World Bank Staff) Working Paper No 715
39. Martin P 2003 Mexico-US Migration (Washington, DC: Institute for International Economics) www.iie.com/publications/chapters_preview/332/08iie3349.pdf
40. Massey D and Espinosa K E 1997 What's driving Mexico-US migration? A theoretical, empirical, and policy analysis Am. J. Sociol. 102 939–99
41. Massey D, Arango J, Hugo G, Kouaouci A, Pellegrino A and Taylor J 1993 Theories of international migration—a review and appraisal Populat. Dev. Rev. 19 431–66
42. Massey D S, Rafael A, Jorge D and Humberto G 1987 Return to Aztlan: The Social Process of International Migration from Western Mexico (London: University of California Press)
43. Miluka J, Carletto G and Davis B 2010 The vanishing farms? The impact of international migration on Albanian family farming J. Dev. Stud. 46 140–61
44. Myers N 2005 Environmental refugees: an emerging security issue 13th Economic Forum, Organization for Security and Co-operation in Europe (Prague, May 2005)
45. Nitya R personal communication
46. Pandey S, Bhandari H and Hardy B 2007 Economic Costs of Drought and Rice Farmers' Coping Mechanisms: A Cross-Country Comparative Analysis (Manila: International Rice Research Institute) p 203
47. Quinn M A 2009 Estimating the impact of migration and remittances on agricultural technology J. Dev. Areas 43 199–216
48. Radel C and Schmook B 2008 Mexican male transnational migration and its linkages to land use change in a southern Campeche ejido J. Latin Am. Geograph. 7 59–84
49. Reij C, Tappan G and Belemvire A 2005 Changing land management practices and vegetation on the Central Plateau of Burkina Faso (1968-2002) J. Arid Environ. 63 642–59
50. Rhoda R and Burton T 2010 Geo-Mexico: The Geography and Dynamics of Modern Mexico (Ladysmith: Sombrero Books)

51. Rogaly B 2003 Who goes? Who stays back? Seasonal migration and staying put among rural manual workers in eastern India J. Int. Dev. 15 623–32
52. Rogaly B personal communication
53. Rogaly B, Jhuma B, Daniel C, Abdur R, Kumar R and Amrita S 2001 Seasonal migration, social change and migrants' rights: lessons from West Bengal Econ. Polit. Weekly 36 4547–59
54. Romero J 2006 Contract farming and producer participation in the avocado industry in Michoacan MSc Thesis Simon Fraser University
55. Rubenstein H 1992 Migration, development and remittances in rural Mexico Int. Migration 30 127–53
56. Schmook B and Radel C 2008 International labor migration from a tropical development frontier: globalizing households and an incipient forest transition Human Ecol. 36 891–908
57. Scoones I 1998 Sustainable rural livelihoods: a framework for analysis IDS Working Paper No. 72 (Brighton: IDS)
58. Singer A, Durand J and Massey D S 1995 Coming into the country: the social process of border crossing among undocumented Mexican migrants Working Paper (Philadelphia, PA: Population Studies Center, University of Pennsylvania)
59. Stark O 1982 Research on rural-to-urban migration in LDCs: the confusion frontier and why we should pause to rethink afresh World Dev. 10 63–70
60. Stark O and Bloom D E 1985 The new economics of labor migration Am. Econom. Rev. 75 173–8
61. Stuart J and Kearney M 1981 Causes and Effects of Agricultural Labor Migration from the Mixteca of Oaxaca to California (La Jolla, CA: Center for US-Mexican Studies, University of California at San Diego)
62. Tacoli C 2009 Crisis or adaptation? migration and climate change in a context of high mobility Prepared for Expert Group Meeting on Population Dynamics and Climate Change UNFPA and IIED In Collaboration with UN-HABITAT and the Population Division ed G José Miguel (London: UNFA/IIED) pp 104–18
63. Taylor L R 1986 Synoptic dynamics, migration and the Rothamsted insect survey J. Animal Ecol. 551–38
64. Taylor J E 1987 Undocumented Mexico–US migration and the returns to households in rural Mexico Am. J. Agricult. Econ. 69 626–38
65. Taylor E, Arrango J, Hugo G, Kouaouci A, Massey D and Pellegrino A 1996 International migration and community development Population Index 62 397–418
66. Taylor J E and Martin P 2001 Human capital: migration and rural population change Handbook for Agricultural Economics (New York: Elsevier)
67. Taylor E and Yunez-Naude A 1999 Education, migration and productivity: an analytical approach and evidence from rural Mexico Development Centre of the Organizations for Economic Co-operation and Development (OECD) vol 95 pp 901–18 No.5
68. Tiffen M et al 1994 More People, Less Erosion: Environmental Recovery in Kenya (Chichester: Wiley)
69. Tsegai D 2004 Effects of Migration on the Source Communities in the Volta Basin of Ghana—Potential Links of Migration, Remittances, Farm and Non-farm Self-employment Activities
70. Vargas-Lundius M and Villareal R 2008 International Migration, Remittances and Rural Development (Rome: IFAD)
71. Wadood A and Kumari P 2009 Impact of climate change on Jharkhand agriculture: mitigation and adoption Workshop Proceedings: Impact of Climate Change on Agriculture pp 207–10 www.isprs.org/proceedings/XXXVIII/8-W3/B3/B3-5-63.pdf
72. Wang Y-C, Tung W W, Li J L and Waliser D E 2010 Sensitivity of the South Asian summer monsoon circulation to the cloud microphysics and associated radiative effects 29th Conf. on Hurricanes and Tropical Meteorology (Tucson)
73. Wouterse F S 2008 Migration, poverty and inequality, evidence from Burkina Faso IFPRI Discussion Paper 00786

74. Zimmerer K 1993 Soil erosion and labor shortages in the Andes with special reference to Bolivia, 1953–1991: implications for 'conservation-with-development' World Dev. 21 1659–75

CHAPTER 2

Farmers' Perceptions of Livestock, Agriculture, and Natural Resources in the Rural Ethiopian Highlands

Rea Tschopp[1], Abraham Aseffa[2], Esther Schelling[3], and Jakob Zinsstag[4]

[1]Swiss Tropical and Public Health Institute, PO Box 4002, Basel, Switzerland Armauer Hansen Research Institute (AHRI/ALERT), PO Box 1005, Addis Abeba, Ethiopia

[2]Armauer Hansen Research Institute (AHRI/ALERT), PO Box 1005, Addis Abeba, Ethiopia

[3]Swiss Tropical and Public Health Institute, PO Box 4002, Basel, Switzerland

[4]Swiss Tropical and Public Health Institute, PO Box 4002, Basel, Switzerland

ABSTRACT

Increasing human and livestock populations in Ethiopia are leading to a growing demand for food and feed. Cereal cropping is highly prioritized at the cost of the livestock subsector and the environment. Grazing land is decreasing, leading to overstocking and overgrazing of pastures, thus fueling conflicts over scarce resources and exacerbating further land degradation. Two independent surveys were carried out in 4 areas in the Ethiopian Highlands, using questionnaires to investigate livestock husbandry as well as farmers' perceptions and attitudes regarding the relationship among cropping, livestock, and natural resources, in the context of broader reflection on what could help support Ethiopia's human, animal, and environmental needs in a sustainable way. In total, 684 farmers were interviewed in 75 villages. The majority of animals were said to be fed on natural pasture and crop residues; only 1.3% of the respondents purchased supplementary feed. Overall, cropland had increased in the study area at the cost of grazing land, and overstocking of pastures was seen as a major problem. Decreasing grazing land was also considered to be due to drought in Woldia and increased human population in Gurage. No pasture management system was in place at community level in our study sites. Less than 2% of the respondents perceived and understood land degradation and subsequent reduced land fertility to be a constraint for sustainable feed production. Measures and priorities for future livelihoods were perceived differently by farmers from different regions. We discuss strategies for de-stocking cattle herds, the nature of which is intrinsically tied to cereal cropping. This study highlights the lack of understanding amongst farmers of the causes and effects of land degradation and the lack of community-based strategies for conservation agriculture.

Keywords: Ethiopia, agriculture, livestock, land use, conflict, natural resources

INTRODUCTION

Ethiopia, well known for its recurrent famines, is amongst the poorest countries in the world and receives the most voluminous food aid in the world (Berry 2003). Agriculture remains the major economic sector in the country, accounting for 43.8% of the national gross domestic product (GDP), 90% of exports, and 85% of employment (CIA 2009). Over 90% of agriculture—characterized by smallholder mixed farming (crop and livestock)—is practiced in the Ethiopian Highlands, which accounts for only 40% of the total territory but carries more than 80% of the human and 70% of the livestock population, making this the most densely inhabited part of the country (Demeke 2006). The Ethiopian Highlands are jeopardized by severe land degradation (Gete and Hurni 2001; Tadesse 2001; Daba et al 2003; Nyssen et al 2004, 2008; Hurni et al 2005). This has direct impacts on agricultural productivity, affecting both cultivated and pastureland through loss of soil and decreased soil fertility, thus constituting a major hazard to sustainable agriculture and feed resources (Hurni 1990; Yirdaw 1996; Nyssen et al 2009). The loss of agricultural value for the period 2000–2010 has been estimated at US$ 7 billion (Sonneveld and Keyzer 2003).

With an Ethiopian population of over 85 million people growing at 3.2% per year (CIA 2009), pressure on the agricultural sector is constantly increasing. Landholdings in the Ethiopian Highlands have an average cropland of 1.2 ha per household (CSA 2007b), but this is predicted to fall to 0.6 ha per household by 2015 due to population growth (Teketay 2001). Despite an overall increase of cropland (at the cost of grazing land) and cereal production, food availability per capita has decreased in the last decade (Sonneveld and Keyzer 2003; CSA 2007b).

In order to secure food availability and alleviate poverty, the Ethiopian government defined cereal intensification as a priority a decade ago (Byerlee et al 2007). By comparison, the livestock subsector remains marginalized, with little effort to improve animal productivity and animal health and promote better management of pastures and thus animal feed. Thus, this subsector does not contribute its full potential to the national economy (Gebremedhin et al 2004; Ibrahim 2004). Ethiopia has the largest livestock population in Africa, with a cattle population of 43 million head (CSA 2007). Animals are kept for milk, meat, draught power, manure, and economic security. Importantly, livestock keeping is intimately linked to agriculture. Traditional farming practices in the Ethiopian Highlands depend on draught oxen for plowing and threshing (Figure 1) (Goe 1987; Gebregziabher et al 2006). Draught power has been shown to be related to poverty because farmers owning fewer oxen cultivate smaller areas and produce less labor-intensive but cheaper pulses instead of cereals (Astatke and Saleem 1996). The increasing livestock population is forced to graze on a decreasing amount of grazing land, which contributes to further land degradation (Gebremedhin et al 2004), poorly nourished animals characterized by low productivity, and conflicts over natural grazing land (CSA 2001, 2007a, 2007b; Nyssen et al 2009).

Numerous quantitative studies have been done on land-use management, but studies that take account of knowledge about farmers' perceptions of their livelihoods and of land-use management have been rare to date. Qualitative research methodology is ideal to investigate perceptions and provides a greater depth of information on the relevance of the subject from the perspective of a target audience in a context-specific setting (Dahlgren et al 2004). Studies using this methodology strengthen and complement previous quantitative studies, which have a high level of measurement precision and statistical power. Using questionnaire surveys, the present study explored prevailing husbandry practices in 4 study areas in the Ethiopian Highlands. It also recorded farmers' perceptions of the current delicate balance among livestock, cropping, and natural resources, and asked them to outline their livelihood objectives.

Figure 1 Field plowing in the Ethiopian Highlands (Woldia Woreda) using oxen pulling a traditional plow, the *maresha*. (Photo by Rea Tschopp)

Material and methods

We conducted 2 independent farmer household surveys using questionnaires with closed and open questions as part of a larger project assessing bovine tuberculosis in rural Ethiopia. Farmers were randomly selected within the multistage sampling framework of the tuberculosis project, according to their willingness to participate and only after they had given their oral consent. All questionnaires were translated into Amharic and back-translated into English for validation of misunderstandings and mistranslations. Interviews were carried out by a trained enumerator. The main researcher was also present during all interviews to verify the accuracy of questionnaire filling. Only farmers who were fluent in Amharic (speaking and understanding) could participate in the interviewing process. We relied entirely on farmers' information (qualitative research), since the purpose of the study was to describe the situation through the farmers' eyes. The reliability and validity of the information were cross-checked through regular visits to farms during bovine tuberculosis testing of animals over 3 consecutive years, and through discussions with village leaders, village elders, and extension workers (in-depth discussions and/or focus group discussions; data not presented in this article).

The first survey was conducted between 2006 and 2007 in 4 Woredas (districts) of 3 regions: (1) Meskanena Mareko (referred to hereafter as Meskan), a Gurage area in the Rift Valley (Southern Nations, Nationalities and People Region, SNNPR) located at 8°10′N and 38°20′E (1800–2170 m), (2) Woldia (Amhara region) located at 11°55′N and 39°35′E (1460–3490 m), (3) Bako Gazer (SNNPR) located at 5°45′N and 36°40′E (1338–1634 m), and (4) the Bale Mountains (Oromia region), a larger geographical area located between 6°50′N and 7°10′N and between 39°40′E and 40°20′E (2120–3500 m). In the latter zone, we regrouped 3 neighboring sites, Woredas, Dinsho, Robe and Goro, into 1 study site (Figure 2). The study sites covered 2 typical agro-ecological zones in the Ethiopian Highlands: *woina dega* between 1500–2300 m and *dega* above 2300 m. The questionnaires for this first survey included general questions on farm size, livestock husbandry (grazing system, fodder, farm input, keeping of other livestock), herd structure and herd turnover (exit due to death/sale and entry due to birth/purchase/gift), use of manure, and off-farm jobs.

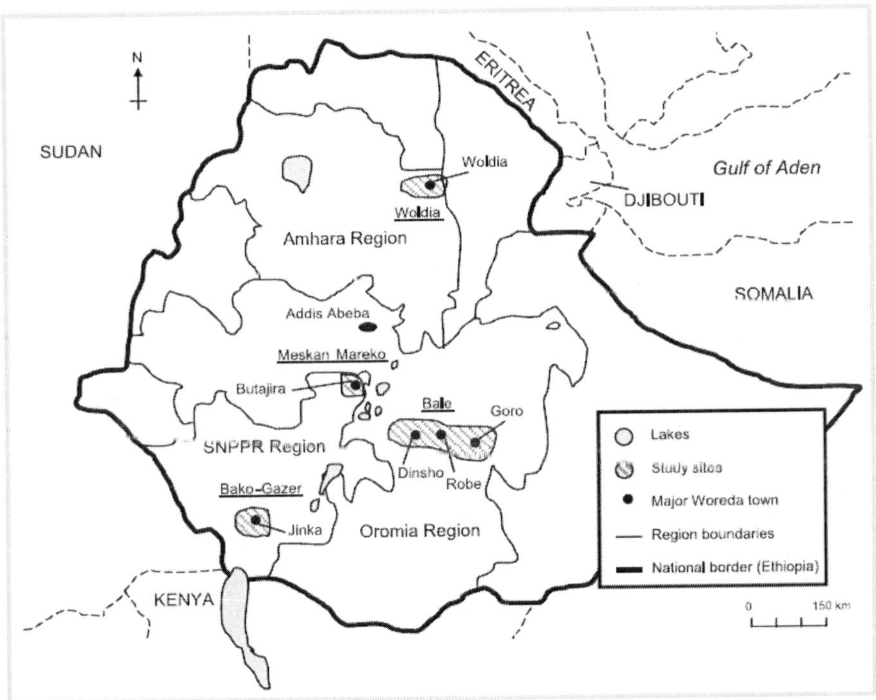

Figure 2 Map of Ethiopia showing the different study sites. (Map by Rea Tschopp)

The second survey was conducted between 2007 and 2008 in only 2 of the Woredas, the eastern part of Woldia (between 1400 and 2000 m) and Meskan (1800–2170 m). This questionnaire focused on questions related to land use, livestock, and interaction between livestock and natural resources, and changes over time as perceived by farmers. Farmers were also asked about occurrences of conflicts over natural resources and their objectives regarding livestock and land availability for the future.

Although the interviewed farmers in both surveys were from various ethnic, religious, and cultural backgrounds, they were all traditional livestock smallholders in the Ethiopian Highlands, involved in both cropping and livestock husbandry with similar farm management.

Additional demographic data at region and zone level were collected from the Central Statistical Agency (CSA), Addis Ababa, and the Ministry of Agriculture and Rural Development, Addis Ababa. All data were double entered in Access and validated for entry errors with the statistical software package Epi Info (version 3.3.2). Analysis was done using the statistical software package STATA 9.1 (StataCorp, Texas, USA) and Microsoft Excel® 2002.

The study received ethical approval from the institutional AHRI/ALERT (AAERC) and the national (NERC) ethical review committees.

RESULTS

Livestock-keeping survey

In total, 536 farmers were interviewed in 4 Woredas, which included 24 Kebeles (smallest administrative unit) and 75 villages. According to their replies, 58% percent of farmers grazed their animals on communal land, whereas the other animals grazed on farmers' own land. Animal feed mainly consisted of forage from natural pastures (free and uncontrolled grazing) and crop residues after harvest, with purchased feed such as oil-cakes and molasses accounting only for 1.3% of the total feed. Veterinary services varied, with farmers stating that 56% of their cattle were regularly vaccinated and only 33% regularly dewormed. Seventy percent of farmers also mentioned that they kept other livestock in addition to cattle.

Cattle herd structure is shown in Table 1. Adult uncastrated and castrated males (37%) exceeded the number of breeding cows kept by interviewees (30%). Nearly a quarter of the herds consisted of oxen (22%). Regardless of age, total males accounted for 52% of herds. With regard to breeding animals, there were only twice as many cows (30%) as bulls (15%), thus 1 bull for 2 cows.

Table 1 Overall herd structure of the interviewed farms.

Cattle categories	Number	Percent of total number
Calves (<1 year)	546	14.0
Juveniles (1–3 years)	730	18.7
Breeding cows	1183	30.4
Breeding bulls	581	14.9
Oxen	850	22.0
Total cattle	3890	100

Regarding herd turnover, 38% of farmers mentioned having purchased at least 1 animal during the previous year, which was more or less equal to the number of farmers (36%) having sold at least 1 animal in the same period. Births were recorded twice as often as deaths, with 63% of farms having had at least 1 calving during the previous year and only 29% of farms recording at least 1 death.

Eighty percent of farmers mentioned holding oxen as draught animals, accounting for 98% of all oxen (Table 2). The remaining 2% of oxen were used for fattening. Twenty-three percent of all breeding males were also used for draught power. Females were rarely used as draught animals (0.4%), and 79% of all draught animals were said to work more than 6 months per year.

Seventy-four percent of respondents valorized manure: 21% used it as fertilizer, but the majority (79%) used it as a source of fuel in the household and sold the remaining unused manure. Overall, 38% of respondents invested in farm improvement. As shown in Table 3, these inputs varied a great deal by region. Veterinary services were viewed by farmers as the most important husbandry input to give, with the exception of Bale: Between 62% (Gurage) and 94% (Woldia) were seeking veterinary care, whereas improvement of breed genetics and improvement of feed, both assets contributing to increasing animal productivity, were perceived as less important, with a maximum of 11% respondents in favor of breed improvement and 34% favoring feed improvement. No husbandry improvements were observed in the Bale Mountains. However, only 40 farmers were interviewed, and thus these figures do not necessarily reflect the reality in that region. Only 8% of farmers mentioned having alternative off-farm income source.

Table 2 Number of oxen kept per household and percentage of households with few to several oxen.

Number of oxen	Number of households keeping oxen	Percent of households keeping oxen
0	113	21
1	148	27
2	211	39
3	32	6
4 and more	37	7
Total	536	100

Table 3 Farm input by region in percent of respondents.

Farm input	Gurage region ($n = 172$)	Woldia ($n = 189$)	Bako Gazer ($n = 135$)	Bale Mountains ($n = 40$)
Veterinary care	62	94	78	0
Improvement of cattle housing	11	4.6	0	0
Improvement of breeds	11	7	0	0
Improvement of feed	34	6	2	0

Survey of farmers' perceptions and objectives

This survey included 69 interviews in 16 villages from 5 Kebeles in the Gurage region (Meskan) and 79 interviews in 22 villages from 6 Kebeles in Woldia (total questionnaires: 148).

Cattle population:

Change in herd size over time differed between the 2 regions: 41% of respondents in Meskan answered that their herd was now larger than in the previous 10 years, due to purchase, birth, and gifts from nongovernmental organizations (NGOs) and the government. Only 20% of farmers reported smaller herds than in the last 10 years. By contrast, 44% of Woldia farmers stated that they had to decrease their herd size over the years due to diseases, drought, and severe feed shortage. Only 18% had more cattle by comparison with past years. The majority of farmers in both regions (73% in Woldia, 56% in Meskan) stated that stocking density on communal land was too high, leading to overgrazing and degradation of pastures.

Pastures:

No pasture management system was in place in Woldia and Meskan. Figure 3A shows the answers of respondents from both regions concerning the current, past, and future availability of pasture forage for their livestock. The majority of farmers in both regions acknowledged a current lack of grazing land, which would dramatically worsen in the future and overstocking/overgrazing of pastures. The reasons for pasture shortage as perceived by farmers are given in Figure 3B. The major reason given by over 60% of respondents from both regions was clearly increased cropland. All interviewed farmers said that land used for crops was greater in surface area than the grazing land they could access. Farmers were not able to provide absolute figures for surfaces, and the statements relied on their perception of changes in land use. Most Woldia farmers (91%) mentioned that they had cropland 4 times and more the size of pastures, while only 62% of Gurage farmers mentioned having this proportion. Only 16% of farmers in Meskan and 3.7% in Woldia thought that a greater cattle population was a limitation to available pastures. In Woldia, drought was said to be an important source of shortage of grazing land (31.5% of respondents). Decreased land fertility was not seen as a reason for shortage of grazing land in Woldia, and only 1.6% of Gurage respondents observed a decrease in land fertility.

Problems and benefits of communal grazing were assessed with farmers from Meskan only (Figure 4A, B). Shortage of forage and animal emaciation were seen as the main drawbacks of communal grazing by a quarter of the respondents. One third perceived easy herding as a major benefit of communal grazing.

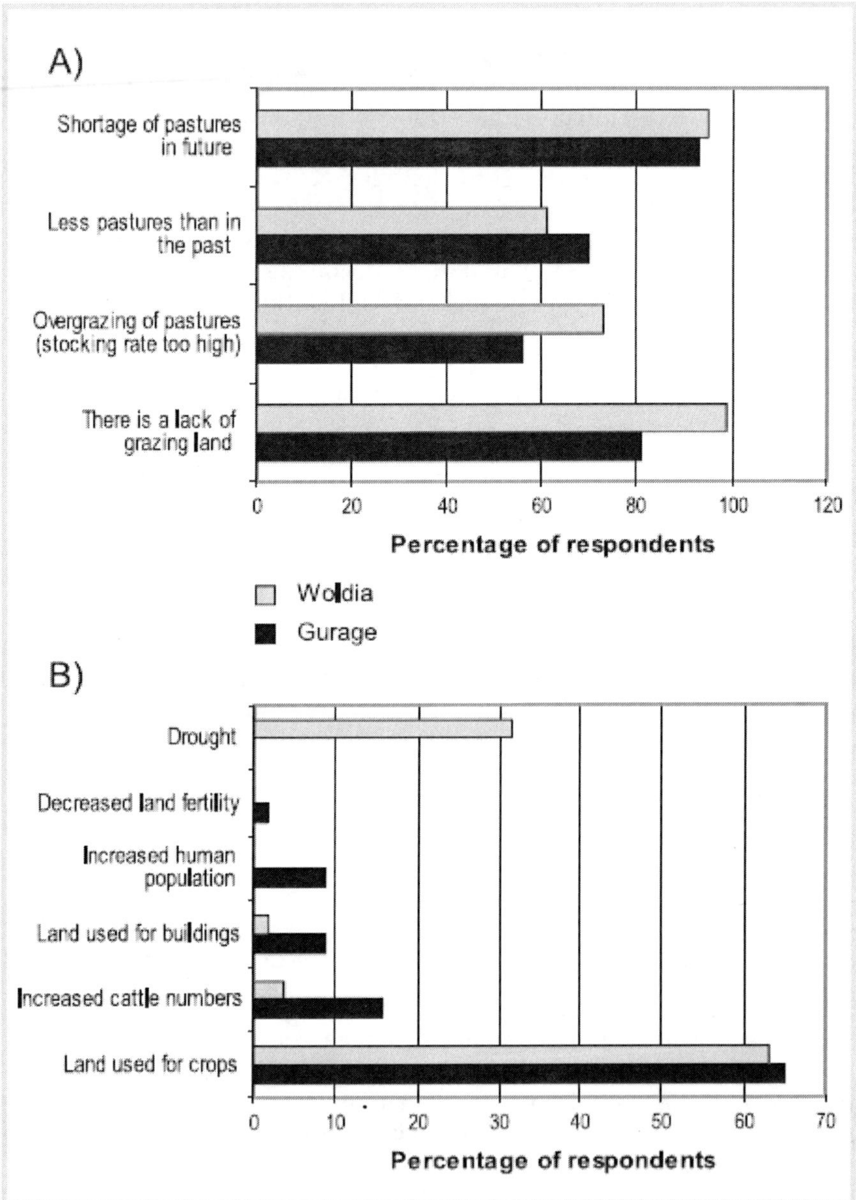

Figure 3(A) Grazing availability for livestock and (B) reasons for grazing shortage, as perceived by Gurage and Woldia farmers.

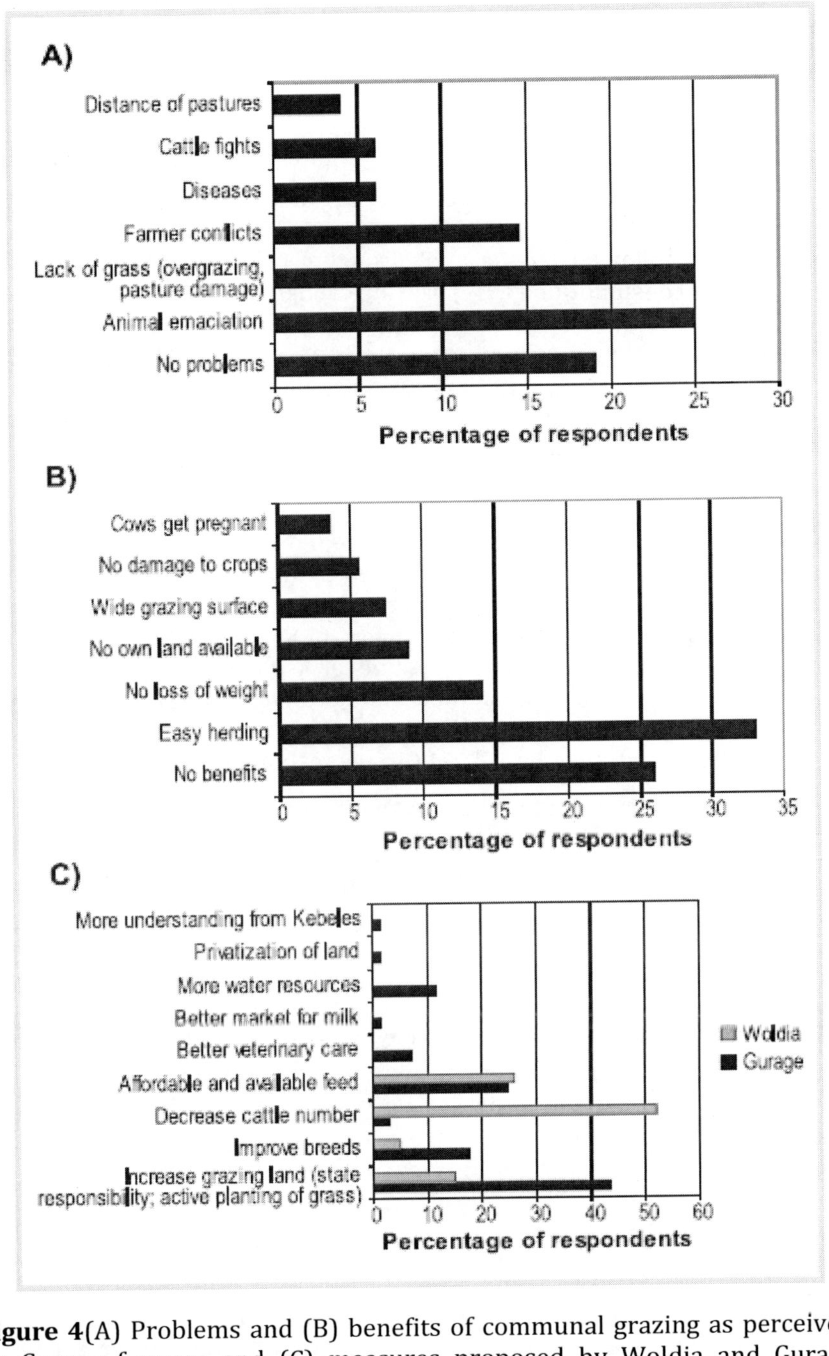

Figure 4(A) Problems and (B) benefits of communal grazing as perceived by Gurage farmers and (C) measures proposed by Woldia and Gurage farmers (multiple answers possible).

Water resources:

Rivers and lakes were mentioned as the main water sources for all Woldia farmers and 60% of Gurage farmers. Because of the often long walk to watering sources, 57% of herds in Meskan and 75% of herds in Woldia were watered only once a day. Forty-four percent of Gurage farmers complained about regular shortage of water. The main complaint was that water was diverted by richer farmers for field irrigation (37.5% of respondents), followed by seasonal droughts (25% of respondents) and human water consumption (19%). During drought periods, water access for livestock was said to have been restricted for 16% of the interviewed households. Problems related to common watering were reportedly infestation of cattle by leeches (25% of respondents) during the dry season and long walking distances to water sources (21% of respondents), whereas disease and injuries from fights were rare (4%).

Conflicts:

Twenty-two percent of farmers in Meskan and 44.3% in Woldia declared that communal grazing land was also used by farmers from other villages. Conflicts over grazing land were mentioned by 23% of respondents in Meskan but only by 4% of Woldia farmers. In Meskan, half of the conflicts involved fellow farmers from the same or from other villages sharing natural resources, and half of the conflicts involved Kebele authorities. Farmers said they reacted with oral complaints. Friction over water resources was mentioned by 18% of Gurage respondents.

Priorities and measures for the future proposed by farmers:

Figure 4C shows farmers' priorities and objectives for the future. These varied between the regions. Less than half of the farmers wished to have more pastures available. In Meskan region, 44% of Gurage farmers considered that additional grazing land should simply be provided by the government. The next 2 prioritized managerial improvements were improved breeds (17% of respondents) and having access to available and affordable supplementary feed (25%). Decreasing herd size was not seen as a major option by Gurage farmers (only 3% of respondents), whereas half of the Woldia farmers saw this as the major measure to overcome land degradation.

DISCUSSION

Ethiopia is a country dependent on its agricultural sector, which is characterized by low productivity. Traditional production systems have led to severe land degradation, resulting in decreased feed resources for both human and livestock. Currently, human and livestock populations are simultaneously increasing rapidly, using more resources than in the past and putting increasing pressure on land (Figure 5) (Nyssen et al 2009).

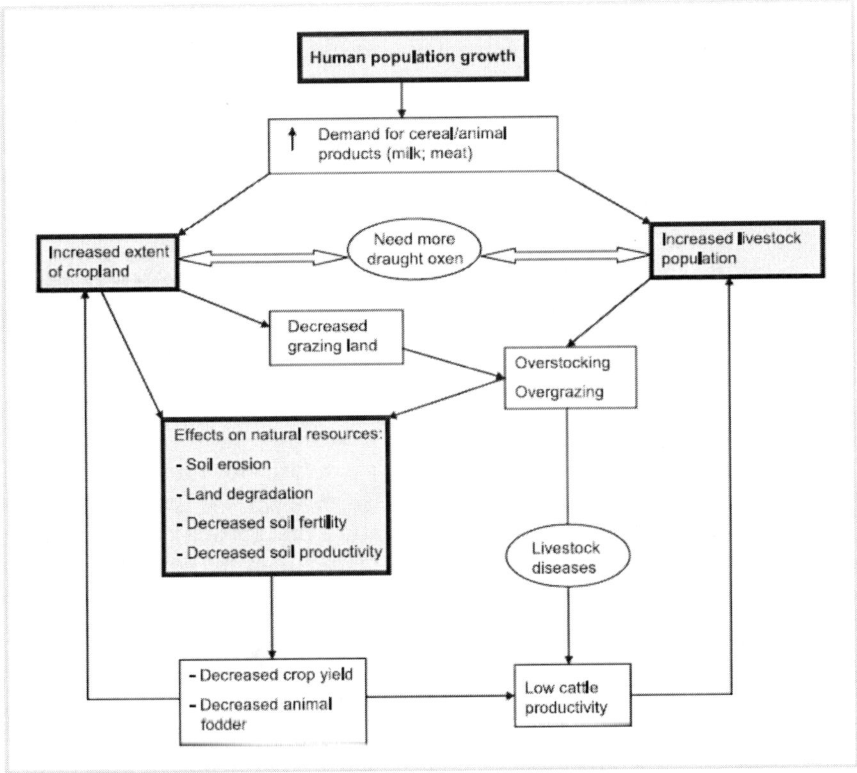

Figure 5 Simplified flow chart summarizing the traditional low-intensity crop–livestock system in Ethiopia and its relationship to natural resource dynamics. (Flow chart by Rea Tschopp)

Data from the Central Statistical Agency (CSA) have shown that over the last decade cropland has increased at the expense of grazing land (CSA, 1996a, 2001, 2007b). Ever-decreasing grazing land combined with a rapidly growing livestock population of over 90 million head (CSA 2007a) is likely to lead to massive overstocking and overgrazing of available pastures and increased land degradation. This national situation correlates with data from our study: 41% of interviewed Gurage farmers stated that while their herd size was now larger than 10 years ago, their grazing land had decreased massively in favor of cropland, thus further increasing the stocking rate. All interviewed farmers, regardless of region, stated that they needed to prioritize cropland to feed their growing families.

The majority of the Gurage and Woldia farmers complained about the current situation regarding lack of pastures to accommodate their animals and about overstocking/overgrazing problems. Farmers' perceptions regarding other reasons for lack of grazing land and forage in the future—besides increasing cropland—differed by region. Woldia farmers considered drought to be a major constraint on both grazing land and herd size, whereas Gurage farmers considered the increasing human population to be a major constraint, since it is linked in their view with increasing livestock and more land needed to build infrastructure.

None of the farmers in Gurage or Woldia perceived land degradation and subsequent decreased land fertility as a problem. Research has shown, however, that overstocking damages land, but kept in optimal numbers, livestock contribute through accumulation of manure to increased biomass production on grazing land (Tadesse et al 2002, 2003). In our study, manure was collected from the fields as a primary source of fuel and/or income and thus was used only partially for soil fertilization (only 21% of respondents used it as fertilizer). Less than 2% of the farmers interviewed understood decreased soil fertility of pastures to be one reason for feed shortage. Water shortage was perceived as a much bigger problem than shortage of grazing land.

Overall, farmers relied almost entirely on forage from natural pastures to feed their animals, with only 1.3% of respondents purchasing feed. Communal grazing was perceived by the interviewees as providing more benefits than problems despite the associated lack of forage. Poor nutritional status contributes to low animal productivity. To compensate for low individual productivity, farmers tend to increase their herd size, as shown in the Gurage study, which in turn puts more pressure on pastures. Only external factors such as severe drought, as shown in the Woldia case, lead to herd depletion through sale and death. Such events may in turn force some of the farmers to use communal land for cereal cropping in order to survive, thus further decreasing available grazing land, as well as soil fertility due to lack of manure.

Farmers' stated solutions to the overstocking/overgrazing problem differed depending on the region: Woldia farmers said they wished to have fewer but more productive animals, which in turn would reduce the need for grazing land. On the other hand, Gurage farmers were asking local authorities for more pastures, as if land was an expandable commodity. They also considered more productive breeds as a priority but were not willing to reduce their herd size. Agricultural extensification rather than intensification was clearly the preference of these farmers. Finally, they wished to have greater access to water resources. Both Woldia and Gurage farmers also saw the need for increased and affordable supplementary feed.

No household-level or community-based land-use management was reported in the 2 study areas. Our study also highlighted the fact that interviewed farmers did not fully perceive the limitations of existing natural resources, that is, the complete cause–effect chain, and the full extent and implications of overgrazing, land degradation, future availability of feed, and sustainability of natural resources. This contrasts with the attitude of rural communities in other parts of Ethiopia (eg pastoralist communities in Afar or Borana, in Tigray), who have a long tradition of restrictive regulations of grazing areas at village level and an understanding of the limitations of existing natural resources and consequently the need for managing them (Gebremedhin et al 2004; Abule et al 2005; Edossa et al 2005).

In Tigray, a region with extreme land degradation, agricultural intensification and conservation agriculture have proved to be one solution to the aforementioned agricultural trap, as revealed by participatory research (Astatke et al 2003; Nyssen et al 2009). Conservation agriculture, which combines social and economic benefits from integrating production and protection of the environment, is a relatively new concept, especially in Africa (Dumanski et al 2006). The goal of conservation agriculture is sustainability in farming systems and thus food production and security of livelihoods (Govaerts et al 2008). Rockstrom et al (2009) showed that conservation agriculture practices in Ethiopia increased crop yields when compared to traditional practices. These yields additionally increased when fertilizer was added. Further examples in Tigray include increased field irrigation systems (Nyssen et al 2009), terracing and stone bunds on slopes (Nyssen et al 2007), catching and storing runoff water in ponds (Fekadu et al 2007), and rehabilitation of degraded land by exclosures and thus limitation of uncontrolled and free grazing that further damages land (Mekuria et al 2007). However, conservation agriculture, being a complex multicomponent technology, requires adaptation to local farming systems prevailing in different regions of a country, and/or in different agro-ecological areas (Govaerts et al 2008).

Decreased feed availability on overstocked/degraded communal grazing land as seen in this study is likely to have direct impacts on animal health; this is reflected by animals showing poor body condition, decreased productivity, and decreased resistance to diseases (Pandey et al 1993; Mishra et al 2001). Communal grazing also directly intensifies the risk of disease transmission between animals and increases the parasitic load on pastures (Lefèvre and Blancou 2003). A quarter of the interviewed farmers perceived communal grazing to be associated with animal emaciation, but only 6% of them saw a possible link with diseases. The increasing livestock population also strains the scarce veterinary services available in the country. Official figures show that only 2 million cattle (4.6%) were vaccinated nationwide in 2007 (CSA 2007a). In contrast, half of the interviewed farmers in our study stated that they regularly vaccinated their animals. This discrepancy can be explained by the fact that the latter farmers had better access to veterinary care services than farmers from other regions, such as pastoralist zones. Their statement may also have only included valuable animals, such as oxen, and not their entire herd. The need for increased veterinary care was clearly one of the major priorities for husbandry improvement given by most respondents (62% of Gurage and 94% of Woldia farmers).

Friction over scarce natural resources exists in most of Ethiopia. In some areas, the different players manage to reach agreements (Nyssen et al 2009), whereas in others conflict and violent clashes occur, such as in the Awash River basin (Edossa et al 2005) and the Gambella region (Sewonet 2003; Reuters 2008). Verbal friction was described by 23% of Gurage respondents in our study, half of which involved the relevant Kebele authorities, who were said to have not taken their needs seriously. Farmers' attitudes in this study site stress the lack of collective action at village level to cooperate in resource management and the fact that farmers rely on the authorities to improve the situation.

Ethiopia has the largest livestock population in Africa, with 1 head of cattle for less than 2 people. Furthermore, within the cattle population, males account for half of the total herd structure at the national level (CSA 2007a) and in our study. Destocking cattle herds seems to be the logical solution to decrease the economic and ecological burden on the agricultural sector. In urban areas, farmers have switched to more productive and intensive systems, including highly productive exotic dairy breeds in order to supply big cities with the increasing demand for milk. However, highly productive exotic breeds and their crosses currently account for only 1% of the total cattle population, and demand for milk is still much higher than the supply; moreover, animal feed remains unaffordable, and grazing areas are lacking (CSA 2007a, 2007b; SNV 2008).

In rural areas, traditional production systems rely entirely on animal power. Increased cropland will in turn lead to a need for more draught animals and thus to stagnation in the agricultural system and further exacerbation of pressure on land, implying further degradation (Figure 5). Taking into account animal losses and the minimum age at which animals can be used as draught animals, currently each household realistically needs to maintain a minimum herd size of 8–10 animals to permanently secure at least 2 oxen for plowing (Sandford 1982). The majority of farmers in our study possessed 2 oxen, and 12.5% of the respondents even had 3 and more oxen. Draught animals worked more than 6 months per year (plowing and threshing), making it difficult to share working animals among farmers and thus decrease the number of oxen in a village. Cows are primarily kept to produce the next male generation for draught power rather than for milk production. Improvement of breed genetics, as encouraged in urban and peri-urban areas to increase productivity, and thus decrease animal numbers, is not a solution for these essentially ox-production systems. This is also reflected in the farmers' attitudes in our study. Regardless of fodder availability, only 11% of Gurage and 7% of Woldia farmers said they would like to have improved breeds with better productivity. Moreover, our study showed that overall no efforts were made for livestock housing and feed supplementation as alternatives to natural grazing.

However, even in these traditional production systems, research has shown that strategies can be used to decrease the oxen population necessary for production. Astatke et al (2003) showed that modification of the traditional plow reduced tillage and soil erosion. These authors reported 50% less draught animals required compared to the traditional plowing system. Aune et al (2001) have also discussed potential alternatives to ox-plowing in Ethiopia, such as hoe tillage, zero tillage, and cow traction.

Finally, increasing the export capacity for live animals and meat may help decrease the livestock surplus. Around 350,000 cattle and 1.2 million small ruminants are exported annually (FAO 2007). However, the export industry is still underdeveloped in Ethiopia, and most trade is informal (cross-border) from pastoralist areas (Little 2005). These areas are also the main suppliers of animals to export abattoirs, and exporters help destock their herds during times of crisis (eg severe drought).

CONCLUSIONS

Livestock remain the most important component in Ethiopian farmers' lives for daily survival and economic security. However, cereal cropping is highly prioritized at government and farm level at the cost of the livestock subsector, the environment, and natural resources. The introduction of conservation agriculture and rehabilitation of degraded land in parts of the country has shown benefits in economic and ecological terms compared to the traditional agricultural system. However, these strategies work best when they are community- or village-driven. This study showed that the perceptions and attitudes of farmers toward agriculture (cropping and livestock) and natural resources in our study sites, as well as their priorities for future livelihoods, diverged greatly depending on the region. It also highlighted the lack of understanding among these farmers regarding the causes and consequences of land degradation, and their lack of insight into the current unsustainability of their farming system, which jeopardizes future sources of food for both humans and livestock. We thus conclude that there is a need for increased community-based awareness and participatory trials with conservation agriculture in the study areas.

ACKNOWLEDGMENTS

We are very grateful to the Wellcome Trust (UK) for funding this study, which was done in collaboration with Swiss National Centre of Competence in Research (NCCR) North-South. We thank Armauer Hansen Research Institute (AHRI/ALERT) (Addis Abeba) for logistic and data management support. We also thank Lina Gazu, Mesgebu Asmro, Bamlaku Tilahun, and Alemayehu Kifle for their valuable help and support during fieldwork. Our thanks also go to all the villagers who were willing to participate in this study. We are grateful to Anne Zimmerman for her help in editing the manuscript, as well as the anonymous reviewers for their valuable comments.

REFERENCES

1. Abule, E., H. A. Snyman, and G. N. Smit. 2005. Comparisons of pastoralist's perceptions about rangeland resource utilization in the Middle Awash Valley of Ethiopia. *Journal of Environmental Management* 75:21–35.
2. Astatke, A., M. Jabbar, and D. Tanner. 2003. Participatory conservation tillage research: An experience with minimum tillage on an Ethiopian Highland Vertisol. *Agriculture, Ecosystems and Environment* 95:401–415.
3. Astatke, A. and M. A. Saleem. 1996. Draught animal power for land-use intensification in the Ethiopian Highlands. *World Animal Review (FAO)* 86:3–11.
4. Aune, J. B., M. T. Bussa, F. G. Asfaw, and A. A. Ayele. 2001. The ox ploughing system in Ethiopia: Can it be sustained? *Outlook on Agriculture* 30:275–280.
5. Berry, L. 2003. Land Degradation in Ethiopia: Its Extent and Impact Study commissioned by the Global Mechanism with support from the World Bank ftp://ftp.fao.org/agl/agll/ladadocs/ETHIOPIA_LD_CASE_STUDIES.doc; accessed on 25 June 2008.
6. Byerlee, D., D. J. Spielman, D. Alemu, and M. Gautam. 2007. Policies to Promote Cereal Intensification in Ethiopia: A Review of Evidence and Experience. Development Strategy and Governance Division, Discussion Paper 00707, June 2007. Washington, DC: International Food Policy Research Institute (IFPRI).

7. CIA [Central Intelligence Agency] 2009. CIA factsheet. The World Factbook: Ethiopia https://www.cia.gov/library/publications/the-world-factbook/geos/et.html; accessed on 15 April 2010.
8. CSA [Central Statistical Agency] 1996a. Agricultural Sample Survey for 1995/96. (CD-ROM). Addis Abeba, Ethiopia: CSA.
9. CSA [Central Statistical Agency] 1996b. Report on Livestock, Poultry and Beehives. Population and Number by Size of Holdings 1995/1996. Vol VI. Statistical Bulletin 152. Addis Abeba, Ethiopia: CSA.
10. CSA [Central Statistical Agency] 2001. Agricultural Sample Survey 2000/01. Vol II: Report on Livestock and Livestock Characteristics. Statistical Bulletin 245. Addis Abeba, Ethiopia: CSA.
11. CSA [Central Statistical Agency] 2007a. Agricultural Sample Survey 2006/07. Vol II: Report on Livestock and Livestock Characteristics. Statistical Bulletin 388. Addis Abeba, Ethiopia: CSA.
12. CSA [Central Statistical Agency] 2007b. Agricultural Sample Survey 2006/07. Vol IV: Report on Land Utilization. Statistical Bulletin 388. Addis Abeba, Ethiopia: CSA.
13. Daba, S., W. Rieger, and P. Strauss. 2003. Assessment of gully erosion in eastern Ethiopia using photogrammetric techniques. *Catena* 50 (2):273–291.
14. Dahlgren, L., M. Emmelin, and A. Winkvist. 2004. Qualitative Methodology for International Public Health. Umea, Sweden: Print och Media, Umea University.
15. Demeke, S. 2006. Contribution of Jimma University, College of Agriculture and Veterinary Medicine (JUCAVM) to Ethiopian Agriculture Seminar paper. Seminar held between 8 July and 14 August 2006 at Nova Scotia Agricultural College (NSAC). http://www.nsac.ns.ca/international/International_Projects/Current_Projects/Ethiopia/SolomanPaper.pdf; accessed on 20 April 2008.
16. Dumanski, J., R. Peiretti, J. Benetis, D. McGarry, and C. Pieri. 2006. The paradigm of conservation tillage.*Proceedings of the World Association of Soil and Water Conservation* P1:58–64.
17. Edossa, D. C., M. S. Babel, A. D. Gupta, and S. B. Awulachew. 2005. Indigenous systems of conflict resolution in Oromia, Ethiopia. In: Proceedings of the International Workshop on “African Water Laws: Plural Legislative Frameworks for Rural Water Management in Africa.” Johannesburg, South African Department of Water Affairs and Forestry; http://www.nri.org/projects /waterlaw /AWLworkshop/DESALEGN-CE.pdf; accessed in June 2008.
18. Fekadu, Wondumagegnehu, Tsegay Alemtsehay, Ashebir Dereje, Tekle Hailemariam, Gebre Addisu, Kiros Mewael, S. Geerts, D. Raes, J. Nyssen, and J. Deckers. 2007. Household Water Harvesting Structures in Geba Catchment. Tigray Livelihood Papers No. 5. Mekelle, Ethiopia: VLIR–Mekelle University IUC Programme.
19. FAO [Food and Agriculture Organization] 2007. FAOSTAT Database http://faostat.fao.org; accessed on 20 April 2010.
20. Gebregziabher, S., A. M. Mouazen, H. Van Brussel, H. Ramin, J. Nyssen, H. Verplancke, M. Behailu, J. Deckers, and J. De Baerdemaeker. 2006. A review: Animal drawn tillage, the Ethiopian ard plough, maresha. *Soil and Tillage* 89:129–143.
21. Gebremedhin, B., J. Pender, and G. Tesfay. 2004. Collective action for grazing land management in crop–livestock mixed systems the Highlands of Northern Ethiopia. *Agricultural Systems* 82:273–290.
22. Gete, Zeleke and H. Hurni. 2001. Implications of land use and land use cover dynamics for mountain resource degradation in the Northwestern Ethiopian Highlands. *Mountain Research and Development* 21 (2):184–191.
23. Goe, M. R. 1987. Animal Traction on Smallholder Farms in the Ethiopian Highlands [PhD thesis]. Ithaca, NY:Department of Animal Science, Cornell University.
24. Govaerts, B., N. Verhulst, A. Castellanos, K. D. Sayre, J. Dixon, and L. Dendooven. 2008. Conservation agriculture and soil carbon sequestration: Between myth and farmer reality. *Critical Reviews in Plant Science* 28:97–122.
25. Hurni, H. 1990. Degradation and conservation of soil resources in the Ethiopian Highlands. In: Messerli, B. and H. Hurni. editors. African Mountains and Highlands:

Problems and Prospective. Marceline, MO: Walsworth Press for the African Mountains Association (AMA). pp. 51–63.

26. Hurni, H., K. Tato, and G. Zeleke. 2005. The implications of changes in population, land use, and land management for surface runoff in the Upper Nile Basin area of Ethiopia. *Mountain Research and Development* 25 (2):147–154.

27. Ibrahim, M. 2004. Extension experiences in Ethiopia. Paper presented at the Ministry of Agriculture and Rural Development Planning Workshop. Addis Ababa, Ethiopia. Available from the first author of the present paper.

28. Lefèvre, P. C. and J. Blancou. 2003. Particularités épidémiologiques des régions chaudes. In: Lefèvre, P. C., J. Blancou, and R. Chermette. editors. Principales maladies infectieuses et parasitaires du bétail. Europe et régions chaudes. Tome 1. Généralités, Maladies virales. Paris, France: Lavoisier. pp. 47–54.

29. Little, P. 2005. Unofficial Trade When States Are Weak. The Case of Cross-Border Commerce in the Horn of Africa. Research Paper No. 2005/13. Helsinki: EGDI (Expert Group on Development Issues) and UNU-WIDER (United Nations University–World Institute for Development Economics Research).

30. Mekuria, W., E. Veldkamp, M. Haile, J. Nyssen, B. Muysd, and K. Gebrehiwot. 2007. Effectiveness of exclosures to restore degraded soils as a result of overgrazing in Tigray, Ethiopia. *Journal of Arid Environments* 69:270–284.

31. Mishra, C., H. H. T. Prins, and S. E. Van Wieren. 2001. Overstocking in the trans-Himalayan rangelands of India. *Environmental Conservation* 28 (3):279–283.

32. Nyssen, J., J. Poesen, D. Gebremichael, K. Vancampenhout, M. D'aes, G. Yihdego, G. Govers, H. Leirs, J. Moeyersons, J. Naudts, N. Haregeweyn, M. Haile, and J. Deckers. 2007. Interdisciplinary on-site evaluation of stone bunds to control soil erosion on cropland in Northern Ethiopia. *Soil and Tillage Research* 94:151–163.

33. Nyssen, J., J. Poesen, N. Haregeweyn, and T. Parsons. 2008. Environmental change, geomorphic processes and land degradation in tropical highlands. *Catena* 75:1–4.

34. Nyssen, J., J. Poesen, J. Moeyersons, J. Deckers, H. Mitiku, and A. Lang. 2004. Human impact on the environment in the Ethiopian and Eritrean Highlands—A state of the art review. *Earth-Science Reviews* 64:273–320.

35. Nyssen, J., G. Simegn, and N. Taha. 2009. An upland farming system under transformation: Proximate causes of land use change in Bela-Welleh catchment (Wag, Northern Ethiopian Highlands). *Soil and Tillage Research* 103:231–238.

36. Pandey, V. S., F. Chitate, and T. M. Nyanzunda. 1993. Epidemiological observations on gastro-intestinal nematodes in communal land cattle from the highveld of Zimbabwe. *Veterinary Parasitology* 51:99–106.

37. Reuters 2008. Twenty Killed in Western Ethiopia Land Clashes http://in.reuters.com /article/worldNews/idINIndia-33747520080524; accessed on 20 May 2008.

38. Rockstrom, J., P. Kaumbutho, J. Mwalley, A. W. Nzabi, M. Temesgen, L. Mawenya, J. Barron, J. Mutua, and S. Damgaard-Larsen. 2009. Conservation farming strategies in East and Southern Africa: Yields and rain water productivity from on-farm action research. *Soil and Tillage Research* 103:23–32.

39. Sandford, S. 1982. Livestock in the Communal Areas of Zimbabwe. A report prepared for the Ministry of Lands, Resettlement and Rural Development. London, United Kingdom: Overseas Development Institute (ODI).

40. Sewonet, A. 2003. Ethiopia: Breaking the Cycle of Conflict in Gambella Region United Nations Country Team in Ethiopia (UNCT). http://www.who.int/disasters /repo/8684.pdf; accessed on 20 May 2008.

41. SNV Netherlands Development Organisation. 2008. Study on dairy investment opportunities in Ethiopia. July 2008, Addis Ababa, Ethiopia. http://www.snvworld.org/en/Documents/Dairy%20Investment%20Opportunities%20in%20Ethiopia.pdf; accessed in October 2010.

42. Sonneveld, B. G. J. S. and M. A. Keyzer. 2003. Land under pressure: Soil conservation concerns and opportunities for Ethiopia. *Land Degradation Development* 14:5–23.

43. Tadesse, G. 2001. Land degradation: A challenge to Ethiopia. *Environmental Management* 27 (6):815–824.
44. Tadesse, G., D. Peden, A. Astatke, and A. Wagnew. 2003. Effect of manure on grazing lands in Ethiopia, East African highlands. *Mountain Research and Development* 23 (2):156–160.
45. Tadesse, G., M. Saleem, A. Abyie, and A. Wagnew. 2002. Impact of grazing on plant species richness, plant biomass, plant attribute and soil physical and hydrological properties of Vertisol in East African Highlands.*Environmental Management* 29 (2):279–289.
46. Teketay, D. 2001. Deforestation, wood famine, and environmental degradation in Ethiopia's Highland ecosystems: Urgent need for action. *Northeast African Studies* 8 (1):53–76.
47. Yirdaw, E. 1996. Deforestation and forest plantations in Ethiopia. In: Palo, P. M. and G. Mery. editors. Sustainable Forestry Challenges for Developing Countries. Dordrecht, The Netherlands: Kluwer Academic Publishers. pp. 327–357.

CHAPTER 3

How Much Meteorological Information Is Necessary to Achieve Reliable Accuracy for Rainfall Estimations?

Mohammad Valipour

Young Researchers and Elite Club, Kermanshah Branch, Islamic Azad University, Kermanshah, Iran

ABSTRACT

This paper reports the study of the effect of the length of the recorded data used for monthly rainfall forecasting. Monthly rainfall data for three periods of 5, 10, and 49 years were collected from Kermanshah, Mashhad, Ahvaz, and Babolsar stations and used for calibration time series models. Then, the accuracy of the forecasting models was investigated by the following year's data. The following was concluded: In temperate and semi-arid climates, 60 observation data is sufficient for the following year's rainfall forecasting. The accuracy of the time series models increased with increasing amounts of observation data of arid and humid climates. Time series models are appropriate tools for forecasting monthly rainfall forecasting in semi-arid climates. Determining the most critical rainfall month in each climate condition for agriculture schedules is a recommended aim for future studies.

Keywords: Iran; weather data; precipitation

1. INTRODUCTION

More accurate forecasting of monthly rainfall is significantly important in water resource management and crop pattern design. In this study, time series models were used to forecast monthly rainfall in four different climates. After publishing the paper of Box and Jenkins [1], Box–Jenkins models became a general time series model of hydrological forecasting. These models include the Auto Regressive Integrated Moving Average (ARIMA), the AR Moving Average (ARMA), the Auto Regressive (AR), and the Moving Average (MA). Access to basic information requires integration from the series (for a continuous series) or calculating all differences in the series (for a continuous series). Since the constant of integration in derivation or differences are deleted, the probability of using accurate amount in this process is not possible. Therefore, ARIMA models are non-static and cannot be used to reconstruct the missing data. However, these models are very useful for forecasting changes in hydrological processes considering agricultural water management, evapotranspiration, and water crisis issues [2,3,4,5,6,7, 8,9,10,11,12,13,14,15,16,17]. Models of time series analysis (Box–Jenkins models) in various fields of hydrology and rainfall forecasting in irrigation schedules are widely applied, some of which will be described in what follows.

Serinaldi and Kilsby [18] presented a modular class of multisite monthly rainfall generators for water resource management and impact studies. The results of the case study show that the model can capture several characteristics of the rainfall series. In particular, it enables the simulation of low and high rainfall scenarios more extreme than those observed as well as the reproduction of the distribution of the annual accumulated rainfall, and of the relationship between the rainfall and circulation indices such as the North Atlantic Oscillation (NAO) and the Sea Surface Temperature (SST), thus proving the framework to be well-suited for sensitivity analysis under alternative climate scenarios and additional forcing variables. Luc et al. [19] studied an application of artificial neural networks (ANNs) for rainfall forecasting successfully. Wei et al. [20] used weather satellite imagery to forecast rainfall in Taiwan. Andrieu et al. [21] studied the adaptation and application of a quantitative rainfall forecasting model in a mountainous region. This work showed that a limit on forecast lead-time may be related to the response time of the precipitating cloud system. Burlando et al. [22], using ARMA models, forecasted short-term rainfall. The hourly rainfall parameter from two gaging stations in Colorado, USA, and from several stations in Central Italy, was used. Results showed that the event-based estimation approach yields better forecasts. Hu et al. [23] studied rainfall, mosquito density, and the transmission of the Ross River virus using a time-series forecasting model. Their results showed that both rainfall and mosquito density were strong predictors of Ross River virus transmission in simple models. Ramírez et al. [24] used an ANN technique for rainfall forecasting applied to the São Paulo region. The results showed that ANN forecasts were superior to the ones obtained by the linear regression model, thus revealing a great potential for an operational suite. Han et al. [25] successfully forecasted drought based on remote sensing data using ARIMA model. Chattopadhyay and Chattopadhyay [26] compared ARIMA and ARNN models using univariate modeling of summer-monsoon rainfall time series. Anctil et al. [27] studied the impact of the length of observed records on the performance of ANN and of conceptual parsimonious rainfall-runoff forecasting models. The results showed that the best performance about evenly for 3- and 5-year training sets, but multiple-layer perceptrons (MLPs) performed better whenever the training set was dominated by wet weather. The MLPs continued to improve for input vectors of 9 years and more, which was not the case of the conceptual model. Jia and Culver [28] using bootstrapped ANNs suggested that even a small set of periodic instantaneous observations of stage from a staff gauge, which can easily be collected by volunteers, can be a useful data set for effective hydrological modeling. M. Baareh et al. [29] applied the ANN and AR models to the river flow forecasting problem. A comparative study of both ANN and the AR conventional model networks indicated that the ANNs performed better than the AR model. They showed that ANN models can be used to train and forecast the daily flows of the Black Water River near Dendron in Virginia and the Gila River near Clifton in Arizona. Xiong and O'Connor [30] used four different error-forecast updating models—AR, AR-threshold (AR-TS), fuzzy AR-threshold (FU-AR-TS), and an ANN—for real-time river flow forecasting. They found that all of these four updating models are very successful in improving the flow forecast accuracy. Chenoweth et al. [31] estimated the ARMA model parameters using neural networks. Their results showed that the ability of neural networks to accurately identify the order of an ARMA model was much lower than reported by previous researchers, and is especially low for time series with fewer than 100 observations. Using forecasting of hydrologic time series with ridge regression in feature space, Yu and Liong [32] showed that the training speed in the data mining method was much faster than the ARIMA model. See and Abrahart [33] used data fusion for hydrological forecasting. Their results showed that, using data fusion methodologies for ANN, fuzzy logic, and

ARMA models, accuracy of forecasting increased. Using hybrid approaches, Srinivas and Srinivasan [34] improved the accuracy of AR model parameters for annual streamflows. Using Fourier coefficients, Ludlow and Enders [35] estimated the ARMA model parameters with relatively good accuracy. Chenoweth et al. [36] estimated the ARMA model parameters using Hilbert coefficients. Their results showed that Hilbert coefficients are a useful tool for estimating ARMA model parameters. Balaguer et al. [37] used the time delay neural network (TDNN) and ARMA model in support centers for crisis management. The obtained correlation results for TDNN model and ARMA were 0.88 and 0.97, respectively. This study confirmed the superiority of the ARMA model to the TDNN. Toth et al. [38] used the ANN and ARMA models to forecast rainfall. The results show the success of both short-term rainfall-forecasting models for forecasting floods in real time. Mohammadi et al. [39] forecasted Karaj reservoir inflow using data of melting snow and ANN and ARMA methods, as well as regression analysis. Sixty percent of inflow in dams occurs between April and June, so forecasting the inflow in this season is very important for a dam's performance. The highest inflows were in the spring due to the snow melt caused by draining in threshold winter. The results showed that an ANN, compared with other methods, has lower significant errors. Mohammadi et al. [40] in another study estimated the parameters of an ARMA model for river flow forecasting using goal programming. Their results showed that the goal programming is a precise and effective method for estimating ARMA model parameters for forecasting inflow. Other researchers [41,42,43,44,45,46,47,48] estimated the parameters of the ARMA and ARIMA models and compared their ability for inflow forecasting. By comparing root mean square errors of the model, it was determined that the ARIMA model can forecast inflow to the Dez reservoir from 12 months with a lower error than can the ARMA model.

Therefore, considering the above-mentioned research, the efficacy of time series models forecasting field and hydrologic sampling can be determined. The effect of recorded data in ANN forecasting has been previously studied. However, concurrent use of time series models to determine the number of required observation data has not been performed. This study aims to forecast monthly rainfall using time series models and determine appropriate observation data according to different climate conditions.

2. MATERIAL AND METHODS

In this study, to investigate the effect of climate conditions on the amount of appropriate observation data, four synoptic stations with different climate conditions in Iran, including temperate, humid, arid, and semi-arid were selected.

Table 1 displays information on the synoptic stations used in this study. In order to forecast rainfall on a monthly basis, the rainfall data period from 1951 to 2000 was gathered. In fact, the used data involved 2400 data (all stations) that began in January 1951 and ended in December 2000. Figure 1 shows the locations of the synoptic stations. In this study, time series models were used to forecast monthly rainfall. These models include ARIMA, ARMA, AR, and MA. For this purpose, MINITAB software was used to run all time series models. Finally, via three methods, the ability of time series models to forecast monthly rainfall was surveyed. In the first method, 60 data (1995–1999) regarding the amount of rainfall forecasted for the year 2000 was used. In the second method, 10 years of monthly rainfall data (1990–1999) regarding the amount of rainfall forecasted for the year 2000 was used. In the third method, 588 data (1951–1999) regarding the amount of rainfall forecasted for the year 2000 was used.

Table 1. Information of synoptic stations used in this research.

Station	Climate	Elevation Related to Sea Level (m)	Longitude	Altitude
Kermanshah	Temperate	999.2	59°38′ E	36°16′ N
Mashhad	Semi-arid	1322.0	47°7′ E	34°17′ N
Ahvaz	Arid	22.5	48°40′ E	31°20′ N
Babolsar	Humid	−21.0	52°39′ E	36°43′ N

3. RESULTS AND DISCUSSION

Table 2 shows the obtained R^2 for four different stations according to the amount of observation data. Figure 2, Figure 3, Figure 4 and Figure 5 show the amount of forecasted rainfall for Kermanshah, Mashhad, Ahvaz, and Babolsar synoptic stations, respectively.

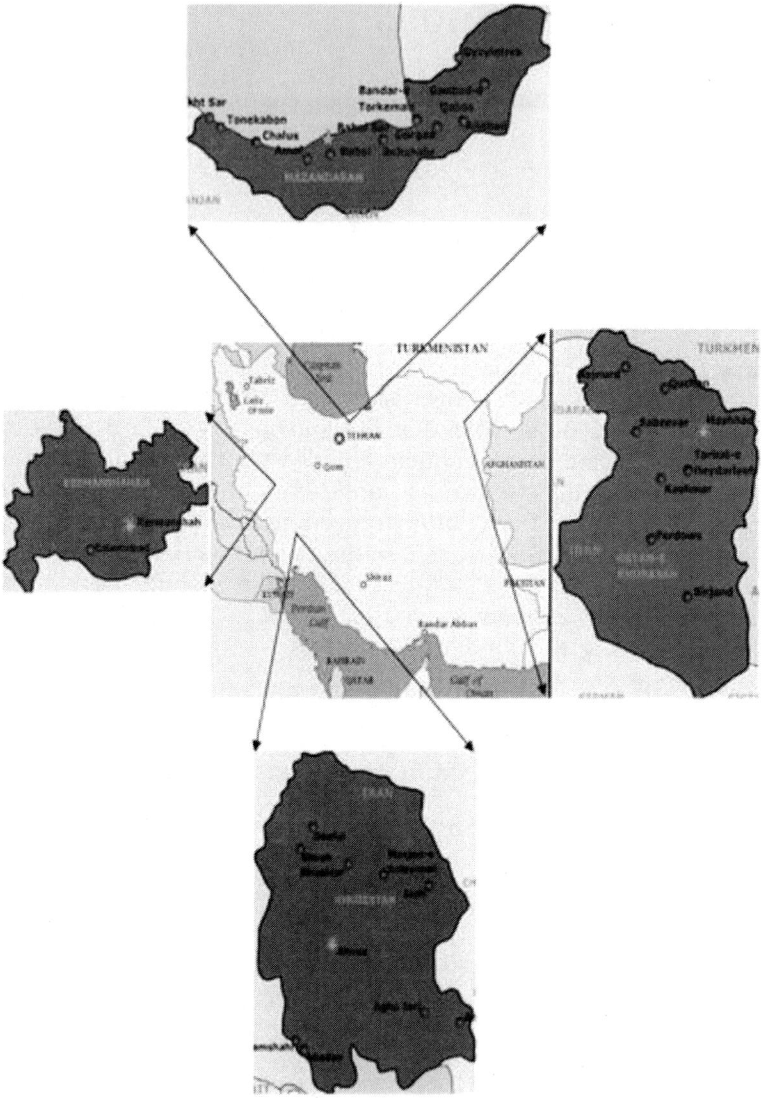

Figure 1. Position of synoptic stations in Iran.

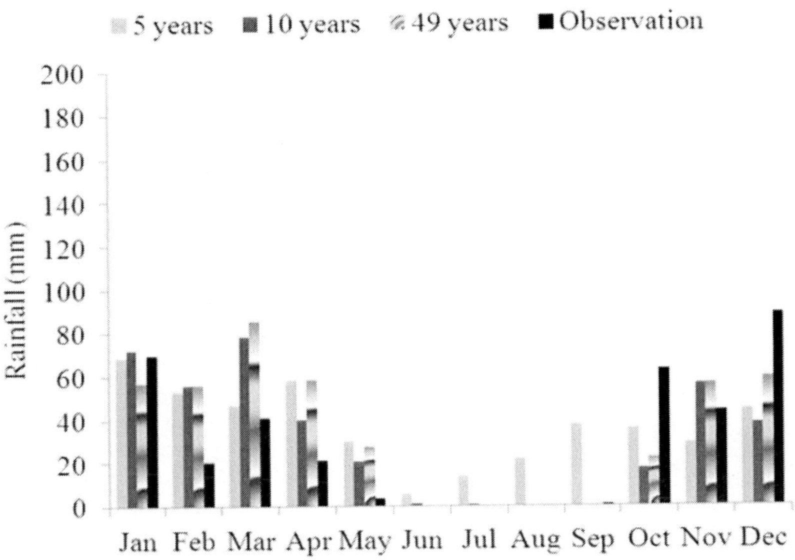

Figure 2. Amount of forecasted rainfall for the Kermanshah synoptic station.

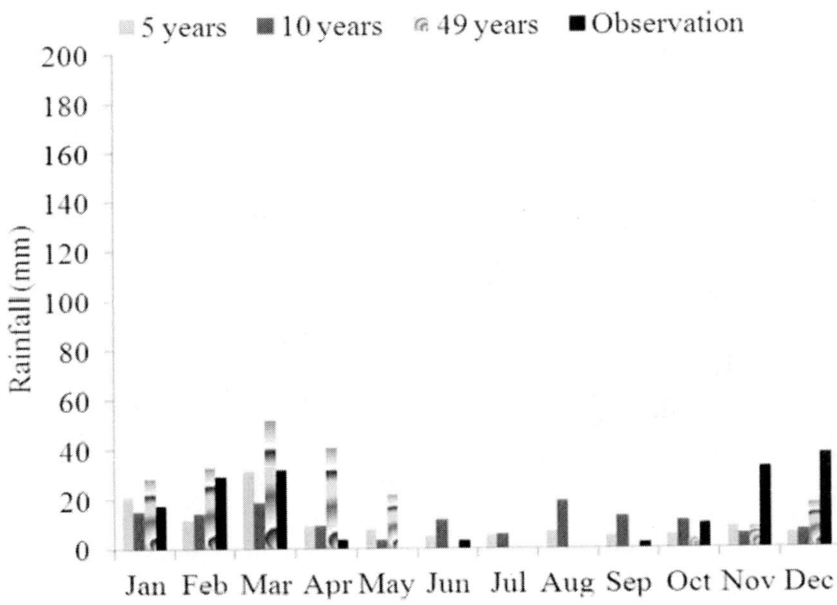

Figure 3. Amount of forecasted rainfall for the Mashhad synoptic station.

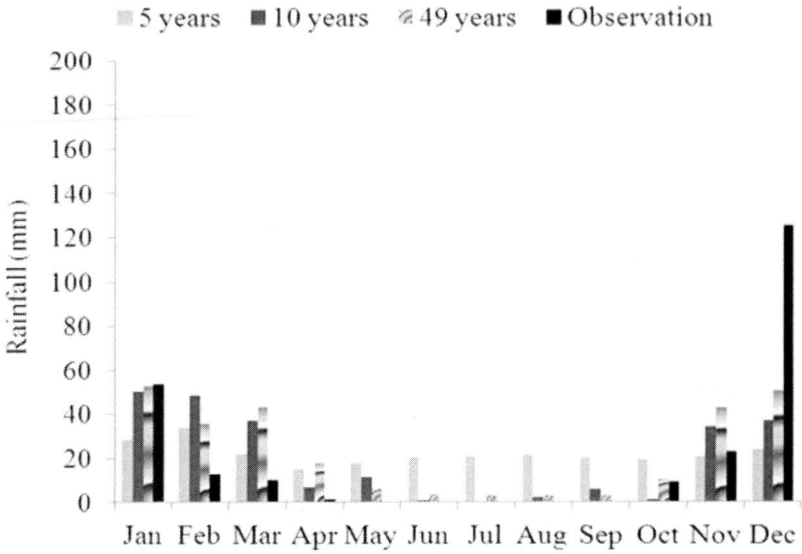

Figure 4. Amount of forecasted rainfall for the Ahvaz synoptic station.

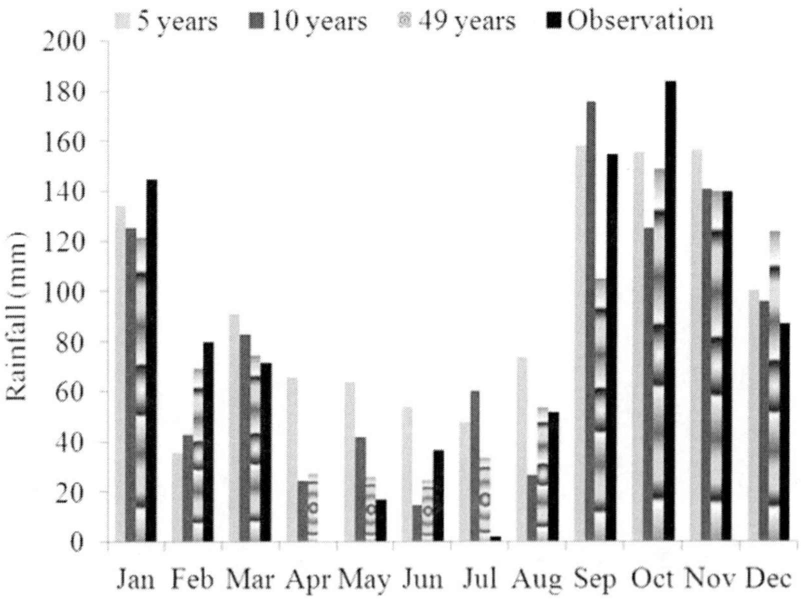

Figure 5. Amount of forecasted rainfall for the Babolsar synoptic station.

Table 2. Obtained R^2 for four different stations according to the amount of observation data.

Station	5 Years (60 Data)	10 Years (120 Data)	49 Years (588 Data)
Kermanshah	0.81	0.81	0.81
Mashhad	0.96	0.96	0.96
Ahvaz	0.77	0.83	0.87
Babolsar	0.70	0.73	0.82

According to the obtained results for R^2 in Table 2, in a temperate climate (Kermanshah), due to low changes of monthly rainfall over the long term, 60 observation data was sufficient for forecasting. In this station, the R^2 was 0.81, and an increase in observation data is not recommend.

In a semi-arid climate (Mashhad), due to the low monthly rainfall relative to the Kermanshah and Babolsar stations and the lack of torrent precipitation in arid climates (for example, December in Figure 4), the highest R^2 was obtained.

In the Mashhad station, data from 5 years prior was sufficient for forecasting, and this is a good result for deficit data situations. However, according to Table 2 and Figure 4 and Figure 5, with 60 observation data, the accuracy of forecasted rainfall was not ideal.

The Ahvaz station, with an arid climate due to drought periods and torrent precipitation, needs at least 588 data for accurate forecasting ($R^2 = 0.87$ rather than $R^2 = 0.77$)

The Babolsar station, with a humid climate due to flood periods in some years, depends on an increase in observation data if rainfall forecasting is to improve in accuracy. A difference in R^2 from 49 years of observation data accrued in the Babolsar synoptic station, compared with only 5 years of observation data ($R^2 = 0.82$ rather than $R^2 = 0.70$) was observed.

4. CONCLUSIONS

This paper reports the study of the effect of the length of the recorded data for monthly rainfall forecasting. Monthly rainfall data for three periods of 5, 10, and 49 years were collected from Kermanshah, Mashhad, Ahvaz, and Babolsar stations and used for calibration time series models. Then, the accuracy of forecasting models was investigated with the following year's data. The following was concluded:

1. In temperate and semi-arid climates, 60 observation data is sufficient for the following year's rainfall forecasting.
2. The accuracy of the time series models increased with increasing amounts of observation data of arid and humid climates.
3. Time series models are appropriate tools for forecasting monthly rainfall forecasting in semi-arid climates.
4. Determining the most critical rainfall month in each climate condition for agriculture schedules is a recommended aim for future studies.

CONFLICTS OF INTEREST

The authors declare no conflict of interest.

REFERENCES

1. Box, G.E.P.; Jenkins, G.M. Series Analysis Forecasting and Control; Prentice-Hall Inc.: London, UK, 1976.
2. Khoshravesh, M.; Gholami Sefidkouhi, M.A.; Valipour, M. Estimation of reference evapotranspiration using multivariate fractional polynomial, Bayesian regression, and robust regression models in three arid environments. Appl. Water Sci. 2015.
3. Valipour, M.; Singh, V.P. Global Experiences on Wastewater Irrigation: Challenges and Prospects. In Balanced Urban Development: Options and Strategies for Liveable Cities; Maheshwari, B., Singh, V.P., Thoradeniya, B., Eds.; Springer: Bern, Switzerland, 2016; pp. 289–327.
4. Valipour, M.; Gholami Sefidkouhi, M.A.; Raeini-Sarjaz, M. Selecting the best model to estimate potential evapotranspiration with respect to climate change and magnitudes of extreme events. Agric. Water Manag. 2017.
5. Valipour, M.; Gholami Sefidkouhi, M.A.; Khoshravesh, M. Estimation and trend evaluation of reference evapotranspiration in a humid region. Ital. J. Agrometeorol. 2017, in press.
6. Valipour, M. Future of agricultural water management in Africa. Arch. Agron. Soil Sci. 2015, 61, 907–927.
7. Valipour, M. Calibration of mass transfer-based models to predict reference crop evapotranspiration. Appl. Water Sci.2015.
8. Valipour, M. Analysis of potential evapotranspiration using limited weather data. Appl. Water Sci. 2015.
9. Valipour, M. Increasing irrigation efficiency by management strategies: Cutback and surge irrigation. ARPN J. Agric. Biol. Sci. 2013, 8, 35–43.
10. Valipour, M. Necessity of Irrigated and Rainfed Agriculture in the World. Irrig. Drain. Syst. Eng. 2013, S9, e001.
11. Valipour, M. Evolution of Irrigation-Equipped Areas as Share of Cultivated Areas. Irrig. Drain. Syst. Eng. 2013, 2, e114.
12. Valipour, M. Use of surface water supply index to assessing of water resources management in Colorado and Oregon, US. Adv. Agric. Sci. Eng. Res. 2013, 3, 631–640.

13. Valipour, M. Hydro-Module Determination for Vanaei Village in Eslam Abad Gharb, Iran. ARPN J. Agric. Biol. Sci.2012, 7, 968–976.
14. Valipour, M. Ability of Box–Jenkins Models to Estimate of Reference Potential Evapotranspiration (A Case Study: Mehrabad Synoptic Station, Tehran, Iran). IOSR J. Agric. Vet. Sci. 2012, 1, 1–11.
15. Valipour, M. A Comparison between Horizontal and Vertical Drainage Systems (Include Pipe Drainage, Open Ditch Drainage, and Pumped Wells) in Anisotropic Soils. IOSR J. Mech. Civil Eng. 2012, 4, 7–12.
16. Valipour, M. Application of new mass transfer formulae for computation of evapotranspiration. J. Appl. Water Eng. Res. 2014, 2, 33–46.
17. Valipour, M.; Montazar, A.A. An Evaluation of SWDC and WinSRFR Models to Optimize of Infiltration Parameters in Furrow Irrigation. Am. J. Sci. Res. 2012, 69, 128–142.
18. Serinaldi, F.; Kilsb, C.G. A modular class of multisite monthly rainfall generators for water resource management and impact studies. J. Hydrol. 2012, 464, 528–540.
19. Luc, K.C.; Ball, J.E.; Sharma, A. An application of artificial neural networks for rainfall forecasting. Math. Comput. Model. 2001, 33, 683–693.
20. Ludlow, J.; Enders, W. Estimating non-linear ARMA models using Fourier Coefficients. Int. J. Forecast. 2000, 16, 333–347.
21. Baareh, A.K.M.; Sheta, A.F.; Khnaifes, K.A. Forecasting River Flow in the USA: A Comparison between Auto Regressive and Neural Network Non-Parametric Models. J. Comput. Sci. 2006, 2, 775–780.
22. Mohammadi, K.; Eslami, H.R.; Dardashti, S.D. Comparison of Regression ARIMA and ANN Models for Reservoir Inflow Forecasting using Snowmelt Equivalent (A Case Study of Karaj). J. Agric. Sci. Technol. 2005, 7, 17–30.
23. Mohammadi, K.; Eslami, H.R.; Kahawita, R. Parameter estimation of an ARMA model for river flow forecasting using goal programming. J. Hydrol. 2006, 331, 293–299.
24. Ramírez, M.C.V.; de Campos Velho, H.F.; Ferreira, N.J. Artificial neural network technique for rainfall forecasting applied to the São Paulo region. J. Hydrol. 2005, 301, 146–162.
25. Rezaei, M.; Valipour, M.; Valipur, M. Modelling evapotranspiration to increase the accuracy of the estimations based on the climatic parameters. Water Conserv. Sci. Eng. 2016.
26. See, L.; Abrahat, R.J. Multi-model data fusion for hydrological forecasting. Comput. Geosci. 2001, 27, 987–994.
27. Srinivas, V.V.; Srinivasan, K. Post-blackening approach for modeling dependent annual streamflows. J. Hydrol. 2000, 230, 86–126.
28. Toth, E.; Brath, A.; Montanari, A. Comparison of short-term rainfall predication models for real-time flood forecasting. J. Hydrol. 2000, 239, 132–147.
29. Valipour, M. Long-term runoff study using SARIMA and ARIMA models in the United States. Meteorol. Appl. 2015, 22, 592–598.
30. Valipour, M. Temperature analysis of reference evapotranspiration models. Meteorol. Appl. 2015, 22, 385–394.
31. Valipour, M. Use of average data of 181 synoptic stations for estimation of reference crop evapotranspiration by temperature-based methods. Water Resour. Manag. 2014, 28, 4237–4255.
32. Valipour, M. Evaluation of radiation methods to study potential evapotranspiration of 31 provinces. Meteorol. Atmos. Phys. 2015, 127, 289–303.
33. Valipour, M.; Eslamian, S. Analysis of potential evapotranspiration using 11 modified temperature-based models. Int. J. Hydrol. Sci. Technol. 2014, 4, 192–207.
34. Anctil, F.; Perrin, C.; Andréassian, V. Impact of the length of observed records on the performance of ANN and of conceptual parsimonious rainfall-runoff forecasting models. Environ. Model. Softw. 2004, 19, 357–368.
35. Andrieu, H.; French, M.N.; Thauvin, V.; Krajewski, W.F. Adaptation and application of a quantitative rainfall forecasting model in a mountainous region. J. Hydrol. 1996, 184, 243–259.

36. Balaguer, E.; Palomares, A.; Sorie, E.; Martin-Guerrero, J.D. Predicting service request in support centers based on nonlinear dynamics, ARMA modeling and neural networks. Expert Syst. Appl. 2008, 34, 665–672.
37. Burlando, C.; Rosso, R.; Cadavid, L.G.; Salas, J.D. Forecasting of short-term rainfall using ARMA models. J. Hydrol.1993, 144, 193–211.
38. Chattopadhyay, S.; Chattopadhyay, G. Univariate modelling of summer-monsoon rainfall time series: Comparison between ARIMA and ARNN. C. R. Geosci. 2010, 342, 100–107.
39. Chenoweth, T.; Dowling, K.; Hubata, R.; Louis, R. Automatic ARMA identification using neural networks and the extended sample autocorrelation function: A reevaluation. Decis. Support Syst. 2004, 29, 21–30.
40. Chenoweth, T.; Dowling, K.; Hubata, R.; Louis, R. Distance and prediction error variance constraints for ARMA model portfolios. Int. J. Forecast. 2004, 20, 41–52.
41. Han, P.; Wang, P.X.; Zhang, S.Y.; Zhu, D.H. Drought forecasting based on the remote sensing data using ARIMA models. Math. Comput. Model. 2010, 51, 1398–1403.
42. Hu, W.; Tong, S.; Mengersen, K.; Oldenburg, B. Rainfall, mosquito density and the transmission of Ross River virus: A time-series forecasting model. Ecol. Model. 2006, 196, 505–514.
43. Jia, Y.; Culver, T.B. Bootstrapped artificial neural networks for synthetic flow generation with a small data sample. J. Hydrol. 2006, 331, 580–590.
44. Karamouz, M.; Araghinejad, S.H. Advance Hydrology; Amirkabir University of Technology Press: Tehran, Iran, 2012.
45. Wei, C.; Hung, W.C.; Cheng, K.S. A multi-spectral spatial convolution approach of rainfall forecasting using weather satellite imagery. Adv. Space Res. 2006, 37, 747–753.
46. Xiong, L.; O'Connor, K.M. Comparison of four updating models for real-time river flow forecasting. Hydrol. Sci. J.2002, 47, 621–639.
47. Yannopoulos, S.I.; Lyberatos, G.; Theodossiou, N.; Li, W.; Valipour, M.; Tamburrino, A.; Angelakis, A.N. Evolution of Water Lifting Devices (Pumps) over the Centuries Worldwide. Water 2015, 7, 5031–5060.
48. Yu, X.; Liong, S. Forecasting of hydrologic time series with ridge regression in feature space. J. Hydrol. 2006, 332, 290–302.

CHAPTER 4

Sustainable Water Resources Management for Food Security in Kenya: Case of Bwathonaro Catchment

Chris A. Shisanya*, Simon M. Onywere, Joy A. Obando

Chris A. Shisanya, Department of Geography, Kenyatta University, Nairobi, Kenya

ABSTRACT

Food security is important in Kenya today. Bwathonaro sub-catchment in Tana catchment is experiencing food insecurity and severe water related challenges due to high population pressure and over-reliance on irrigated agriculture for livelihood. Bwathonaro river supports irrigated agriculture, which is a major source of livelihood for people in the watershed. These challenges include: water pollution, inappropriate solid waste disposal, illegal water abstraction and/or over-abstraction of water, inefficient irrigation practices, encroachment on and drainage of wetlands as well as inappropriate use of agrochemicals in wetlands. The specific objectives of the study were to 1) document current status of water resource use; 2) propose a strategy for enhancing subsistence food crop farming within the existing khat (miraa) farming system; 3) promote efficient irrigation systems that reduce water use conflict and enhance equitable distribution, and 4) work with community to improve water quality and quantity. The study adopted a four-pronged approach in the data collection procedures, namely: socio-economic surveys, participatory approaches, biophysical data and geospatial tools. Results showed that water use conflicts are persistent in the catchment. However, one Water Resource User Association (WRUA)-BWARUA, has shown consistent progress towards managing water resources for food security. The WRUA has undertaken a number of activities including ensuring that water abstractors observe permit conditions, undertaking surveillance of polluters, awareness creation on protection of wetlands, springs and river banks among other activities in the catchment. The impact of these activities has been positive and includes: reduction of conflict over water, reduction in pollution, more water flowing downstream and more water availability in the catchment for diverse uses including irrigation for agriculture, livestock and wildlife. Such practices could be used in other sub-catchments of Kenya as showcases to facilitate the sustainable use of scarce water resources for the benefit of all stakeholders.

Keywords: Bwathonaro, Food Security, Water Resources, Watershed Management

1. INTRODUCTION

The [1] states explicitly that the world has a water crisis. Poverty is seen as both a symptom and a cause of the water crisis. One of the Millennium Development Goals (MDGs) was to halve the proportion of people without access to clean and safe drinking water. Africa holds 11% out of the available fresh water which supports 13% of the world population in the continent. Water is a key driver of sustainable growth and poverty alleviation, and an essential input to production-agriculture, industry, energy, transport and is a basic requirement for healthy people and ecosystems [2] . Fighting global poverty and reducing hunger are at the height of the global development effort, as seen in the MDGs. High population growth occurs in developing countries which are home to close to one billion malnourished people [3] . In the developing countries, 70% of the poor live in rural areas where productions rely on rain-fed agriculture. Freshwater availability is seen as a limiting factor in food production and livelihood improvement. Meeting the MDGs particularly in poverty-stricken, water scarcity prone regions such as the semi-arid and dry sub-humid savannah regions in Africa were challenging.

In Kenya and since independence, poverty alleviation and reduction of inequalities have been some of the Government's major development objectives with the government putting in place a number of policies and programmes, which emphasize rapid economic growth. This is reflected in various govern- ment policy documents such as development plans, sessional papers and strategy papers. After four decades of trying to implement the development plans, the social and economic inequalities still persist. Poverty has increased from 3.7 million people affected in 1973 to 15 million people at the turn of this century. The elaborate 1997 welfare and monitoring survey in Kenya indicated that 53% of people in rural areas were categorized as overall poor and 51% as food poor [4] . The poor are clustered into certain socio-economic categories that include small-scale farmers, pastoralists, agricultural and casual labourers, unskilled and semiskilled workers, female-headed rural households, the physically handi-capped, HIV/AIDS orphans and street children [4] . Most people in these clusters are landless and lack of formal education.

Poverty situation in Kenya has spawned a wide gap between the rich and the poor. Statistics from Central Bureau of Statistics (CBS) show that income and social services are skewed in favour of the rich [5] . Inequalities are manifested in many forms with the country's 10% of the population controlling 42% of the total wealth while the bottom 10% control less than 1% [5] . Further, regional inequalities within the country are extreme with for example the difference in life expectancy between Central and Nyanza Provinces being a staggering 19 years [6] .

Research has identified a strong link between poverty in Kenya and lack of access to improved water supply and sanitation (WSS). Various key documents, notably the Participatory Poverty Assessment Study [7] ; the National Poverty Eradication Plan [8] ; the National Policy on Water Resources Management [9] and the Human Development Report on Addressing Social and Economic Disparities [10] have articulated this link and have set ambitious targets aimed at increasing access to resources. The planning approach adopted by the Poverty Reduction Strategy Plan [4] is that of bottom up with full involvement and participation of the stakeholders at the grassroots level. The reports and documents outline the scope of poverty and set targets for poverty reduction. They recognise that two-thirds Kenya is arid and semi arid and as a result, only about 160,000 km^2 of land, most of which is situated in the wetter southwest area, is suitable for crop production that supports the current population of approximately 33 million. Future projections show that by the year 2025, per capita water availability will drop to 235 m^3 from the current 650 m^3 as a result of population growth [11] [12] . The level of water scarcity in some regions of the country has become a serious limiting factor for development activities. Consequently and since 2002, there has been a major reform in the water sector with clear roles for the different actors involved in the decentralized institutional framework that separates policy formulation from regulation and water services provision [11] . Most of the interventions measures put in place by the Government are however donor driven and are not integrated into the country's long-term budgetary and development goals, and consequently, they suffer from discontinuity and inconsistency.

Kenya with an estimated population of 32 million (in 2003) and a projected population of 43 million by the year 2015 faces enormous challenges in the management of its limited water resources. These challenges comprise:

- Water scarcity and effects of climate variability;
- Catchment degradation and underdevelopment of the available water;
- Poorly managed water resources assessment and monitoring programmes;
- Pollution and degradation of water resources;
- Trans-boundary water resources management;
- HIV/AIDS and underdevelopment of water resources.

The magnitude of the above issues and challenges and the severity of the water crisis that face Kenya, impact across most sectors of the economy, making water resources management a high priority area for the government. In recognition of the above issues and challenges, the water sector is being transformed in line with the national policies as outlined in the Water Act 2002 [13] ; the Sessional Paper No. 1 of 1999 on National Water Policy on Water Resources Management and Development [8] ; the National Poverty Reduction Strategy Paper of 2003 [14] ; and the Economic Strategy Paper for Wealth and Employment Creation [14] . The Catchment Management Strategy (CMS) is a tool that creates the framework for the management of the water resources and related land resources in the catchment and it outlines how the concept of integrated water resources management can be implemented at the catchment level.

In Bwathonaro watershed, the typical challenges that this project sort to address can be summarised as:

- Water pollution from washing, bathing and watering of animals directly in the river;
- Pollution of water from point sources (abattoirs, toilets, market centres, etc.);
- Inappropriate solid waste disposal in market centres;
- Illegal water abstraction and over-abstraction of water;
- Inefficient irrigation practices relaying on open furrow system that lead to water loss to underling porous formations;

- Encroachment on and drainage of wetlands/Inappropriate use of agro-ch- emicals within the wetlands;
- Introduction of eucalyptus species that are "high water spenders";
- Severe soil erosion on the farms, footpaths and roadsides;
- Lack of information on the water resource (water quantity, quality, rainfall, water use, sediment yield);
- Periodic human and wildlife conflicts;
- Dependence on a single external source of water.

With the above challenges this study aimed to assure availability of water resources for the well being of communities in the Bwathonaro watershed through catchment protection and to ensure availability of good quality water and sustainable management of the environment. The study carried out an integrated watershed management plan with BWARUA towards an environmentally and socio-economically healthy watershed that benefits the communities living within the Bwathonaro watershed. The goal of this research was to enhance water resource management as a national food security strategy in Bwathonaro watershed, Meru County. The objectives guiding this study were: 1) to document the current status of water resource use in Bwathonaro watershed, 2) to propose a strategy for enhancing subsistence food crops within the existing khat (miraa) farming system in the Bwathonaro watershed, 3) to promote efficient irrigation systems that will reduce conflict over water resources and increase equitable distribution in the watershed, and 4) to work with the community and put in place mechanisms to improve water quality and quantity.

2. MATERIALS AND METHODS

2.1. The Study Area

Geographically, the Bwathonaro watershed is bounded by latitude 37°53'00"E and 38°05'00"E and 0°12'00"N and 0°21'00"N with the Nyambene Hills as its catchment (Figure 1). The population is estimated at 46,000 people, who are mainly agriculturalists. The altitude ranges from 2060 m a.s.l. in Ngawa Hills to 700 m a.s.l. at the park area. The climatic conditions of the catchment range from humid to semi-arid. Rainfall is bimodal coming during the long rainy season of March to May and the short rainy season from October to December and varies from about 1300 mm at the hill tops to 600 - 700 mm in the plains around Amwathi [15] . The months of January to February and August to October are dry and very challenging with the area experiencing decreased water availability.

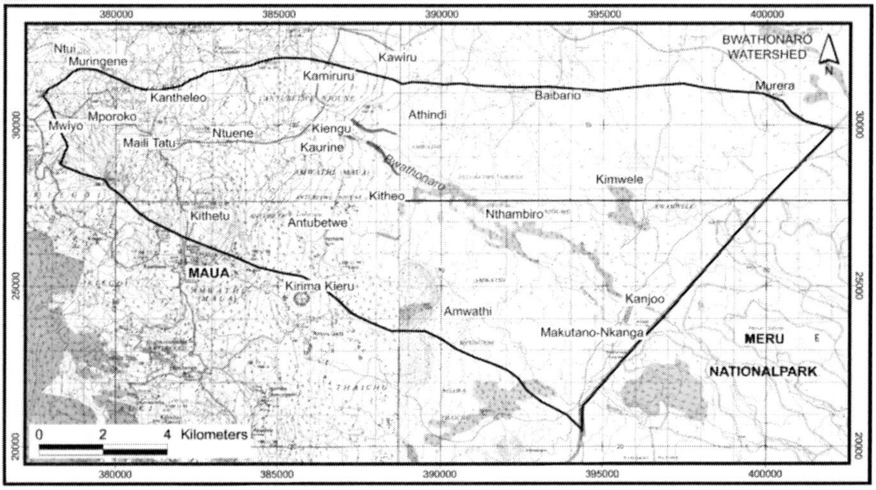

Figure 1. Bwathonaro watershed.

The geology of the catchment comprises mainly of basaltic volcanic rocks with tuffs and pyroclastics of Pleistocene Age. They are young geologically and of clastic nature and have poorly developed soils. Because of climatic variation with altitude, variable soils have developed. The soils are poorly consolidated and with the combination of steep slopes are highly susceptible to erosion. The sources of the drainage into the Bwathonaro catchment contain the porous rock surfaces which allow water to percolate into the deeper horizons underground. Because of this fact irrigation activities loose water from the open furrow structures developed for that purpose. The ground water table is very deep.

Livestock rearing is another important activity in the catchment and includes the keeping of cattle, goat, sheep, donkey and chicken. Donkeys are also an important form of transport, particularly for firewood, water and other goods to the market within the catchment. Market centres such as Muringene, Nthambiro, Kiengu and Kanjoo have grown mainly through miraa trade. The main source of energy for cooking in the homesteads within the catchment is firewood. The demand for firewood has lead to depletion of indigenous tree species, scarcity of wood resources, deforestation, high prices and introduction of eucalyptus species. Additionally, women are forced to walk longer distances to collect firewood.

The data in Table 1 provides a summary of Bwathonaro catchment characteristics.

2.2. Methods

2.2.1. Primary and Data Collection

The research adopted a four pronged approach in the data collection procedures. These included socio-economic surveys, participatory approaches, biophysical data and geospatial data. A socio-economic baseline survey was being undertaken in order to document the current status of water resource use in Bwathonaro watershed. This was linked to the existing livelihood strategies and the interventions to food security in the project area. The major concern was the management of the water resources in relation to other natural resources that impact on food production. The local stakeholders were involved in the identification of constraints, the definition and the selection of potential solutions concerning the water resource use in the watershed. The local community were involved in the study through participatory research methods, which included key informants, in-depth and household interviews and focus group discussions (FGDs). These participatory approaches helped to capture the issues related to water resource use and food insecurity from the participants view points.

Modern geospatial tools of Remote Sensing and Geographical Information Systems were used in mapping and assessing the resources in the watershed which impact food security. This was done in consultation with the Regional Centre for Mapping of Resources for Development at Kasarani, Nairobi. Site surveys were carried out on the physiography, geology, geomorphology, soils, hydrology and climate. In addition, information was obtained from relevant government departments. A geographical positioning system (GPS) was used in locating the various natural resources particularly water within a spatial context.

2.2.2. Secondary Data Collection

Secondary data supplemented the primary data. These were derived from various sources such as desk study, internet, FBO, CBO, NGO and government policy and reports. Other secondary data were obtained from the aerial photographs, topographic, geological, hydrological maps as well as satellite imagery.

Table 1. Summary of the key features of the Bwathonaro watershed.

Catchment area	149 km^2
Annual average rainfall	~900 mm
Surface water abstractions rates.	Not known
Groundwater abstractions rates.	Not known
Average borehole yield	Not known
Borehole specific capacity	Not known
Hydropower production	Nil
Average discharge during low flow	Not known
Altitude	750 - 1800 m a.s.l.
Average temperature	18°C
Water demand	Not known
Estimated water demand by 2030	Not known
Population	46,000
Population density	350 persons/km^2

3. RESULTS AND DISCUSSIONS

3.1. Socio-Economic Activities

The main economic activities in the catchment are subsistence and commercial farming. The commercial farming mainly involves the growing of khat (miraa), macadamia nuts, coffee, tea and horticulture. Subsistence farming involves the growing of maize, beans, potatoes, arrow roots, bananas, sugar cane, sweet potatoes, cassava, yams, and kales. Sorghum, millet and pigeon peas are grown mainly in the Lower Zone. In this area miraa is grown through irrigation during the dry season. Since the Lower Bwathonaro catchment is the only area where cultivation of miraa in the dry season is possible, market prices for miraa during the months of July, August, September and October are very high, making it a very profitable activity. This leads to conflicts among the miraa growers and other water users.

Livestock rearing is another important activity in the catchment and includes cattle, goat, sheep, donkey and chicken. Donkeys are also an important form of transport, particularly for firewood, water and other goods for the market. Markets such as Muringene, Nthambiro, Kiengu and Kanjoo have been developed mainly for the trading of miraa. The main source of energy for cooking in the homesteads within the catchment is firewood. The demand for firewood has lead to depletion of indigenous tree species, scarcity of wood resources, deforestation, high prices and introduction of eucalyptus species. Additionally, women are forced to walk longer distances to collect firewood. Due to water scarcity in parts of the catchment, women and children are forced to walk long distances (up to 5 km) to fetch water. This has implications on other productive sectors of the local economy due to loss of precious time for other activities. It also has health implications on the young girls, which often develop contracted pelvis which affect their reproductive systems.

3.2. Challenges in Bwathonaro Catchment

The following water management related problems were identified in the Bwathonaro catchment:

1) Water pollution from washing, bathing and watering of animals directly in the river;
2) Pollution of water from point sources (abattoirs, toilets, market centres etc.);
3) Inappropriate solid waste disposal in market centres;
4) Illegal water abstraction and over-abstraction of water;
5) Inefficient irrigation practices;
6) Encroachment on and drainage of wetlands/Inappropriate use of agro-ch- emicals in wetlands;
7) Introduction of eucalyptus species;
8) Soil erosion on the farms, footpaths and roadsides;
9) Lack of water resource information (water quantity, quality, rainfall, water use, sediment yield);
10) Human and wildlife conflicts.

Dependence on a single external source of water.

3.3. Satellite Image Analysis Results

From the assessment of the image data of the project area, one full scene of Landsat Thematic Mapper (TM) and Enhanced Thematic Mapper (ETM) path/ raw p168r060 adequately covered the study area.

The images met Landsat sensor parameters and area coverage (185 × 185 km^2).

- Focus area of study: Subsets of Bwathonaro Catchment;
- Issues of concern: Land use and settlement trends at the catchment, vulnerability of the communities to flood hazards and other impacts, sustainability of the livelihood systems based on land use policy and legislative framework governing the lives of the riparian communities;
- Vision: to utilize existing data for planning for sustainable management of the Bwathonaro watershed and wetland resources to improve the livelihood of the local communities and address the issue of poverty.

Image data sets were acquired from US Geological Survey (USGS) through Regional Centre for Mapping of Resources for Development (RCMRD), Nairobi and used for the interpretation of the status of land cover in Bwathonaro catchment and included:

Year: 1976	Year: 1987	Year: 2005	Year: 2003
MSS L2_p180r60_2m1 9760211	TM L5_p168r60_5t 870225	ETM L7_p168r060_7k2 0000221	ETM L7_1168060_060 20031214

- DTM Data from Active Sensors-Shuttle Radar Topographic Mission;(SRTM);
- Landsat mosaic FCC bands 432 in MrSID format; Mosaic ID Zone N/S-37_loc;
- Scanned 1:50,000 scale topographic maps for the larger Nyambene hills area (sheets 107_2,4 zone 37, 108_1,2,3,4-Zone 37, 109_1,2,3,4-zone 37, 121_2,4-zone 37, 122_1,2,3,4-zone 37, 123_1,3 zone 37);

The Landsat TM (7 bands) and ETM (9 bands) images have spatial resolution of 28 m - 30 m with the thermal band at 80m-resolution and covers 185 × 185 km. The Landsat MSS (4 bands) image has 80m resolution. Currently Landsat 7 is in operation but one of its scanning mirrors has malfunctioned and therefore the current data comes with strip errors that though can be corrected, the images can only be used for visual interpretation.

The Digital Terrain Model (DTM) from Active Sensors-Shuttle Radar Topographic Mission. (SRTM) data is valuable for knowledge of earth surface topography and is of major importance to Earth Sciences, e.g. hydrology, geomorphology. The data was obtained free of charge from:

http://seamless.usgs.gov/ and ftp://edcsgsg.cr.usgs.gov/pub/data/srtm/. The current status of SRTM DTM for Africa is the first continuous large-scale homogeneous product and at 90-meter vertical resolution with a relative vertical accuracy of less than 10 meters, the data offers a great improvement compared to other existing global DTM of 1 km horizontal resolution from USGS.

All the Landsat satellite data had all the image bands and are in Geotiff Format. The reference projection for the study site is UTM zone 37, Datum: WGS84, Map Units: Meters, Central Meridian : 39°, Reference Latitude: Equator (0°), Scale Factor at Origin: 0.9996, False Easting 500,000, False Northing: 0. Preprocessing and identification of the band combinations that can give the best visual interpretation of the image subsets has been done. An example of the various image band combinations for the larger environment of the Bwathonaro catchment subset of scene p168r060 is shown in Figure 2.

3.4. Geo-Referencing

The satellite image of 1976, 1987, 2000 and 2003 geo-referenced against scanned topographic map at a scale of 1/50,000 using a number of control points. The data was written to the Universal Transverse Mercator (UTM) with WGS84 as datum projection. Second order polynomial transformation and nearest neigh- bour re-sampling method was be used for this process.

3.5. Image to Image Registration for the Year 1987, 2000, and 2003 Images

The satellite images of 1987, 2000, and 2003 were co-registered to one another. This process, called image to image registration, is usually used for time-series image analysis. It ensures that the pixel grids of the images of the 1987, 2000, and 2003 (all with 30 m spatial resolution) conform to one another hence enabling pixel by pixel comparison of the images. Since the Landsat Multi-spectral Scanner (MSS) image of 1976 image has a resolution of 80m it could not be utilized for similar registration but never the less the data was used to assess the status of land cover in that period. In this study, the satellite images were composited to a false colour composites FCC using band 4 (red), band 3 (green) and band 2 (blue). In this band combination the red colour represents dense vegetation, whilst the green/blue colour will mean very scattered vegetation cover or no vegetation cover or bare ground, in a Pseudo-colour composite of bands.

The Bwathonaro Watershed Landsat image 11th Feb 1976

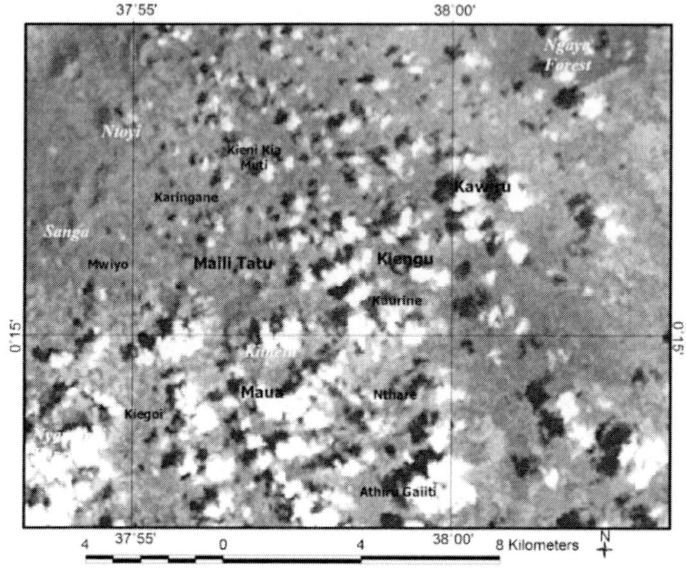

(a)

The Bwathonaro Watershed Landsat image 25th Feb 1987

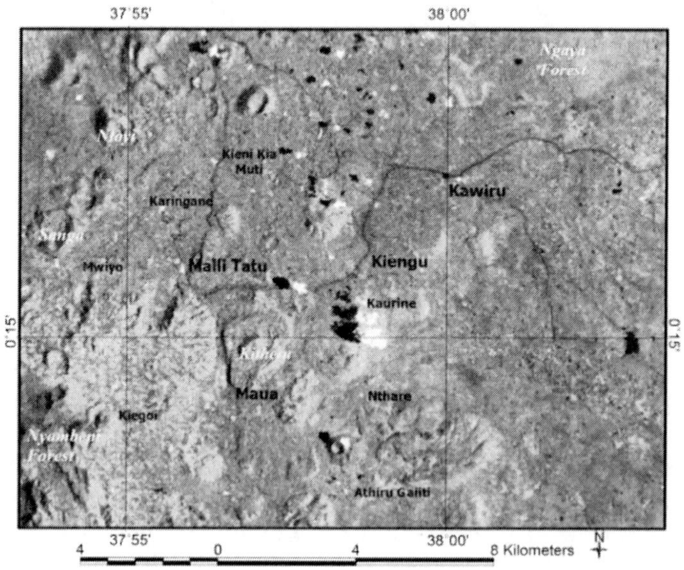

(b)

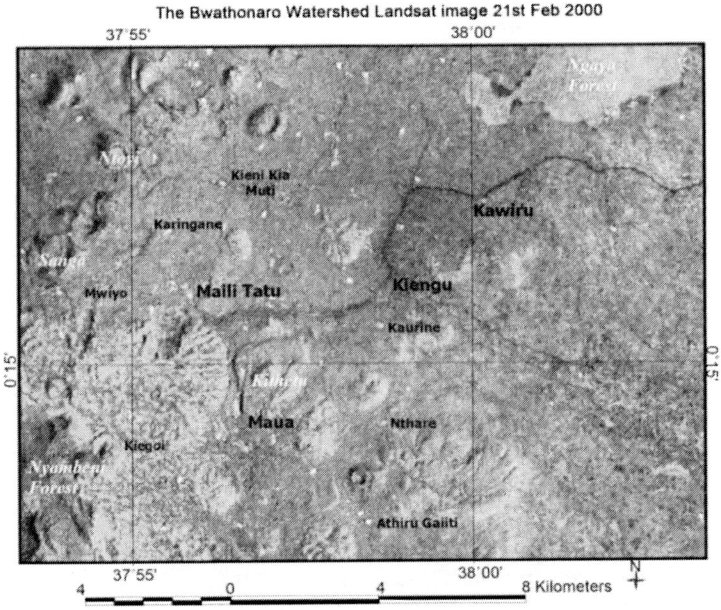

(c)

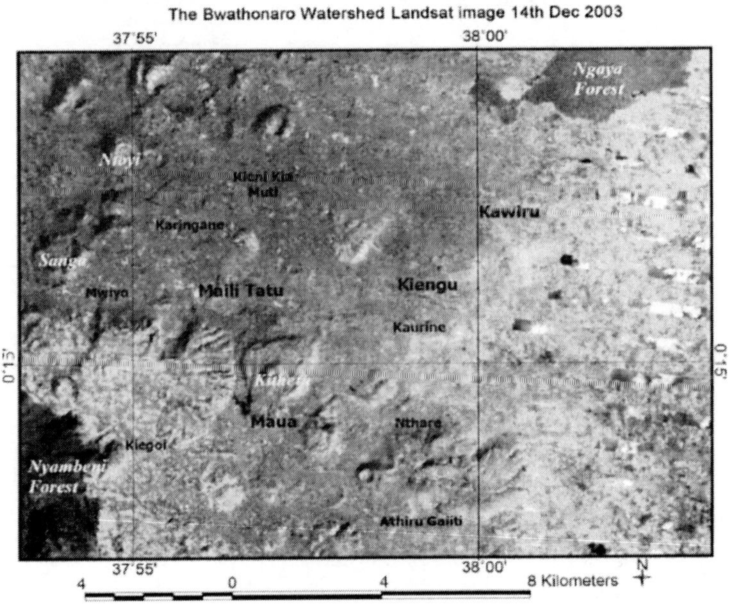

(d)

Figure 2. (a)-(d) the result of different band combinations of Landsat ETM data of the 2000 image and displayed using red, green blue colour redentions.

3.6. Image Interpretation

The key areas of the image interpretation included mapping forest cover and its changes and assessment of impacts of the forest loss on the local rainfall patterns and stream discharge. Remotely sensed data based on the pseudo-colour image composited using bands 742 RGB where vegetation appeared in green was used for interpretation because the colour composite is understood by the Bwathonaro community as it represents the colour of the natural vegetation. The FCC was used to map the forest cover and its changes between 1976 and 2008.

The image analysis was done at two levels: mapping forest cover and its changes and assessment of impacts of forest cover loss on the hydrological processes between 1976 and 2008. Satellite images of 1976, 1987, 2000, and 2003 from Landsat were used for detection of changes in the forest cover. The results are shown in Figure 3 and Table 2.

Table 2. Land cover types in Bwathonaro watershed.

Land cover classes in Bwathonaro Sub-catchment	Area in Km^2	Percentage of the total area (%)
Bare ground	7.31	1.11
Closed grassland	20.682	3.15
Closed shrubs	194.779	29.70
Deforested	2.665	0.40
Forest	35.419	5.40
Grassland	97.284	14.83
Indigenous forest	45.434	6.93
Mountain vegetation	1.633	0.25
Open Shrubs with soil	1.062	0.16
Open grassland	34.599	5.28
Open shrubs	135.066	20.60
Plantation forest	0.791	0.12
Riverine trees	12.154	1.85
Rocks	2.588	0.39
Scattered shrubs	19.171	2.92
Shrubs	15.115	2.30
Urban area	2.583	0.39
Very open shrubs	22.182	3.38
Wetland	5.255	0.80
Total area in Km^2	**655.774**	**100**

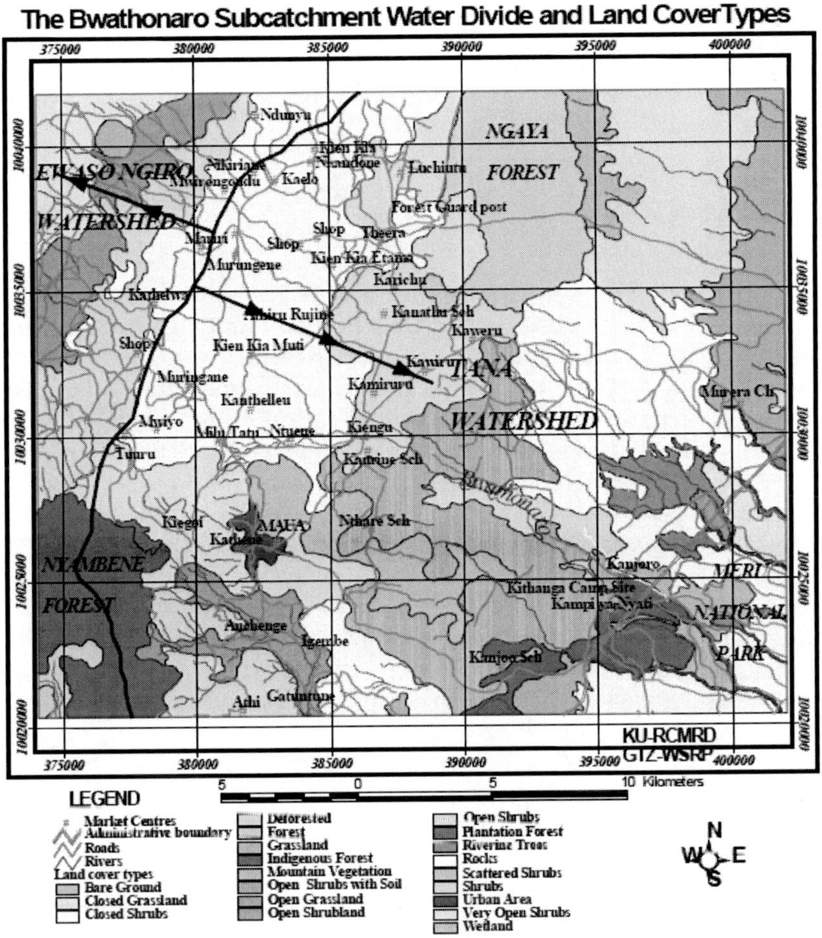

Figure 3. Land cover types in Bwathonaro watershed.

Change detection will be enhanced by conducting field visits to the area to make ground truthing. Rainfall and stream flow data available from existing measurements within the catchment area (such as at weather stations and stream gauges) will be used to illustrate hydrological alterations which will be correlated to forest cover change to establish the type or nature of relationship and the strength of relationship.

The destruction of the forest cover is reflected in the reduced green vegetation cover over the period of the images considered: 1976, 1987, 2000 and 2003. This is threatening the hydrological functions, endangering water supplies for the community in the Bwathonaro watershed. The exposed soils may also cause severe siltation, flooding, affecting dry season flows and disrupting socio-eco- nomic activities such as irrigation. The downstream habitat quality may also be affected considering the increased settlement and encroachment into the riparian environment.

4. CONCLUSION

So far, one Water Resource User Association (WRUA)-BWARUA, that is a partner in this project, established in Bwathonaro Sub-Catchment in Meru Sub-Re- gion, Tana catchment, has shown consistent progress towards managing water resources for food security since its inception in 2005. The WRUA has undertaken a number of activities including ensuring that water abstractors observe permit conditions, undertaking surveillance of polluters of River Bwathonaro, awareness creation on protection of wetlands, springs and river banks among other activities in the catchment. The impact of these activities has been quite positive and they include: reduction of conflict over water, reduction in pollution, more water flowing downstream and more water availability in the catchment for diverse uses including irrigation for agriculture, livestock and wildlife. Such practices could be used in other sub-catchments as showcases to facilitate the sustainable use of scarce water resources for the benefit of all stakeholders.

5. RECOMMENDATIONS

This study recommends the following:

1) With regard to illegal and over abstraction of surface water, there is need to enforce the rules and regulations on water abstraction. This will include empowering Bwathonaro watershed communities to support enforcement of the law, renewal of permits, formation of water user groups and metering and paying for abstracted water.

2) Irrigation will continue to play a major role in the livelihoods of the Bwathonaro watershed communities. Inefficient irrigation practices such as use of unlined open furrows, leaking pipes and uncontrolled watering of Miraa trees is common within the watershed. Farmers will be encouraged to adopt more efficient irrigation practices such as piped conveyance, lined furrows and sprinkler and drip irrigation. Key will be establishment of water structures incentives, regulations and restrictions that will help, guide, influence and co-ordinate the farmers' behaviour for the efficient use of water for irrigation.

3) The wetlands serve to regulate the flow in the rivers and to trap sediment. During the dry season low flows are sustained by wetlands and in the rain season, flash floods are reduced and sediments are trapped. The wetlands in the catchment are threatened by encroachment for farming purpose, drainage and use of agrochemicals. The community should be sensitised on the importance of wetlands and discouraged from utilising them for agricultural activities.

4) Eucalyptus, which has been introduced in the catchment, is known to have high rate of water uptake resulting in reduction of water resources. It also causes deterioration of the soil structure, loss of fertility and hinders growth of other vegetation under its canopy. The introduction of a number and diversity of low water consuming tree species should be encouraged, especially environmentally friendly indigenous trees.

5) Severe soil erosion is prevalent in the catchment leading to deep gullies on the hillsides and roadsides. Farmlands especially on steep slopes suffer from severe erosion. Infrastructures such as roads are damaged, thus limiting access to markets. Proposed interventions to curb erosion will include proper land use, soil and water management and appropriate engineering works. Implementation of riparian reserve requirements should be done according to recommendations in the Agriculture Act.

6) There is a general lack of data for effective water resources management. Instruments for collecting water related data should be installed in the catchment to improve existing ones. Kenyatta University, BWARUA and WRMA will work together to develop a plan for data collection and management.

Monitoring of water resources (rainfall, stream flow, groundwater, abstractions, and water quality) and catchment conditions should be done regularly in order to provide data that can be used to assess the impact of undertaking the above interventions. This should be coordinated by WRMA in collaboration with BWARUA.

ACKNOWLEDGEMENTS

The Commission for Higher Education (CHE) of Kenya is thanked for funding the study. Bwathonaro Water User Association (BWARUA) hosted the researchers in the watershed and provided very useful primary data.

CITE THIS PAPER

Shisanya, C.A., Onywere, S.M. and Obando, J.A. (2017) Sustainable Water Resources Management for Food Security in Kenya: Case of Bwatho- naro Catchment. Open Access Library Jour- nal, 4: e3524. https://doi.org/ 10.4236/oalib.1103524

REFERENCES

1. UN World Water Report (2003) Water for People, Water for Life. The United Nations World and Water Development Report, New York.
2. International Research Institute (IRI) (2006) A Gap Analysis for the Implementation of the Global Climate Observing System Programme in Africa. Final Report.
3. Swedish International Water Institute (SIWI) (2007) Rain: The Neglected Resources: Embracing Green Water Management Solutions, Stockholm, Sweden.
4. Government of Kenya (GoK) (2001) Poverty Reduction Strategy Paper for the Period 2001-2004. Ministry of Planning and National Development, Nairobi Kenya.
5. Central Bureau of Statistics—CBS (2000) Economic Survey of Kenya, 2000.
6. Society for International Development—SID (2004) Pulling Apart, Facts and Figures on Inequality in Kenya, SID, Nairobi.
7. Government of Kenya (GoK) (1997) The Second Participatory Poverty Assessment Study—Kenya. A Report Prepared by AMREF and the Human Social Services of the Office of the President and Ministry of Planning and National Development. Government Printer, Nairobi, Kenya.
8. Government of Kenya (GoK) (1999) National Poverty Eradication Plan 1999-2015. Department of Development Co-Ordination, Ministry of Planning and National Development, Nairobi, Kenya. Government Printer.
9. Government of Kenya (GoK) (1999) Sessional Paper No.1 of 1999 on National Policy on Water Resources Management. Government Printer, Nairobi, Kenya.
10. UNDP (2002) Addressing Social and Economic Disparities. Human Development Report United Nations Development Programme, Nairobi, Kenya.
11. Water Resource Management Authority (WRMA) (2006) Catchment Management Strategy. Final Draft.
12. Ngige, A. and Macharia, F. (2006) Kenya Water Policy Overview Paper, IT Power East Africa.
13. Government of Kenya (GoK) (2002) The Water Act 2002 (Cap 372) Government Printer, Nairobi, Kenya.
14. Government of Kenya (GoK) (2003) Kenya Participatory Impact Monitoring (KePIM); Perspectives of the Poor on the Credit and Extension Policies. Ministry of Finance and Planning, Nairobi, Government Printer.
15. Jaetzold, R., Hornetz, B. and Shisanya, C.A. (2007) Farm Management Handbook of Kenya Vol II: Natural Conditions and Farm Management Information, Part B, Eastern Kenya. Ministry of Agriculture, Nairobi.

CHAPTER 5

Adoption and Scaling-Up of Conservation Agriculture in Tanzania: Case of Arusha and Dodoma Regions

Frederick Cassian Kahimba[1], Khamaldin Daud Mutabazi[1], Siza Donald Tumbo[1], Kenneth Francis Masuki[2*], Winfred Baptist Mbungu[1]

[1]*Soil-Water Management Research Group, Sokoine University of Agriculture, Morogoro, Tanzania*

[2]*Association for Strengthening Agricultural Research in Eastern and Central Africa (ASARECA), Entebbe, Uganda*

ABSTRACT

A study was conducted to assess the adoption and scaling-up of Conservation Agriculture in Arumeru district, Arusha region, northern Tanzania and Chamwino and Dodoma Urban districts in Dodoma region, Central Tanzania. The study employed structured questionnaire survey and key informant interviews as the main data collection methods. Data analysis was done descriptively to determine factors that influence the adoption. Results showed that farmers in Arusha have highly adopted terraces, minimum tillage and cover cropping whereas their counterparts in Dodoma have highly adopted large planting pits, minimum tillage, and rippers. The intensity of adoption in Arusha is higher for the wealthy compared to the poor while in Dodoma the wealth status was not a factor that influenced adoption of the Conservation Agriculture (CA) technologies. The process of technology adoption should include the use of various participatory methods including farmer field schools, experimental plots, farmer exchange visits, and training of trainers among others. Use of farmer groups, incentives, and support instruments such as Savings and Credit Co-operative Society (SACCOS) or warehouse receipts systems are also important to ensure that farmers realize some acceptable profits from their efforts. Involvement of various stakeholders is also very important including local governments and agricultural change agents at national, regional, district, ward and village levels. Hence, the intention to promote CA technologies should not only look at the economic importance, but also its socio-economic importance to the local people in the area. Their desire to adopt and out-scale a technology should be among the most important investment factors that the government and development partners should consider.

Keywords: Social Capital; Diffusion; Water System Innovation; Conservation Agriculture

1. INTRODUCTION

Conservation agriculture (CA) as a concept for natural resource saving strives to achieve acceptable profits with high and sustained production levels while concurrently conserving the environment [1] -[3] . It is an approach that advocates the concept of sustainable intensification of production [1] . It is a combination of tested technologies and/or principles in agricultural production and is gaining acceptance as an alternative to both conventional agriculture and organic agriculture as a means of ensuring sustainability [1] . One principle of CA is maintenance of adequate soil cover, especially achieved by leaving crop residues on the soil surface. This is the key to the success of CA systems, and the yield of crops direct-seeded into bare soil is often considerably lower than that of crops sown with conventional tillage practices [4] [5] . CA attempts to maintain a permanent or semi-permanent organic soil cover and promotes the concept of optimizing yields and profits while ensuring provision of local and global environmental benefits and services [6] . A good CA system is required to have at least 30% of soil surface cover with crop residue at planting for optimum effects of the system [7] . However, at the level of small-scale farmers, CA practice of reduced tillage is constrained with the availability of sufficient crop residues for mulch. This is commonly reported as a problem in sub-Saharan Africa and other many regions [7] -[10] . Delving on this observation, Giller et al. [11] , asserted that CA is inappropriate for majority of smallholder farmers because its adoption implies a direct competition for a scarce resource as some crop residues like maize, sorghum or millet straws which are often used for cooking fuel in the scarcity of firewood or other fuels. In more marginal environments, crop productivity is lower and therefore crop residues are scarcer and competition for them greater. Nevertheless, in most cases CA combines modern technologies, with traditional knowledge on soil husbandry. In Tanzania, CA is available and practiced in different forms, the most common ones being: terracing, conservation tillage, pit and trench farming and micro-catchment water harvesting systems. This study was conducted to identify and recommend investment areas in CA with the greatest potential to improve incomes and food security for poor farmers. It specifically focused on determining the extent of adoption and importance of various CA technologies in the case study areas and the estimated spread of the same in Tanzania; assessing the factors that contribute to adoption; investigating the constraints that limit adoption of CA technologies; and recommending steps to be used in promoting adoption and up-scaling of CA technologies in Tanzania.

2. METHODOLOGY

2.1. The Study Location

The study was conducted in Arumeru district in Arusha region, and Chamwino and Dodoma Urban districts in Dodoma region (Figure 1). Arumeru district is located between longitudes 35°E and 37°E and latitude 3°S. The district covers an area of 2966 km^2, which is 3.5% of Arusha Region [12] . Despite its proximity to the Equator, Arusha's elevation of 1400 m amsl on the southern slopes of Mount Meru keeps temperatures relatively low and alleviates humidity. Cool dry air is prevalent for much of the year. The temperature ranges between 13 and 30 degrees Celsius, with an average of around 25 degrees (http://en.wikipedia.org/wiki/Arusha). It has distinct wet and dry seasons with average annual rainfall of 654 mm (http://www.meoweather.com/).

Dodoma is located at latitude 6°S and longitude 35°E, in the centre of the country. The town is 486 kilometers (302 mi) west of the business city of Dar es Salaam, and 441 kilometers south of the tourist city of Arusha. It covers an area of 2669 square kilometers. Dodoma features a semi-arid climate with relatively warm temperatures throughout the year. The region averages 570 mm of precipitation per year, the bulk of which occurs during its short wet season between December and March. The remainder of the year comprises dry season.

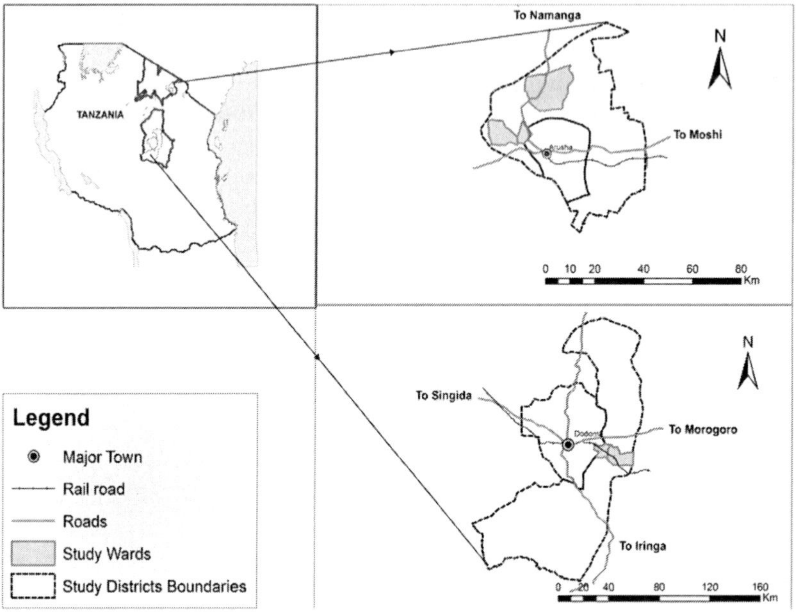

Figure 1. Map of Tanzania showing case study wards and villages in Arusha and Dodoma.

2.2. Sampling Techniques and Sample Size

Five wards whose villages practice CA technologies were purposely chosen in this study. Three wards, Musa (Likamba village), Olturumet (Ilkuishen and Ekenywa villages) and Oldonyosambu (Oldonyoas village) were chosen in Arumeru district, while Kikombo ward (Kikombo and Chololo villages) was chosen in Dodoma urban district and Msamalo ward (Mnase and Mgunga villages)was chosen in Chamwino district. A total of 200 randomly selected farmers were interviewed from the 8 villages.

2.3. Data Collection and Analysis

Structured questionnaire survey and key informant interviews were mainly used for data collection in this study. A structured questionnaire with both open ended and closed question was used to obtain quantitative data from the sampled respondents. The survey was used to collect demographic, crop production and marketing, socioeconomic, livelihood, and food security factors from the sampled respondents. The questionnaire was pre-tested and corrections were made accordingly in one village in Dodoma, and one village in Arusha and adjustments were made before its final administration. A multi-stage sampling technique was employed to get households from different villages. A total of 25 respondents were used from each village to make a total of 100 households from 4 villages for each district.

Key informants interviews were conducted to fill gaps from the questionnaire survey and verify the results. The study involved a wide range of stakeholders from farmers, key village and ward leaders and officials, district leaders and officials, and members of Non-Governmental Organizations (NGOs) that operate in the study areas.

Descriptive analyses were used to determine factors that influence the adoption. The questionnaire survey data was analysed using SPSS statistical package and MS Excel software. Results from the analysis were interpreted and the gaps were also addressed through interviews with key informants.

3. RESULTS AND DISCUSSION

3.1. Adoption of Conservation Agriculture and Its Importance to Food Security

3.1.1. Roles of CA Technologies in Improving Yields and Food Sufficiency

Analysis of farmers' responses on losses they would incur if the technologies were not introduced was used to assess their perception on the role of a particular technology in sustaining their livelihood. Results from Arusha and Dodoma regions (Figures 2(a) and (b)) imply that the adoption of CA technologies has helped farmers increase yield. Farmers felt that the yields would be affected without CA technologies. The yield losses range from less than 25% to 100% with majority of farmers indicating yield losses of about 50%. Farmers indicated that terraces, minimum tillage, large pits and cover cropping have had impacts on crop yields.

Figures 3(a) and (b) show households food self-sufficiency in terms of production and their purchasing power. Results show that majority of respondents in Arusha are capable to producing between 50% and 75% of their household requirements and to some extent they can also afford to purchase food in case of some deficits. This group is followed by that capable of producing 75% to 100% of their food and with good purchasing power in case of some deficit or crop failure. However, this is contrary to respondents in Dodoma, whom the majority can produce between 25% and 75%, and majority can afford to some extent to purchase food in case of deficit or crop failure. This information signifies the importance of improving agricultural practices that are geared at increasing production, such as the use of CA technologies, especially in Dodoma where majority are not producing enough and have weak purchasing power.

3.1.2. Crops grown and Their Yields under Different CA Technologies

Results of CA technologies with corresponding crops are shown in Tables 1 and 2. In Arusha maize crop was planted across all the CA technologies, due to its importance as both food and cash crop in the region (Table 1). On average about 55% of all the farms had maize planted, followed by beans 29%, Lablab 6%. The farmers find maize as the crop with a good return to investment compared to others, and would employ the most trusted technology for the production of maize. The high numbers of those who used terraces and minimum tillage and the combination of the two technologies to grow maize confirms that as supported key informants. Beans, though has a good market value, but its harvest has been declining in recent years.

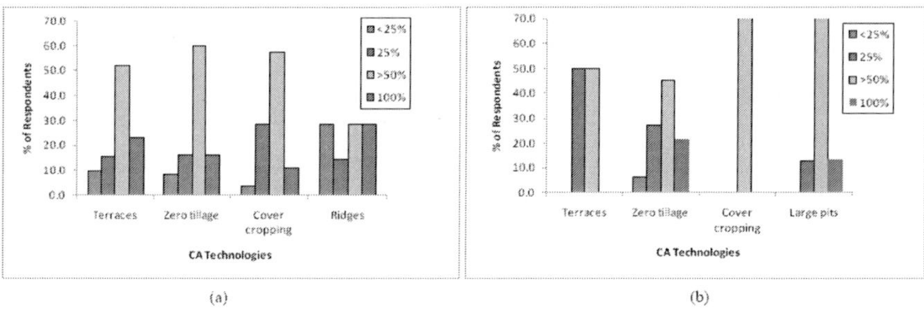

Figure 2. (a) Loss in yield due to lack of technology in Arusha; (b) Loss in yield due to lack of technology in Dodoma.

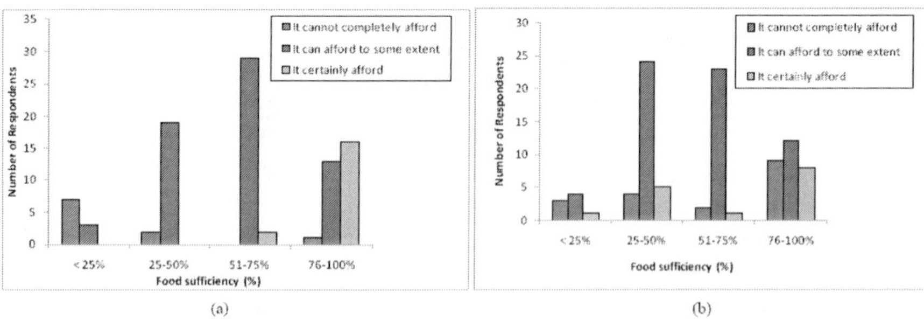

Figure 3. (a) Household food sufficiency in Arusha; (b) Household food sufficiency in Dodoma.

Table 1. Proportion of crop types grown under different CA technologies in Arusha.

CA Technology	Maize	Beans	Vegetables	Lablab	Other	Total
Terraces	60.0	32.7	1.8	1.8	3.6	100
Minimum tillage	51.1	31.9	8.5	4.3	4.3	100
Cover cropping	47.2	33.3	0.0	2.8	16.7	100
Ridges	55.6	27.8	11.1	0.0	5.6	100
Terraces & min. tillage	54.9	33.3	2.0	2.0	7.8	100
Terraces & cover crop	60.0	13.3	0.0	26.7	0.0	100
Average	**54.8.**	**28.7**	**3.9**	**6.3**	**6.3**	**100**

Table 2. Proportion of crop types grown under different CA technologies in Dodoma.

CA Technology	Maize	Sorghum	Groundnuts	Lablab	Others	Total
Terraces/Ridges	15.4	**46.2**	15.4	23.1	0.0	100
Minimum tillage	14.1	**46.9**	21.9	12.5	4.7	100
Cover cropping	27.8	11.1	**44.4**	11.1	5.6	100
Large pits	17.5	**75.0**	7.5	0.0	0.0	100
Terraces & Min. tillage	15.6	23.3	**36.7**	18.9	5.6	100
Terraces & Cover crop	9.3	25.6	30.2	**34.9**	0.0	100
Average	**16.6**	**38.0**	**26.0**	**16.8**	**2.7**	**100**

One farmer Simon Kutingala in Ekenywa village (Kilimapunda CA group) in Arusha during interviews said that:

"Maize has a good market value because it is used by the majority as a food crop, but crop like lablab is mainly bought by traders from the neighbouring country. Beans have a good market value, but the yield has been decreasing in recent years".

In Dodoma region sorghum is the dominant food crop, highly cultivated across all the technologies (38% of all the crops, Table 2). This crop is a drought tolerant suited to the semi-arid climate of Dodoma (300 - 500 mm annual rains). For the large pits practice, about 75% of the crop grown is sorghum. Also, terraces and minimum tillage are employed in the cultivation of sorghum. Cover cropping and a combination of terraces and cover cropping or minimum tillage are mainly employed in groundnuts cultivation.

The yields for different crops in the study area are generally lower, typical of dryland areas in sub-Saharan Africa. This underscores the rationale of promoting CA technologies to improve productivity. On average, terrace performed better (1.3 ton/ha) than other technologies with regard to maize yield (Table 3). Other CA technologies under which at least a ton of maize per hectare was realized were large pits and ridges with an average yield range of 1.1 and 1.0 tons/ha respectively. However, at median its only terraces which attained the yield of one ton per hectare. As the median not affected by outliers compared to the mean it stands to be an ideal statistic presenting a typical situation in a particular population.

Ridge technology had much better results on yield of beans. However, only a few farmers practiced ridging in beans production. The yield levels of sorghum, groundnuts, and lablab were typically low across the CA technologies. These four are among the crops promoted in the dryland farming areas of Tanzania including the study sites. Sorghum and groundnuts are common crops grown in Dodoma and across other regions in the semi-arid central part of Tanzania. Lablab is widely grown across the semi-arid northern Tanzania mainly for export to neighbouring country.

Table 3. Yields of crops under different CA technologies (ton/ha).

Crop	Statistics	CA technology						
		Cover Crop	Terrace	Min. Till	Large pits	Ridges	Terrace + Min. Till	Terrace + Cover Crop
Maize	N	16	34	24	7	11	42	11
	Mean	0.8	**1.3**	0.8	**1.1**	**1.0**	0.9	1.0
	Std. Dev.	0.6	1.2	0.8	1.0	0.7	0.9	0.7
	Median	0.8	**1.0**	0.6	0.4	**0.8**	0.6	1.0
Sorghum	N	2	2	24	30	3	18	9
	Mean	0.3	0.5	**0.5**	**0.4**	0.3	0.5	0.6
	Std. Dev.	0.1	0.0	0.4	0.5	0.2	0.5	0.4
	Median	0.3	0.5	0.3	0.3	0.3	0.3	0.8
Beans	N	12	17	8	-	5	17	2
	Mean	0.4	0.2	**0.8**	-	**1.5**	0.5	**0.7**
	Std. Dev.	0.3	0.3	0.6	-	1.3	0.4	0.3
	Median	0.3	0.1	**0.6**	-	**1.3**	0.3	**0.7**
Groundnuts	N	7	-	11	3	-	27	8
	Mean	0.1	-	0.3	0.1	-	0.2	0.2
	Std. Dev.	0.1	-	0.2	0.1	-	0.3	0.1
	Median	0.1	-	**0.3**	0.1	-	0.1	0.2
Lablab	N	2	-	7	-	2	17	18
	Mean	0.9	-	0.3	-	0.3	**0.6**	0.4
	Std. Dev.	0.9	-	0.2	-	0.0	0.5	0.4
	Median	0.9	-	0.3	-	0.3	**0.5**	0.5

N = number of observations (valid cases). Min. Till. = Minimum Tillage.

The relatively better yield (at median) of sorghum was attained under terracing. However, considering minimum tillage as a control, the typical increment in yield gain (0.2 ton/ha) may not likely cover the costs of terracing in terms of monetary investment and labour. Either, there is no difference in using large planting pits or minimum tillage in sorghum cultivation. However, farmers have indicated that they are likely to loose up to 50% of the crop if they don't practice CA technology as already indicated in Figure 2. It has been found that in a good year, there is significant difference in yield between conservation tillage and conventional tillage [13].

With regard to lablab, it is a cover crop and hence is self-provisioning in terms of benefits of cover crop use. The results indicate that other CA technologies did not result into yields of lablab higher than just growing the crop itself as a cover crop.

3.1.3. The Extent of Adoption of CA Technologies

Survey results indicate that the number of farmers adopting different CA techniques varied from place to place depending on biophysical characteristics of the farm areas.

In Arusha region majority of respondents (52.1%) adopted terraces, followed by almost 23.7% of the respondents adopting less land disturbing technologies such as zero tillage, ripping, and minimum tillage and about 16.6% adopting cover crops technology and 6.5% adopting ridges. However, there was little adoption of such technologies as mulching and jaruba with a combined percentage of 1.1% (Figure 4(a)). This pattern is likely to happen in Arumeru district due to topographic nature of the study area. Terraces in Arusha are practiced in sloping lands, and in areas where land degradation was high. The key informants confirmed that the intervention of SCAPA program tremendously impacted into their adoption aiming at reducing soil degradation, improving soil fertility through improving water retention, and increasing soil organic matter.

In Dodoma majority of the farmers (29.1%) have adopted conservation tillage followed by large planting pits (hereafter referred to as large pits and locally known as Chololo[1] pits) (26.2%) (Figure 4(b)). The wider practice of conservation tillage is due to the use of ox-drawn rippers, a farming technique that is widely used in the area. Large pits are gaining more popularity due to their ability to conserve soil moisture when the rains are little and the slope is steep for other cultivation methods. Cover crop technique has been adopted by only 8% of the respondents this is due to that fact that this area practice free grazing mode of livestock keeping where crop residues serve as animal feeds.

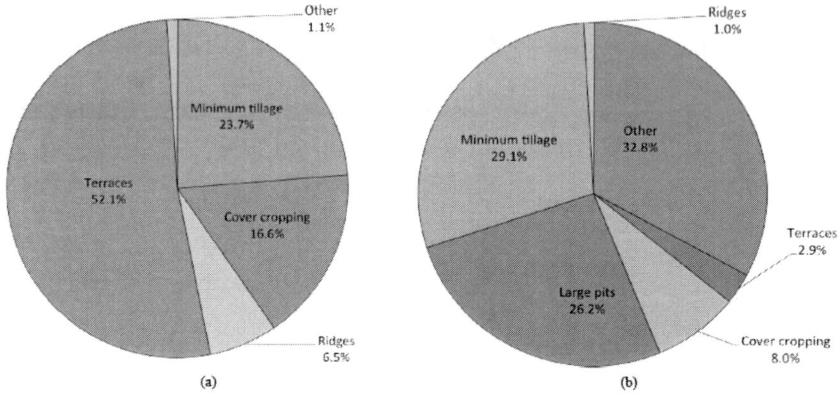

Figure 4. (a) Adopters of different technologies in Arusha; (b) Adopters of different technologies in Dodoma.

This poses a competition between animals and soil cover, the case that has been identified as typical problem of CA system in Sub Saharan Africa (SSA) region [8] [10] . Ridge cultivation (1.0%) and terracing (2.9%) are not popular technologies in Dodoma region since they cannot easily be made using ox-ploughs, which is a popular cultivating method in the region.

3.1.4. Types of Farmers Who Adopted Conservation Techniques

Results of analysis to identify type of farmers who adopted conservation technologies based on wealth status and gender indicate that wealth status influence the decision to adopt conservation technologies. Three wealth groups namely poor, average and wealthier were identified. The indicators used for ranking of wealth status included roofing, floor and wall materials of the main house, and possession of means of transport (motorbike, and bicycle) and cell phone.

In Arusha, majority of farmers that are wealthier have invested on CA technologies in three or more farms (Figure 5(a)). Most of those who have average wealth and the poor have two technologies in their farms. This is contrary to farmers in Dodoma whereby most of them have adopted three technologies regardless of their wealth status (Figure 5(b)). The reason for this pattern is due to the fact that Dodoma being a drier area, farmers had to adopt various technologies for capturing and retaining water.

Results on gender role in household decisions to adopt conservation agriculture technologies are presented in Figures 6(a) and (b). In Arusha, husbands make adoption decision with regard to ridges and terraces (>50%

of respondents). For the minimum tillage and cover cropping both husbands and their spouses make the decisions to adopt equally.

In Dodoma over 60% of the respondents indicated that husbands make the decision to adopt large pits and minimum tillage. However, the decision to adopt cover cropping is made by both husbands and wives equally. This shows that men have an upper edge to decide to adopt technologies that are labour intensive than those, which do not require too much labour such as conservation tillage and cover cropping.

3.1.5. Estimates of the Spread the CA Technologies in Tanzania

The spread of CA technologies in the study areas has been facilitated through groups, where farmers form groups through which different stakeholder change agents conduct training. In Arusha, key informants recall that groups were formed in villages and farmers were given training on CA technologies.

"Our group started with 19 members and this is a third year we are practicing CA, many farmers in the village have adopted and the number of those practicing CA has reached more than 50 households in the village, even a primary school within the village has started practicing CA in their farms" said Godwin Sanare a farmer in Ilkuishen village.

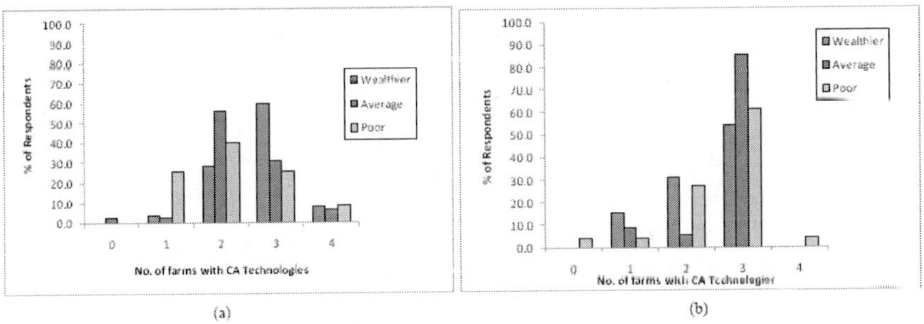

Figure 5. (a) Farmers adoption of CA technologies based on wealthy status in Arusha; (b) Farmers adoption of CA technologies based on wealthy status in Dodoma.

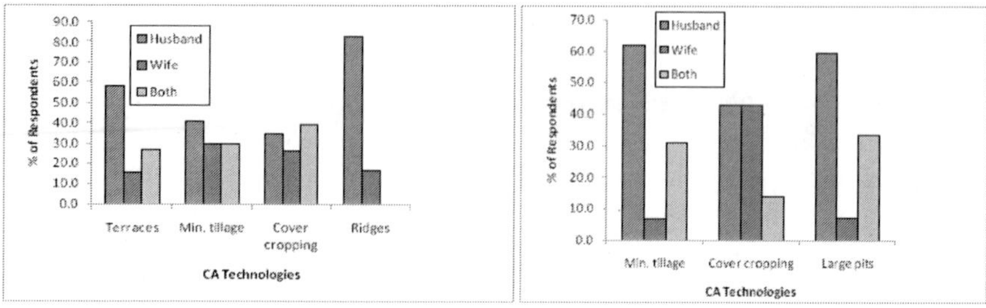

(a) (b)

Figure 6. (a) Gender influence on a decision to adopt CA technologies in Arusha; (b) Gender influence on a decision to adopt CA technologies in Dodoma.

CA technologies started with one group in the Olturumet ward in 2005, but now four CA groups (Merikinoi, Kilimapunda, Tuamke and Olorishi village) have been formed. Another farmer from Ekenywa ward reported that from a group of 26 farmers CA has continued to be adopted by other farmers to more than 80 farmers and many more are adopting. It was also reported that about 300 households in the village from 1200 households in the village have adopted CA technologies. There are also CA groups in other areas such as Nduruma and Enyorata.

Conservation agriculture technologies have been adopted by farmers in places where interventions were promoted. Change agents played a great role in the adoption and widespread of CA technologies. In Arusha, early efforts are credited to Soils Conservation and Agroforestry Project Arusha (SCAPA), which started in the late 1990s. Efforts by NGOs such as Research, Community and Organizational Development Associates (RECODA), Women in Agricultural Development and Environmental Conservation (WADEC), and World Vision in collaboration with Selian Agricultural Research Institute (SARI) through project funded by The Food and Agricultural Organization (FAO) and the district have promoted the adoption of CA in most wards in the district. Interviews with key informants from NGOs and extension agents revealed that CA technologies have spread to 11 out of 21 wards of Arumeru District. The wards are Olturumet, OldonyoSambu, Musa, Namayala, Mlangarini, Maroroni, Kisongo, Nduruma, Manyire, Oljoro, and Oltoroto. Apart from villages and wards in Arumeru district, there have been successes of spreading CA in other districts of Arusha and Manyara regions. Key informant interview revealed that Karatu and Babati Districts were one of the early adopters of CA technologies through the Conservation Agriculture for Sustainable Agriculture and Rural Development (CASARD) project. The farmers, extension agents and NGOs who were interviewed reported that different groups were taken to Karatu and Babati on study tours and field visits to learn about CA.

The primary focus for Dodoma case study was to investigate the adoption and diffusion of Chololo pits. According to key informant interview, at the moment a farmer group with 25 members is practicing the Chololo pits farming and about 45 - 50 other farmers that are not in the group are also practicing it in the studied villages. The limitation of the technology is that it can only be practiced in heavy soils and sloppy areas. Majority of the village farm-lands are located in flat loamy soils that are not suitable for Chololo pits.

The Chololo pits were invented by inventor farmer from Chololo villagein 1997. Thereafter they were promoted by various NGOs such as Inades Formation Tanzania (IFTz), Lay Volunteer International Association (LVIA), and Muunganowa Vikundivya Wakulima Tanzania [MVIWATA (Union of Farmers groups)]. In Dodoma urban and Chamwino districts the technology has spread in the wards of Handali, Msamalo, Kikombo, Ng'ong'ona, and Makulu. In these wards 60 innovator farmers were trained by IFTz on the Chololo pits. Each of the trained farmers formed their own group of farmers who practiced the technology. The conservation tillage technology (using ox-ploughing) is commonly practiced in Msamalo, Ikoa, Igandu, Handali, Kikombo, and Buigiri wards.

The Chololo pits technology got more publicity when it was involved into competition of best farmer innovations within east Africa region by the UNDP and won a third place. The District Agricultural Development Officer (DALDO) of Chamwino district and Municipal Agricultural Development Officer (MALDO) of Dodoma Urban, have reported that the Chololo pits technology is now being practiced in six districts of Dodoma region namely Dodoma urban, Kongwa, Kondoa, Chamwino, Bahi, and Mpwapwa. The Inades Formation Tanzania officials indicated that the main source of spread is through farmers who visited areas where the technology is being practiced. Also the Inades Formation Tanzania has been proposing adoption of the Chololo pits in the respective districts.

Inades Formation Tanzania officials indicated that semi arid areas of Morogoro, Singida, and Shinyanga, Kilimanjaro, Mwanza, and Manyara regions are potential regions for promotion of Chololo pits as these regions have highlands with steep slope and heavy soils suited for the technology. The CA technologies are currently promoted in other parts of the country like Kilimanjaro and Tanga regions by RECODA; in Morogoro region and in Kilindi district in Tanga region by CARE International; and in the Southern Highlands regions of Mbeya and Njombe by Agricultural Research Institute (ARI) Uyole.

3.2. Factors Contributing to Adoption and Up-Scaling of CA Technologies in Tanzania

3.2.1. Technological and Economic Factors

Various factors have contributed to the adoption of CA technologies by farmers (Figures 7(a) and (b)). The ability of the CA technology to conserve soil moisture was the main factor for adopting terraces and conservation tillage in Arusha and large pits in Dodoma. Technologies with less labour intensiveness were also favoured in Dodoma region as is the case with minimum tillage and cover cropping. Presence of external incentives such as provision of subsidized farm inputs (e.g. lablab and pigeon pea seeds provided by RECODA and SARI in Arusha), and provision of training on the use of draught animal power at subsidized costs in Dodoma were also contributing factors for farmers to adopt new technologies.

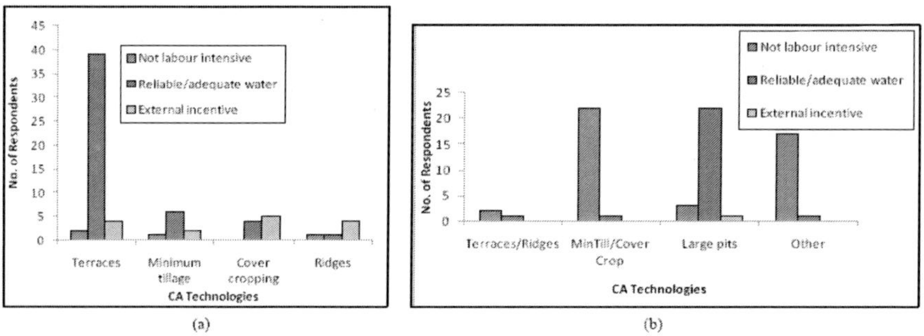

Figure 7. (a) Reasons for adoption of CA technologies in Arusha; (b) Reasons for adoption of CA technologies in Dodoma.

3.2.2. The Roles of Change Agents

This study hypothesized that the intervention of change agents and the messages that they carry are also a factor that brings wider adoption of technology. Respondents were asked on the number of visits made by the various change agents and the type of messages they passed over. Figure 8 shows the frequency at which change agents were visiting farmers. It is quite clear that extension agents were more in contact with farmers than the other three groups (researchers, NGOs and private sector). The intervention by extension agents is relatively higher in Arusha than Dodoma with exception of one case by the NGOs, which indicates that the "more than twice" visit by NGOs in Dodoma was higher than in Arusha.

The efforts by the government to increase the number of extension workers to the village level are clear as nearly every village has at least one extension officer. The key informants interviews also revealed that apart from the government efforts to increase the number of extension officers, there is a lack of motivation to the staff due to lack of transport to meet the demands and needs of people in the villages. Extension officers themselves expressed the difficult conditions they have to endure to travel from one point to the other and the housing conditions in villages and the lack of training package in CA. Despite having bylaws that prohibit grazing of animals on farms, enforcement of such laws at village level has been hindered by traditions to the detriment of CA adoption efforts.

The advices from the change agents delivered to farmers included agronomic practices, pests and diseases, livestock husbandry, CA technologies and others. In the study area in Arusha, 53.8% of the advices were on CA technologies while advices on agronomic practices and livestock husbandry were 21.3% and 17.3%, respectively (Figure 9), indicating higher level of CA technologies promotion compared to other technologies. In Dodoma, the advices higher percent compared to others was on agronomic practices (45.7). Advices on CA technologies came third at 20.0%. This implies that farmers in Arusha were receiving more advices on CA technologies than in Dodoma.

The key informants interviews revealed that the promotion of CA technologies in Arusha started way back 1980's by SCAPA. This project provided training, formation of groups, demonstration of plots and other supports. However, farmers in Mnase Ward in Dodoma, reported that the government was not supportive of their efforts in adopting CA technologies and especially Chololo pits.

3.2.3. Inputs and Equipment Suppliers

The inputs and equipment suppliers have helped to bring agricultural inputs and equipment close to the farmers thereby facilitating the adoption and up scaling of the CA technologies. Ward extension officers indicated that in Dodoma region an NGO called LVIA successfully promoted conservation tillage using ox-drawn rippers by conducting training and issuing a set of oxen and ox plough at a subsidized price. Farmers were supposed to pay in three installments after they have sold their harvest. The government also through village extension workers has been distributing improved seeds to farmers at a subsidized price so that majority of the farmers could afford and adopt the used of improved seeds. The farmers also revealed that farmer organizations such as Savings and Credit Co-operative Society (SACCOS) have helped farmers by bringing farm inputs such as fertilizers, pesticides and seeds in store located within villages thereby making them easily accessible to farmers.

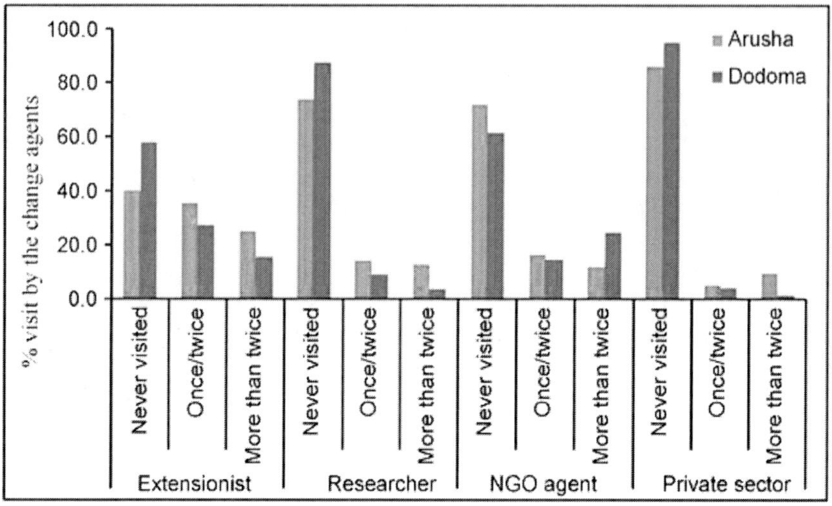

Figure 8. Frequency of contact of farmers by the change agents.

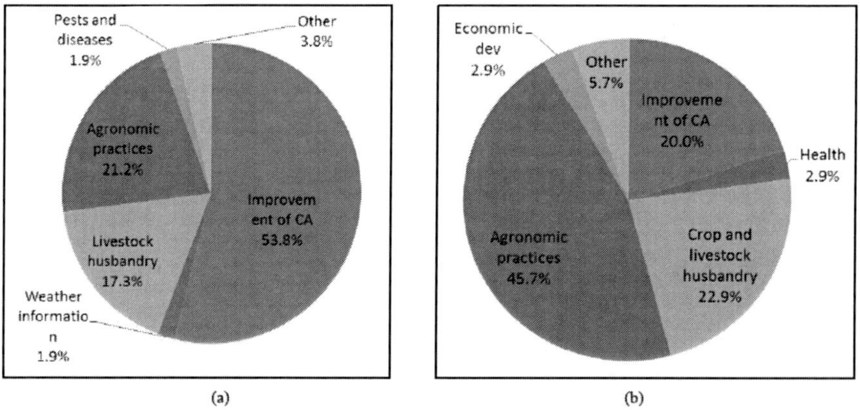

Figure 9. (a) Advices from change agents in Arusha; (b) Advices from change agents in Dodoma.

3.2.4. Market Interventions and Communication Infrastructure

Analysis of places where farmers sell their produce indicated that in Arusha farmers sell mostly dried maize and beans in a nearby township compared to the sales within the village (Figures 10(a) and (b)). On the other hand, sales within the village were the leading market outlets in Dodoma than to neighbouring villages and nearby townships. The leading crops in Dodoma are Sorghum/millet, groundnuts, and Marcia (improved sorghum variety).

Small traders lead the process of buying farm produce in both regions (Figures 11(a) and (b)). Large traders are not common in the villages because most of the farmers are subsistent and they harvest only enough for their household consumption leaving little amount for selling.

With regard to transport and communication infrastructure, the study villages in Arusha do not have year round passable roads with exception of Oldonyo Sambu. However, they are well connected in terms of telecommunication. In Dodoma, the roads are well maintained. There is also a rail service in Kikombo and Msamalo wards. All villages are well connected and served with various mobile telephone networks.

3.2.5. Policies and Bylaws at National and Local Scale

Policies and bylaws were mentioned to play a vital role in the adoption and spread of CA technologies. The farmers and change agents noted a number of government policies and local by-laws that influence the adoption of CA. The change agents indicated that the government is keen and supportive on CA.

Formulation, mainstreaming and implementation of proper policy interventions; enforcement of the existing rules and regulations in the areas of land use planning, land tenure, developing smart market and fertilizers subsidies at local, national and sub-regional levels. The National Agricultural Input Voucher Scheme (NAVIS) was identified as one of success ways of implementing agricultural policy through which the government of Tanzania has managed to equip farmers with inputs to support their efforts of alleviating poverty and reducing hunger. The government supports local by-laws that prohibit grazing on farms after harvest, which was limiting CA adoption. At village level an offender of such by-laws are liable to fines. For instance, in Ilkuishen and Ekenywa villages in Oltrumet ward a fine of TZS. 50,000/= was set for a cattle grazing on a field.

The farmers also noted the provision of farm implements to some of the CA groups as one of the efforts to promote CA. The farmers mentioned that the provision of loans by the government in terms of implements to groups and individuals who practice CA was a major boost. A farmer in Ekenywa acknowledged to have received the loan for a direct seeder and has been able to re-pay the loan according to schedule.

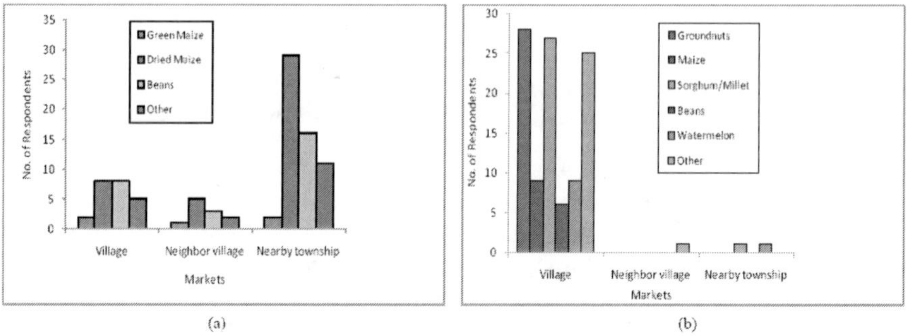

Figure 10. (a) Markets for farmers to sell their produce in Arusha; (b) Markets for farmers to sell their produce in Dodoma.

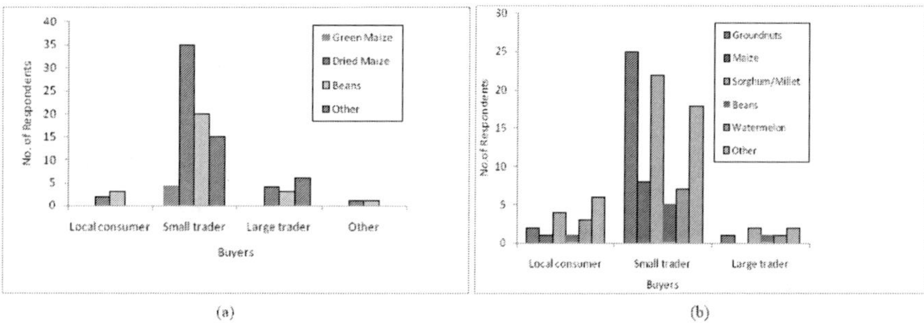

Figure 11. (a) Buyers of farm produce in Arusha; (b) Buyers of farm produce in Dodoma

3.3. Constraints That Limit CA Technologies Adoption and Up-Scaling

3.3.1. Perception of Limitations for CA Adoption

Analysis was carried out to investigate why some farmers were not adopting conservation technologies. Labour intensiveness, lack of training, and lack of capital to invest in new technologies were the main constraints for farmers not to adopt in Arusha (Figure 12(a)). Lack of training, poverty, and land ownership were the main reasons for farmers not to adopt CA technologies in Dodoma Region (Figure 12(b)). Farmers are hesitant to invest in labour on new technologies such as large pits and terraces on hired farms that they are not sure of continuing to farm in the subsequent season.

Other reasons for non-adoption from both regions were lack of interest, lack of incentives, and time constraints. In addition, availability of farm inputs, costly implements, low returns, and lack of land for implementing the technologies were also reasons mentioned by farmers.

3.3.2. Technical, Biophysical, and Economic Constrains

Failure of adoption of the CA technologies was also contributed by some technical, biophysical, and economic constrains. Lack of immediate returns was among the reasons for not adopting as most technologies take more than 2 years for farmers to start realizing results. The key informants also indicated that lack of training to farmers and extension agents was also a major constraint. Interview with farmers indicated that technologies such as the Chololo pits in Dodoma and terraces in Arusha are labour intensive and expensive compared to conventional tillage. Technologies such as Chololo pits are area specific. They are suited to sloping areas with heavy soils. Considering the minimum tillage using ox-drawn equipment, which is a popular technology in Dodoma, most farmers (over 80%) are subsistent and they do not produce enough to have the financial power to adopt the technologies.

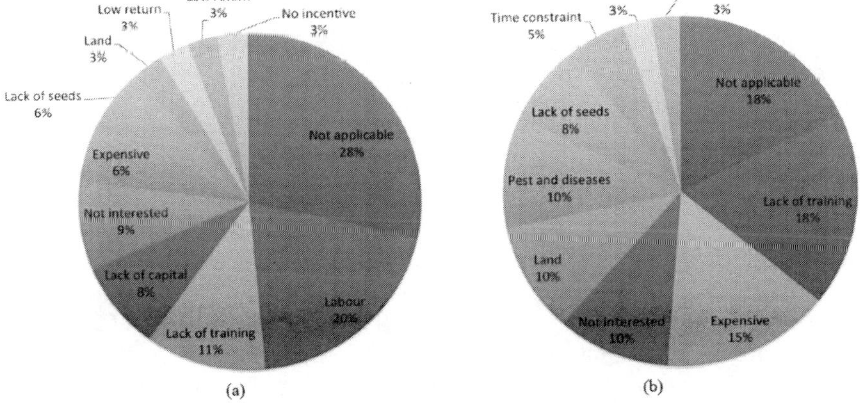

Figure 12. (a) Limitations for adoption of CA technologies in Arusha; (b) Limitations for adoption of CA technologies in Dodoma.

3.3.3. Legal and Institutional Framework

Newer technologies originating from farmers (bottom-up) get little support and attention by the system. Local government continues to support standard and conventional technologies such as bench terracing and minimum tillage using ox-drawn equipment.

Land tenure was also a factor limiting widespread of CA technologies. From the key informants interview many of the younger generations indicated that main reason for failure to invest in CA technologies was lack of ownership of the farms as most of them were hiring from the village elders. Results from the questionnaire survey indicated that 95% of respondents who adopted CA technologies were owners of the farms. CA technologies involve living crop residues for the next cropping season. However, these residues are sometimes eaten by livestock. In Arusha region there is a bylaw for livestock that destroy crops in the field, which is a fine of TShs 50,000/—for every cow that is caught grazing on any farm. But the implementation of the fine is hindered by social relationships as most people tend to seek reconciliation to end the matter without paying fines. Finally, extension agents are not well motivated; there is a need for them to be given proper training on CA, transport, and housing facilities.

3.3.4. Markets and Communication Constraints

The marketing systems in rural areas within the study villages are not well established and coordinated. Farmers and village leaders revealed that there are no permanent and formal markets within their villages. Despite the presence of farmer groups (SACCOS), there is still poor communication among farmer groups and farmers. As a result the producers lack market information and hence fail to have collective marketing strategies. Every farmer sells same commodities with different inflated prices. On the other hand, presence of number of uncoordinated middlemen is a reason as to why farmers selling their products at very low prices, resulting in low household income. Buyers inflate volume-measuring instruments during purchase of crops from farmers (e.g. 25 L bucket for a price of 20 L), and deflate the volumes during selling back (e.g. 16 L for a price of 20 L). In Dodoma Region, the National Food Reserve Agency (NFRA) used to buy only specialized crop (maize). There is a need for the NFRA to buy more crops such as sorghum, millet and sunflowers, since these are the types of crop grown by majority and are suited to the semi-arid climate characteristics of Dodoma.

3.3.5. Socio Economic Constrains

Socio economic issues were among the factors contributing to poor adoption and up-scaling of CA technologies. When different change agents promote the CA technologies in the villages most of the strategies focus on the fewer wealthier farmers (10%) who can afford to invest on the technologies and buy improved seeds. An example is availability of equipment and tools that are affordable to farmers. This has resulted in majority of the poor farmers failing to afford and hence adopt CA technologies. Subsistence farmers have their alternative means of living such as selling labour, preparing charcoal, and doing small business. They also use livestock as their own banks and sell some whenever they get financial problems. Issuance of Government food aids in case of emergency give farmers assurance of government help in case they do not harvest. As a result of this some of the farmers do not see the necessity of adopting CA technologies for conserving soil moisture thereby improving their farm yields.

3.3.6. Credits, Inputs, and Equipment

The availability of credits and agricultural inputs and equipment at affordable cost has been a constraint to the adoption of CA technologies in both Arusha and Dodoma regions. In Arusha for example, CA implements such as rippers and no-till seeders initially imported from Brazil, were expensive to most farmers, and only few implements were provided to few groups at the time. Nandra Engineering (a local manufacturer in Moshi), and Centre for Agriculture Mechanization and Rural Technology(CARMATEC) in Arusha started fabricating some of the implements, and selling to farmers at prices cheaper than the Brazil equipment. However, farmers in Arusha indicated that the local made rippers were less durable compared to the Brazilian rippers. Limited number of implements to satisfy their needs in time, especially during the farming season when every farmer needs to work on the farm, was reported to be a major setback to wider adoption of the CA technologies. Apart from few implements in groups, even farmers who are eager to adopt are kept at bay by the unavailability of implements. Farm inputs such as seeds especially those of lablab and pigeon pea were also mentioned to be major constraints limiting adoption as they are not locally available and farmers depend on change agents to provide inputs.

3.3.7. Limited Promotion and Publicity

The Chololo pits farming technology in Dodoma was invented in 1997 by a farmer in Chololo village. However, there have been very little efforts to promote and publicise it and thus it did not get a due popularity. While extension workers have been allocated to village and ward levels, they are very few to offer their services adequately to reach out to all the farmers. With regard to publicity of the technologies, especially those invented by farmers, there are very little publications or formal research to advise and popularize the technologies among the research community.

3.4. Steps for Promoting Adoption and Up-Scaling of CA Technologies in Tanzania

Key informants identified various steps that are necessary for adoption and up-scaling of CA technologies in Tanzania. Farmers are willing to adopt technologies that have proven to be beneficial in increasing yields, conserving soil moisture, and reducing soil erosion. Also, most farmers are interested in a technology that has a quick and good return to investments.

The first step identified was training on the CA technologies. The advantages obtained due to particular technology should be clearly stipulated to farmers during the training to encourage them to adopt. Apart from training farmers, there is a need to train the change agents who are involved in the promotion of the CA technologies. The second step is formation of groups. The respondents stressed the importance of groups in the adoption of CA technologies. The study revealed that group formation and networking enhanced the adoption and up-scaling processes in Arumeru district.

In general the respondents proposed the steps or actions they would have taken to promote CA technologies if they had their own million dollars as follows:

1) Sensitize extension workers and councilors at District level to gain policy support.
2) Mobilize the community in collaboration with the local government.
3) Establish resource centers (technology incubators) and use them to provide formal training on CA technologies.
4) Establish demonstration plots accompanied with starter packs to assist farmers to easily adopt CA technologies.
5) Promote the trained farmers to be trainer of trainers.
6) Promote research component to come out with improved tools and technologies.

7) Facilitate the trained farmers to establish demo plots on their own farms and later choose to practice the CA technologies they preferred in their farms.

8) Facilitate formation of village committees on various CA technologies.

9) Promote scaling-out through farmers exchange visits and farmer-to-farmer training.

10) Conduct training on soil water conservation, rainfall characteristics, ecology and biodiversity.

11) Mobilize farmers for collective action in marketing get good prices for their produce e.g. establishment of warehouses, SACCOS or a platform for price negotiation with traders.

4. CONCLUSIONS

This study has evaluated factors that facilitate or constrain adoption of adoption and up-scaling of CA technologies in central and northern parts of Tanzania. Many studies of this nature have been done under various names including CA, in-situ rainwater harvesting, soil and water conservation measures, and water system innovations, to mention a few. Therefore, the findings and facts have not really changed drastically over time. The need to continue investing in the promotion of CA technologies among others is still there in Tanzania and in Africa South of the Sahara. Most of the CA technologies are area specific, but they bear similar characteristics regardless of locations.

Factors that facilitate or constrain adoption have also not changed with the main factors being technological and economic ones with quick return from investment being another important factor. The scaling-out process is well established in Tanzania and it has proved to be effective in promoting adoption. Important elements in the process include the interactive participatory process such as the use of farmer field schools (FFS), experimentation plots, and farmer exchange visits.

The study showed that sometimes the socio-economic importance of a technology can override its economic significance. For example, sorghum is socioeconomically more important to the people of rural Dodoma even though it might give negative economic returns. Therefore, the intention to promote CA technologies should not only look at the economic importance, but also its socio-economic importance to the local people in the area. Their desire to adopt and out-scale a technology should be among the most important investment factors that the government and development partners should consider.

ACKNOWLEDGEMENTS

Frederick Cassian Kahimba, Khamaldin Daud Mutabazi, Siza Donald Tumbo, Kenneth Francis Masuki, Winfred Baptist Mbungu (BMGF) through The Agricultural Water Management (AWM) Solutions project, of which this work was part of it.

REFERENCES

1. Food and Agriculture Organization (2009) Scaling-Up Conservation Agriculture in Africa: Strategy and Approaches. FAO Sub-regional Office for Eastern Africa.
2. Keraita, B., Drechsel, P. and Konradsen, F. (2010) Up and Down the Sanitation Ladder: Harmonizing the Treatment and Multiple-Barrier Perspectives on Risk Reduction in Wastewater Irrigated Agriculture. Irrigation and Drainage Systems, 24, 23-35.
3. Bayala, J., Kalinganire, A., Tchoundjeu, Z., Sinclair, F. and Garrity, D. (2011) Conservation Agriculture with Trees in the West African Sahel—A Review. ICRAF Occasional Paper No. 14. World Agroforestry Centre, Nairobi.
4. Wall, P. (1999) Experiences with Crop Residue Cover and Direct Seeding in the Bolivian Highlands. Mountain Research and Development, 19, 313-317.
5. Sayre, K.D., Mezzalama, M. and Martinez, M. (2001) Tillage, Crop Rotation and Crop Residue Management Effects on Maize and Wheat Production for Rainfed Conditions in the Altiplano of Central Mexico. Conservation Agriculture, A Worldwide Challenge. Proceedings of the I World Congress on Conservation Agriculture, Madrid, 1-5 October 2001, 575-585.
6. Dumanski, J., Peiretti, R., Benetis, J., McGarry, D. and Pieri, C. (2006) The Paradigm of Conservation Tillage. Proceedings of the World Association of Soil and Water Conservation, P1, 58-64.
7. Fowler, R. and Rockstrom, J. (2001) Conservation Tillage for Sustainable Agriculture—An Agrarian Revolution Gathers Momentum in Africa. Soil & Tillage Research, 61, 93-107.
8. Bationo, A., Kihara, J., Vanlauwe, B., Waswa, B. and Kimetu, J. (2007) Soil Organic Carbon Dynamics, Functions and Management in West African Agro-Ecosystems. Agricultural Systems, 94, 12-25.
9. Erenstein, O. (2003) Smallholder Conservation Farming in the Tropics and Sub-Tropics: A Guide to the Development and Dissemination of Mulching with Crop Residues and Cover Crops. Agriculture, Ecosystems and Environment, 100, 17-37.
10. Tursunov, M. (2009) Potential of Conservation Agriculture for Irrigated Cotton and Winter Wheat Production in Khorezm, Aral Sea Basin. Inaugural Dissertation, Institutfür Nutzpflanzenwiss, Und Ressourcenschutz Bodenwissenschaften. University of Bonn.
11. Giller, K.E., Witter, E., Corbeels, M. and Tittonell, P. (2009) Conservation Agriculture and Smallholder Farming in Africa: The Heretics' view. Field Crops Research, 114, 23-34.
12. Nyaki, A.S., Lyimo, S.D., Mawenya, L.J., Kalage, K.S. and Antapa, P.L. (1991) Soil and Water Management for Sustainable Smallholder Development: A Field Exercise and Tour Guide in Tanzania. Workshop on Environment and the Poor. WASWC and IIED, Governments of Kenya and Tanzania.

13. Mkoga, Z.J., Tumbo, S.D., Kihupi, N. and Semoka, J. (2010) Extrapolating Effects of Conservation Tillage on Yield, Soil Moisture and Dry Spell Mitigation Using Simulation Modelling. Physics and Chemistry of the Earth, 35, 686-698.www.elsevier.com/locate/pce http://dx.doi.org/10.1016/j.pce.2010.07.036

CHAPTER 6

New Issues on Protected Area Management

David Rodríguez-Rodríguez

World Commission on Protected Areas (IUCN-WCPA), Spain

1. INTRODUCTION

Protected areas (PAs) are the main strategy to face the global deterioration of ecosystem goods and services due to human causes (Mora & Sale, 2011; Sanderson et al., 2002) and especially of biodiversity (Jenkins & Joppa, 2009; Pullin, 2002).

Almost 150 years have passed since the designation of the first "modern" PAs, back in the XIXthcentury. Designation of PAs has growth rapidly since that date and especially since the last third of the XXth century (Jenkins & Joppa, 2009; McDonald & Boucher, 2011). Currently, there are over 120,000 nationally and internationally-designated PAs in virtually all the countries of the world (UNEP-WCMC, 2008). They covered about 31.235.000 km^2, over 21% of the global terrestrial (approximately 17,290,000 km^2) and ice-covered area (some 13,950,000 km^2) by 2010, according to the Wold Database on Protected Areas (WDPA), and continue to expand. Conservation has become one of the main land uses globally, with approximately 13% of land under some form of protection (Jenkins & Joppa, 2009; McDonald & Boucher, 2011). Vast human, material and economic resources are allocated to conserve and use sustainably the natural and cultural resources provided by PAs. And yet biodiversity and the other ecosystem goods and services continue to be lost (Butchart et al.,

2010). As a result, PA management is receiving increasing attention as one of the key aspects for the effective conservation of PAs.

In this chapter, different aspects related to PA management are discussed. It also tries to clarify some controversial concepts and to encourage the discussion on a number of challenging issues of interest for scientists, managers, policy-makers and conservationists.

2. DEFINITION OF PROTECTED AREAS

There are two internationally accepted definitions of PA: the definition given by the Convention on Biological Diversity (CBD, 1992), and the latest definition proposed by the IUCN (Dudley, 2008). Both have subtle conceptual differences that should be scrutinized in detail to know what exactly is being talked about:

a. According to the CBD, a protected area is "a geographically defined area which is designated or regulated and managed to achieve specific conservation objectives" (CBD, 1992). This broad definition could apply to virtually every space with some regulation to achieve proper use and conservation of resources, such as game reserves, managed logging areas,marinereserves, or Biosphere Reserves, for instance. No specific mention to the conservation of biological features is given. As a result, this definition could also apply to non-biological resources meriting conservation, such as physical, geo-morphological or even cultural features, like museums or cathedrals.

b. According to the most updated definition of PAs by the IUCN, a PA is "is a clearly defined geographical space, recognised, dedicated and managed, through legal or other effective means, to achieve the long term conservation of nature with associated ecosystem services and cultural values" (Dudley, 2008). This definition is more precise, and includes three key and advanced conservation concepts: 1) the "long-term" scope of conservation; 2) the specific conservation of "nature"; and 3) the conservation of complementary ecosystem and cultural goods and services.

All six (seven?) PA management categories proposed by the IUCN (Dudley, 2008) share the principles included in this definition. This second definition of PA is more accurate and thus it will be used along this chapter.

3. MULTIPLE USE. SUSTAINABLE DEVELOPMENT IN PROTECTED AREAS?

Currently, almost no one argues that the overarching goal of PAs is conserving biodiversity and that any other management objective should be subject to biodiversity maintenance, improvement or restoration (Múgica & Gómez-Limón, 2002). There is, however, considerable debate on the weight of additional social and economic objectives in PA management, following the desirable yet vague concept of sustainable development (WCED, 1987).

Global agreements are giving PAs a starring role in many aspects other than biodiversity conservation. One extended mission included in its most updated definition is providing while preserving ecosystem goods and services and associated cultural features other than biodiversity. The most relevant goods and services provided by PAs include: raw materials, food, genetic, medicinal and ornamental resources, water purification, air quality regulation, erosion prevention, mitigation of extreme events, pollination, biological control, carbon sequestration, soil formation, primary production, and nutrient cycling (Chape et al., 2008; Millenium Ecosystem Assessment [MEA], 2005; Naughton-Treves et al., 2005). Lately, PAs are also increasingly conceived as optimum testing fields for monitoring global change (Alcaraz-Segura et al., 2009), as well as useful and cost-effective means for climate change adaptation and mitigation (Dudley et al., 2010). As seen, the "environmental targets" of PAs are quite ambitious. Not all PAs are able to provide all those goods and services, although some of them, especially the biggest ones, might.

Nevertheless, following the dominant tri- (or four-) dimensional concept of sustainability (Spangenberg, 2002), PAs are also entrusted a high number of other social and economic roles, multiplying management objectives virtually to the infinite. Some of those roles are: scientific research, recreation and tourism, inspiration for art and culture, spiritual and cognitive development, aesthetic enjoyment, preservation of traditional cultures and practices, improving social welfare, enhancing environmental education and awareness, promoting peace and security, facilitating people's participation and governance, and boosting economic development at multiple scales (Chape et al., 2008; Naughton-Treves et al., 2005).

Whereas attaining all those targets would be ideal for any PA, the challenge for PAs to do so is overwhelming. These goals are so wide and ambitious that, even in the cases where they are not directly opposed (UNESCO, 2002), it looks doubtful that any piece of land in the world could comply harmoniously with all of them.

Acknowledging the role of man within nature and its importance for the conservation of many species, habitats and ecological processes, especially in Europe where cultural landscapes are paramount (Jongman, 2002), multiple use of PAs sometimes becomes a "dogma" for PA planners, managers and policy-makers, irrespective of very different conditions and situations. Most conservation policies try to force sustainable development to happen, often through the zoning of PAs (Naughton-Treves *et al.*, 2005; UNESCO, 2002) or the placing of tourist infrastructures (Farrell & Marion, 2001), regardless of the fact that some activities may conflict with conservational ends. Such conflicts are frequent in densely populated areas between conservation and recreation activities (Tisdell, 2001).

Figure 1. Cultural features and traditions are animportant and often neglected asset of protected areas

Cultural, recreational, educational and economic values should be pursued in PA planning and management, as long as the key aim of PAs (*i.e.*, conserving biodiversity) is never left behind to satisfy other possibly conflicting objectives. If social and economic objectives are compatible with biodiversity conservation or improvement, we should try to integrate those within planning and management. If they are not, their practice attainment should be considered no more than theoretical goodwill, no matter how temping development promises may look like. If poorly conceived, executed and regulated, even the best socioeconomic proposal might turn out to be the deadliest ecological error (Witte et al., 1991).

Therefore, predicted future trends on the prevailing designation of "multiple use" PAs (IUCN categories V and VI) in the years to come (McDonald & Boucher, 2011), although a good piece of news in itself, might entail subtle risks for effective conservation of resources.

The best solution to make most uses of PAs compatible would be designing a territorial model made of different types (regarding their priority function) of large, zoned PAs connected through biodiversity-oriented managed landscapes (Mata et al., 2009; Rodríguez-Rodríguez, 2012b). However, limited resources and high land and visitor pressure, especially in densely populated areas, force us to put the preservation of ecosystem processes first and before any other use or consideration of PAs (Pressey *et al.*, 2007).

4. THE ROLE OF TOURISM IN PROTECTED AREAS

Tourism is an extremely delicate issue regarding PA management. Whereas often seen and promoted as social and economic salvation for local communities, if unregulated it may, on the one hand, lead to the deterioration or destruction of the resources of the PA and, on the other, threaten local culture. If well planned and managed, however, tourism can provide a significant source of revenue for local populations and / or PA administrators, as well as increase visitor education and environmental awareness (Chape et al., 2008).

Nevertheless, management and conservation problems are intrinsic to tourist visitation of PAs. These include: increasing use of resources, including land-use changes for developing tourism infrastructures in or around PAs; soil trampling and compacting; vegetation removal; animal disturbance; soil erosion; littering; noise making; introduction of alien species and varieties; damage to geological features, cultural sites, vegetation or public use infrastructure; increased fire risk; air pollution from transportation; animal road killing; changes in wildlife behavioural patterns due to human habituation, among others impacts (Chape et al., 2008; Farrell, 2002). Some of these impacts can be so serious that they can compromise the long-term conservation of the resources being protected (Barrado, 1999; Rodríguez-Rodríguez, 2009; Tisdell, 2001). Visitor impacts may also result in a reduced quality of recreational experiences and in conflicts among visitors (Farrell 2002; Phillips, 2000).

Figure 2. Visitors may be an intense pressure in protected areas

Additionally, public use management competes for scarce funding with other management and conservation objectives of PAs, as it is a highly resource-demanding activity (VVAA, 2000).

While zoning PAs and building tourist infrastructure might be useful at reducing impacts from tourism in PAs by directing main fluxes towards the least fragile zones of the PA (Farrell, 2002; UNESCO, 2002), they cannot usually prevent invasion of more vulnerable zones (Barrado, 1999). In fact, tourism infrastructure can attract even more visitors to the PA, and may result in its overcrowding and collapse (Morales, 2001).

As a result, although public use in PAs is a desirable and broadly-accepted concept, it should be strictly regulated or even prohibited in the cases when it is incompatible with conservation targets which must prevail in PAs.

5. FREE ACCESS TO PROTECTED AREAS? FINANCIAL AND EQUITY ASPECTS OF ENTRANCE FEES TO PROTECTED AREAS

As we just saw[1] - , impacts of visitation to PAs are numerous and serious (Chape et al., 2008; Rodríguez-Rodríguez, 2009). Whereas there is free access to most PAs in industrialised countries, no matter what land ownership is, entrance fees are common practise in other parts of the world (Emerton et al., 2006).

There are a number of reasons in favour of charging a fee for accessing PAs. First and most important, entrance fees are effective visitor filters (Emerton et al., 2006). On the one hand, they enhance the quality of visitors by discouraging the least interested visitors often having the poorest environmental awareness (Barrado, 1999) from entering the PA, thus helping prevent most unsocial behaviour (McKercher & Weber, 2008). On the other, they potentially reduce the quantity of visitors, thus diminishing the human pressure on species, soil, vegetation, ecosystems, cultural features and infrastructure (Rodríguez-Rodríguez, 2009).

The second reason, as seen at the beginning of this chapter, is that PAs provide multiple social goods and services related to improved health and well-being: recreation, tourism, sport, relax, inspiration, cultural and aesthetic enjoyment, and spiritual and cognitive development (Chape et al., 2008). The fact that most of these goods and services have not market price does not mean their economic value is negligible; rather on the contrary (Emerton et al., 2006; Kerry et al., 2003). Thus, charging an amount of money to enjoy all these benefits from PAs does not seem exaggerate.

Additionally, paying an amount of money makes the visitor conscious of the value of the visited place (Emerton et al., 2006). It gives him also a temporal "membership" to a "selected" group of people (club) with the right to respectfully enjoy the wonders of the PA.

Finally, most PAs are underfunded (Leverington et al., 2010; Nolte et al., 2010) and budgetary restrictions will tend to be the rule more than the exception in the future due to financial constraints in the public sector which currently makes up most PA funding globally (Phillips, 2000). Thus, provided that incomes from visitors are properly invested in PAs (in the same PA they are expended or in other PAs to support underfinanced or undervisited PAs), PA administrators could count on an additional source of financing to enhance management and conservation and to provide new employment opportunities (Emerton et al., 2006; Phillips, 2000).

Fee collection should be carefully planned and implemented. It must consider adequate equity prior to its implementation so none will be deprived of his or her right to enjoy PAs. Thus, discounts should be applied to low-income visitors upon appropriate certification. Dual-pricing policies charging foreign visitors higher than local ones are an equitable and socially acceptable option in developing countries (Walpole et al., 2001).

Well-thought decisions should also be taken about visitors being charged repeatedly during their visits (for instance, for car parking, access to the PA, and / or visitation of public use infrastructure), and about the appropriate quantities to be charged in view of recovery costs, so no additional charges are imposed to the PA administration (Chape et al., 2008).

FIGURE 3. Parking fee in Peñalara Natural Park,Madrid Region, Spain

6. EFFECTIVE MANAGEMENT VS. EFFECTIVE CONSERVATION

There is not consensus on the effectiveness of PAs as a global strategy for biodiversity conservation. There are examples of conservation success and failure, although the latest are rarely reported (Mora & Sale, 2011). Lack of or inappropriate monitoring and assessment activities are common management deficiencies (Leverington et al., 2010) making the assessment of PAs often difficult, arbitrary or meaningless (Parrish et al., 2003).

Therefore, whereas the protection of spaces for biodiversity conservation remains a valid and useful strategy to mitigate current environmental crisis, a general statement in favour of the actual effectiveness of PAs as a tool for biodiversity conservation is simplistic. Rather, PA effectiveness should be analysed case by case[2] - .

It is generally accepted that legal designation plus effective management result in the effective conservation of PAs (Hockings et al., 2006). It is not always the case, however. There are important pressures and threats to the conservation of PAs and their resources that operate at a scale broader than PAs and that may, therefore, spoil brilliant management efforts, as they are outside the scope, means and competence of PA managers (Alcaraz-Segura et al., 2009; Jameson et al., 2002; Parrish et al., 2003).

Impacts of global change are posing additional pressures on the conservation of biodiversity worldwide (Araújo et al., 2011). Regional impacts can also cause severe effects on faraway zones regardless of their degree of protection or management effectiveness. Acid rain on land and oil spills on sea are good examples. As a result, a PA can be legally designated, well-planned and efficiently managed and see its resources being degraded (Mora & Sale, 2011), more so in the marine environment where connectivity is higher than on land (Jameson et al., 2002).

PAs exchange matter, energy and information with their surroundings. These exchanges are vital to ecological processes underpinning biodiversity and other ecosystem goods and services (Múgica et al., 2002). Shifts in nature, extent, direction and intensity of these interactions may also, however, result in significant alteration of the ecosystem structure and function, even if optimum PA management is in place. Therefore, PAs cannot be managed in isolation from the surroundings influencing those (Radeloff et al., 2010). As a result, nearly as much care should be taken when managing zones surrounding PAs as when managing PAs themselves.

Effective management can play an important role in PA conservation. It can also contribute to adaptation and mitigation of some regional or global pressures and threats, such as climate change, through PAs (Dudley et al., 2010). However, the role of management in attaining long-term conservation of resources should not be overemphasized. PA managers can very rarely achieve effective conservation of managed sites on their own. In contrast, the importance of achieving sustainable development at all scales, integrating PAs in a wise and wider territorial planning (Rodríguez-Rodríguez, 2012b) cannot be stressed enough.

7. ASSESSING PROTECTED AREAS. SCALES AND PERSPECTIVES

For too many years, it was assumed that the resources harboured by PAs were effectively conserved just because they were legally protected. Regrettably, empirical and scientific evidences refute the previous assumption (Butchart et al., 2010). Luckily, that vision is currently outdated and much literature has been written on "paper parks", although they still prevail in certain contexts (Bonham et al., 2008; Davis, 2001).

Currently, the question is: what should be assessed? Or: how to measure PA effectiveness?

Multivariate, socio-ecological systems such as PAs are difficult, costly and time-consuming to characterize, monitor and assess. Lack of basic knowledge increases the uncertainty of what to monitor, how and why (Parrish et al., 2003).

As a result, developing a complete, ecologically-sound and meaningful-for-management assessment system has proved extremely complicated, despite several efforts developed worldwide: Hockings et al. (2000; 2006);Ervin (2003a), Parrish et al. (2003), Pomeroy et al. (2005), Gaston et al. (2006); Mallarach et al. (2008); Ioja et al. (2010); Rodríguez-Rodríguez & Martínez-Vega (in press).

To start untangling this issue, a distinction between different concepts which are often used indiscriminately (Ervin, 2003b)should be made. One is "management effectiveness" and the other is "PA effectiveness". Identifying both concepts would be the same as saying that all a PA can do to protect and conserve its resources is due to management. Once more, evidence does not seem to support this statement (Jameson et al., 2002; Mora & Sale, 2011), as we just saw[3] - .

There are unprotected places well conserved due to lack of human visitation, accessibility and exploitation (Sanderson et al., 2002). There are paper parks enjoying a good conservation status for similar reasons. And there are legally-designated, well-managed PAs being degraded due to unsustainable socioeconomic contexts and / or regional or global pressures and threats (Jameson et al., 2002; Radeloff et al., 2010).

Thus, PA effectiveness is not the same as management effectiveness. Stating so would put all the responsibility for the conservation status of PAs on managers with limited means, capacity and competences to cope with many factors outside their control. Thus, on the one hand, this identification is not precise and, on the other, it may neither be fair to PA managers (Rodríguez-Rodríguez & Martínez-Vega, in press).

Additionally, according to the "Management Effectiveness Evaluation Framework" (Hockings et al., 2000, 2006), protected area management effectiveness evaluation (PAME) should be target-driven. Therefore, if considered precisely, PAME could not be applied to most PAs worldwide lacking specific, clear, mensurable management objectives (Bertzky & Stoll-Kleemann, 2009; Pomeroy et al., 2005).

Therefore, "management effectiveness", which is an important part of "PA effectiveness", should only deal with factors directly linked to management which managers can control and address with sufficient knowledge, capacity and resources, while leaving out context variables they cannot manage or even influence.

There are different conceptions of "PA effectiveness": from its identification with "management effectiveness" (Hockings et al., 2000;Ervin, 2003a; Hockings et al., 2006), to "ecological effectiveness" (Gaston et al., 2006), "ecological integrity" (Parrish et al., 2003), or "sustainability"[4] - (Rodríguez-Rodríguez & Martínez-Vega, in press).

Language precision is fundamental in science, but so far none has come up with an ideal, agreed term to designate overall "PA effectiveness" despite the wide literature on the topic: are we measuring effectiveness?, performance?, functionality?, conservation potential?, sustainability?, success?, ecological integrity?, accomplishment? This uncertainty is because the core object being assessed is complex and integrates several different parameters (often environmental and also social and economic) not easily defined in one known-term.

A second question arises that further highlights the complexity of assessing PAs in an integrated and scientific manner: is "effectiveness" a common, "objective" concept? Is it site-specific and thus, "subjective"? Or can it be both? The answer is not straightforward. There are different perspectives on PA effectiveness assessment closely linked to the scales considered and the objectives pursued with the assessment.

Scales are important when defining PA effectiveness. For example, "management effectiveness", "ecological effectiveness" and "ecological integrity" are usually conceived as "site-specific" due to the different ecological conditions, contexts and, therefore, management needs and objectives of every PA (Hockings et al., 2006). In contrast, "sustainability assessment" proposes a common assessment and valuation method for PAs belonging to the same context (administrative, bio-geographical, or socioeconomic). Based on the most updated definition of PA (Dudley, 2008), it assesses the likelihood that a PA can conserve its natural and associated cultural resources and ecosystem services in the long-term on relevant common parameters for management and decision-making. It is also a site assessment, but it allows comparison and prioritization among the individual PAs assessed as assessment is made upon the same parameters (Rodríguez-Rodríguez & Martínez-Vega, in press).

At a broader scale, the effectiveness of conservation networks should also be assessed using landscape metrics and indicators (Burel & Baudry, 2003).

Table 1 summarizes the different complementary approaches to assessing PA effectiveness.

Table 1. Complementary approaches to assessing protected area effectiveness

Assessment	Scale	Object	*Main* target group
Ecological effectiveness	Individual PA (Site-specific)	Ecological parameters: ecosystems, species and the physical-chemical environment	*Scientists*; PA managers
Ecological Integrity	Individual PA (Site-specific)	Threats; Focal species, communities and ecological systems	*PA managers*; decision-makers
Management effectiveness	Individual PA (Site-specific; Comparable)	Management goals, objectives and strategies regarding: context, planning, inputs, processes, outputs & outcomes	*PA managers*; PA network managers; decision-makers
Sustainability / Effectiveness	Individual PA (Comparable)	State of conservation, planning, management, social and economic context, social perception and valuation, and threats to PAs	*PA network managers*; PA managers; decision-makers
PA network	Landscape; territory	Connectivity, biodiversity representativeness, gap analysis, etc.	*PA network managers*; decision-makers; territorial planners

8. PROTECTED AREA OBJECTIVES VIS À VIS MANAGEMENT OBJECTIVES

Clear objectives are fundamental for adaptive management and, therefore, for effective PA conservation (Hockings et al., 2006; Pomeroy et al., 2005). Nevertheless, it is rare that clearly-stated, science-based objectives be specified in any PA norm or management document (Bertzky & Stoll-Kleemann, 2009; Pomeroy et al., 2005) and, when management objectives are set up, they are often vague, inadequate or even contradictory (Naughton-Treves et al., 2005).

Thus, the establishment of clearly-defined objectives at different levels in the long, medium and short-terms remains one the most urgent needs for effective management and conservation of PAs.

PA objectives should be initially specified in the designation norm of each PA, according to its conservation and management characteristics. Although these objectives might change in the medium-term due to the evolution of natural systems, they are mostly stable throughout the time, as they portrait a long-term vision of the PA. They can be similar to or adapted from the general objectives of the different PA management categories proposed by the IUCN (Dudley, 2008).

In contrast with these long-term "stable" objectives, medium-term specific objectives should be developed for each management planning period; *i.e.*, the validity period of an average management plan of 4-10 years (Thomas & Middleton, 2003). These medium-term planning objectives should try to make the long-term objectives of the PA happen considering the different circumstances of the PA throughout the time.

Finally, management plans are often disaggregated into more operational annual work-plans, whereby the medium-term objectives are made specific to the management means and to the reality of the PA through achievable and measurable short-term objectives (Thomas & Middleton, 2003). These short-term management objectives should also foresee (regarding means and probabilities of occurrence) and address sudden events not previously planned against which rapid action is required (Chape et al., 2008).

Figure 1 shows the conceptualization of the different types of objectives in PAs and their main characteristics.

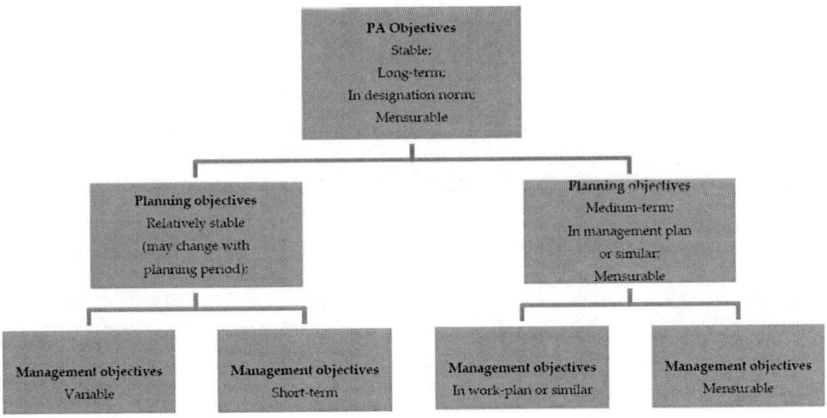

FIGURE 4. Outline of the different levels and objectives of PAs and their main characteristics.

Sensible objective definition leads to adequate planning and, if adequate means are available, to proper management of the PA. However, clear definition and precise forms of measurement of the different objectives are needed at all levels for management planning to be fully effective (Thomas & Middleton, 2003).

9. THE ROLE OF SOCIETY IN PROTECTED AREA MANAGEMENT

PAs are portions of territory where multiple interests coexist and often collide (McKercher & Weber, 2010; Phillips, 2000). Although their main aim is to conserve biodiversity and the other ecosystem goods and services and associated cultural features, they include other environmental, social, economic and governance factors deserving attention. As such, they cannot and should not be managed in a unidirectional manner.

Neither decision-makers nor PA managers can cope with current challenges to PAs alone. An increasingly complex world requires increasingly innovative solutions, also regarding PA management. Thus, they should look for allies in other stakeholders if effective management and affordable enforcement is to be achieved.

Bottom-up, participatory approaches to PA management may be more resource and time-consuming than technical top-down approaches, but they are generally accepted better by all stakeholders and, as a result, they are often more effective and enduring. Consensus results in decisions being more legitimate, and in fewer conflicts (Fraser et al., 2006).

There are two priority groups whose close collaboration should actively be sought: scientists and local populations. Management measures can be more efficient, cost-effective and easily-implemented if they are based on sound science (Chape et al., 2008). Therefore, a stronger, deeper and more trustful relationship between PA managers and scientists is the first priority to enhance effective, participatory management (Zamora, 2010).

Local populations are also key to effective management (Rodríguez-Rodríguez, 2012a). Their views and attitudes towards PAs and their management can determine the success of most management initiatives (Múgica & Gómez-Limón, 2002; Borrini-Feyerabend et al., 2004).

Building a positive attitude towards PAs by local populations consists not only in informing individuals or associations (NGOs, neighbours, etc.) on management issues, but also in consulting them, making joint decisions and seeking their input to actually carry out some management activities (Múgica & Gómez-Limón, 2002). Environmental volunteering can be a very adequate way of participation of local populations into management, and also an effective means of reinforcing the identity of residents with the values, resources and governance of the PA.

Other relevant stakeholders to be considered in PA management are businessmen, politicians and tourists. New opportunities for business and employment related to the PA should be sought, mainly among local enterprises, as long as no negative impacts on the PA or its resources are guaranteed (Phillips, 2000). Therefore, economic activities affecting the PA should be carefully evaluated, planned and regulated in the management plan of each PA. Synergies among business, conservation and management practices should be potentiated; for instance, grazing in grasslands may have better ecological effects on biodiversity when complementing mowing than mowing alone (Metera et al., 2010). Thus, farmers can get free fodder for their animals and make profit by selling farm products associated with the PA, whereas PA managers can save money in ecosystem management.

Politicians should also be incorporated to valuating and managing PAs more actively. Their decisions can determine the future of a PA, including permission for aggressive activities within PAs or their surroundings, diminution or even disappearance of the protection status of a PA. Thus, they should be made aware on the importance of PAs and of their decisions for biodiversity and humans in the long-term. Social groups such as managers, scientists, local populations and NGOs can play an important role in lobbying for the cause of PAs.

Similarly, tourists should be made fully aware of their potential impacts on visited PAs (Hillery et al., 2001) through environmental education and volunteering, and public use infrastructure. Their collaboration with managers should be sought to obtain proper behaviour from them and, where possible, funding for the PA (Rodríguez-Rodríguez, 2012a).

10. KEEPING PACE WITH SCIENCE

PA knowledge and management do not run at the same speed (Nolte et al., 2010; Pullin, 2002). Communication and trust between scientists and managers can also be much improved (Zamora, 2010). As a result, it is rare that sound, updated science be incorporated by managers and decision-makers and applied in everyday management (Pressey et al., 2007). Therefore, common management practices are frequently based just on experience, habit, or at best, on partial scientific knowledge (Pullin, 2002).

Sometimes, lack of scientific knowledge is due to overwork by managers. Too much field and / or administrative work leave them with little or no time or will to get updated on new management issues regularly (Pullin,

2002). Lack of adequate funding, interest or training programmes can also erode appropriate scientific knowledge among managers. Thus, staff training is an overall need in many countries (Leverington et al., 2010; Nolte et al., 2010).

And yet widespread updated scientific knowledge is vital for management and conservation effectiveness (Pressey et al., 2007). Resources devoted to conservation could be more successfully and wisely used if management practices were backed up by sound science (Pullin, 2002). Staff training should be considered an integral part of the management cycle (Kopylova & Danilina, 2011).

However, knowledge-transfer should not be unidirectional, from scientists to managers. The publication or communication of empirical management result and practices, be they successful or a failure, is also very important for advancing towards a more effective management of PAs and can provide science with interesting new research topics and experiences.

Figure 5. Presentations and workshops are useful training options

In conclusion, regular collaboration among scientists, managers and decision-makers is key for effective and efficient management (Pullin, 2002; Rodríguez-Rodríguez & Martínez-Vega, in press). Current disconnection and suspicion should be overcome if more effective ways of conserving PA are to be sought. This collaboration can take the form of regular working-groups or workshops. Ideally, researchers should be incorporated to the regular PA managing staff (Chape et al., 2008), be them restricted to specific PAs, covering a network of PAs, or both. Thus, actual management activities should be done considering both, sound science and best practice.

11. RETHINKING PROTECTION SCALES: FROM PROTECTED AREAS TO PROTECTED TERRITORIES

Current *in situ* strategies for biodiversity conservation are not having enough success to stop or reverse biodiversity loss (Butchart et al., 2010; Mora & Sale, 2011), and thus biodiversity and the other ecosystem goods and services continue to be degraded (MEA, 2005).

Most PAs, especially in Europe, are too small to maintain viable populations within their boundaries in the long-term (Pullin, 2002). Under these circumstances, connectivity among PAs becomes fundamental (Jongman, 2002).

Ecological networks are a conceptual and practical step forward with regard to the initial concept of PA as islands in the territory. However, ecological networks are not without theoretical and practical shortcomings (Boitani et al., 2007). In addition, the current drivers causing environmental degradation are so intense than even ecological networks have been overcome by reality.

Enhanced connectivity of PAs through ecological corridors is thought to have increased PA conservation effectiveness (Beier & Noss, 1998). However, the wide range of scales taking part in ecological processes underpinning ecosystem goods and services (Pressey et al., 2007) make ecological corridors only partially effective, and only under some circumstances (Brunckhorst, 2000; Tzoulas et al., 2007). Moreover, the zones surrounding PAs are becoming increasingly artificial as a result of human activities and are therefore not accomplishing properly their buffer target (Radeloff et al, 2010). As a result, the effective protection and wise management of whole landscapes/seascapes surrounding PAs can be as important as the protection and management of PAs themselves (Jameson et al., 2002).

There is a growing consensus that effective biodiversity conservation cannot take place if complementary conservation scales broader than PA scale are not considered (Mata et al., 2009; Mora & Sale, 2011; Múgica et al.,

2002; Rodríguez-Rodríguez, 2012b). Going from connected islands to connected landscapes is a desirable first step.

National conservation policies must be integrated and coordinated in order to be fully effective. The protection and management of isolated PAs, ecological networks or whole landscapes/seascapes alone is not sufficient for long-term conservation. Additional public conservation policies, mainly related to territorial planning, ecological restoration and environmental impact assessment (of projects, policies, plans and programmes), should complement the establishment and maintenance of PAs at the national level.

Full consideration of ecological processes should also be carefully considered and integrated in other public policies seriously affecting the steady provision of ecosystem goods and services such as housing, mining, fishing, transport or industry. Ecological considerations should always guide (and limit, when necessary) social and economic development policies and strategies if sustainable development is to be achieved (Naredo & Frías, 2005).

However, no matter how effective a national conservation policy is, its effectiveness will always be limited by regional or global trends, making local or national efforts only partially successful (Mora & Sale, 2011).

Future conservation strategies should be regional and global, as the driving forces pressing biodiversity are. Ecologically, it makes little sense planning and managing PAs restricted to local or national administrative boundaries. Thus, deeper and closer international collaboration is needed when planning for conservation.

Focus should change from PAs to protected territories of high conservation value following science-based criteria: species richness, diversity, taxonomic uniqueness, degree of threat, endemism, unusual ecological or evolutionary phenomena, or notable provision of ecosystem goods and services (Pullin, 2002). The ecorregional approach proposed by Olson et al. (2001) seems a promising and adequate scale to plan and manage conservation from an ecological perspective. This is the conservation approach followed by relevant conservation organizations, such as the Nature Conservancy (TNC, 2001).

Management should be much more collaborative and imply different stakeholders from different sovereign authorities. This form of closer international cooperation can have positive impacts in other public policies such as human development or defence (Sandwith et al., 2001).

Even if conservation and management strategies becomeglobal and effective, they will only be a limited solution to current and increasingly unsustainable human development trends (Millenium Ecosystem Assessment [MEA], 2005; Sanderson et al., 2002). Thus, if an enduring solution to the environmental crisis (and thus, sustainable development) is to be achieved, global determined and courageous measures must be taken in order to counter current population, development and consumption trends (Mora & Sale, 2011). All public policies must take into account the bio-physical limits of the Biosphere and they should be planned and implemented accordingly, considering the long-term consequences of today's decisions.

12. FINAL REMARKS: MANAGING THE BIOSPHERE FOR CONSERVATION

The astonishing variety of life on Earth and other vital ecosystem goods and services are severely threatened by human activities well beyond the capacity and competence of PA managers, PA network managers or even decision-makers.

PAs are just part of the solution to the environmental crisis. We should not be excessively optimistic about their ultimate outcome as the main biodiversity conservation strategy. They are a shock therapy for a patient who is turning rapidly to worst.

In situ conservation strategies failed by protecting only some areas on diverse criteria, and leaving the rest of the territory without any form of development control or regulation. We should change (have changed, long ago) our way of thinking and practicing conservation. Conservation should be global, literally. I mean that the entire Biosphere should be protected and, where necessary, appropriately managed, leaving only a number of places of extraordinary social and economic value out of protection for intensive development and human activities, such as mines, cities, transportation corridors, some croplands and plantations.

Some may think this is asking too much, but all signs point to the fact that we are heading towards an unmanageable disaster by surpassing vital ecosystem thresholds (Butchart et al., 2010; MEA, 2005; Sanderson et al., 2002), and that we are running out of time to change our ways. Even the most energetic global initiatives taken today might not be sufficient to prevent the notable impoverishment and eventual collapse of life on Earth.

13. CASE STUDY: MANAGEMENT ASSESSMENT OF THE PROTECTED AREAS OF THE AUTONOMOUS REGION OF MADRID, SPAIN

From 2008 to 2011, we developed the System for the Integrated Assessment of Protected Areas (SIAPA) in the Institute of Economics, Geography and Demography of the Spanish National Research Council (Rodríguez-Rodríguez & Martínez-Vega, in press). We applied the SIAPA to the ten PAs of a paradigmatically unsustainable region: the Autonomous Region of Madrid, Spain (Jiménez et al., 2005). In this epigraph, we show the part of the study's results of the implementation of the SIAPA (Rodríguez-Rodríguez & Martínez-Vega, under review)dealing with management of the ten PAs assessed.

The Autonomous Region of Madrid is a Spanish region of slightly over 8,000 km^2 in the centre of Spain (Figure 6). It has the highest per-capita income in Spain. It also has a rich natural and cultural patrimony that is jeopardized by the implementation of an intensive resource-consuming development model (Mata et al., 2009; Naredo & Frías, 2005), which has led to an increase in economic standards, but also to massive residential and infrastructure developments throughout the region in the past 20 years (Fernández-Muñoz, 2008; Gago et al., 2004).

The protected areas (PAs) of the region of Madrid face numerous pressures, mainly from massive visitor use and intensive land-use transformation (Rodríguez-Rodríguez, 2008) in addition to more general threats arising from global change, such as climate change (Araújo et al., 2011). These unsustainable development dynamics raise concern that effective long-term conservation of the goods and services provided by the region's ecosystems cannot be achieved despite the fact that up to 46 % of its territory is under some kind of protection regime (Mata et al., 2009).

The ten PAs selected for this study are shown in Table 2 and Figure 6.

The 12 management indicators against each PA was assessed are shown in Table 3

Table 2. Main characteristics of the ten protected areas assessed

Protected area	Abbreviation	Area (ha)	Designation year	MAI
Peñalara Natural Park	Peñalara NP	11,637	1990	1.6
Cuenca Alta del Manzanares Regional Park	Cuenca Alta RP	52,796	1985	1.2
Sureste Regional Park	Sureste RP	31,550	1994	1.1
Curso Medium del Río Guadarrama y su entorno Regional Park	Guadarrama RP	22,116	1999	1.1
Pinar de Abantos y Zona de la Herrería Picturesque Landscape	Pinar Abantos y Herrería PL	1,538	1961	0.6
Natural Site of National Interest of Hayedo de Montejo de la Sierra	NSNI Hayedo Montejo	250	1974	1.1
El Regajal-Mar de Ontígola Natural Reserve	Regajal-Ontígola NR	629	1994	0.6
Laguna de San Juan Fauna Refuge	Laguna San Juan FR	47	1991	0.7
Natural Monument of National Interest of Peña del Arcipreste de Hita	NMNI Peña Arcipreste	3	1930	0.0
Preventive Protection Regime of Soto del Henares	PPR Soto Henares	332	2000	0.3

MAI: Value of the Management Index of the SIAPA (in a 0 – 2 point standard scale)

Figure 6. Location of the Autonomous Region of Madrid and the ten protected areas included in the study

Table 3. Management indicators included in the SIAPA

Category	Indicator
Management	Degree of characterization of the protected area
	Degree of fulfillment of management objectives
	Evolution of feature(s) for which the protected area was designated
	Existence of sufficient management staff
	Evolution of investment in the protected area
	Effectiveness of public participation bodies
	Production and distribution of an annual report on activities and outcomes
	Easiness to identify the protected area
	Public use infrastructure
	Existence of environmental education and volunteering activities
	Sanctioning procedures
	Monitoring activities

Despite the importance of management for the effective conservation of PAs (Hockings et al., 2006; Pomeroy et al., 2005), the Management Index (MAI) did not reach a minimum desirable value for the PAs of the Autonomous Region of Madrid: 0.7 / 2 points. Only Peñalara NP had "Adequate" management. Whereas four PAs, including the three regional parks and the NSNI Hayedo Montejo, had a "Moderate" management valuation, the other five PAs had all "Deficient" valuations. The lowest value was for NMNI Peña Arcipreste (MAI = 0), as all its indicators scored 0 or lacked any information to be valuated. The MAI of PPR Soto Henares was also extremely low (MAI = 0.3).

Low scores were found for two PAs which had a director at the moment of the assessment: Regajal-Ontígola NR (MAI = 0.6) and Laguna San Juan FR (MAI = 0.7). This might be due to the scarce attention paid by the administration to these two PAs, and because of the excessive amount of work of the director, who had to make compatible the management of both PAs with the management of one of the biggest and most conflicting PAs: the Sureste RP (Rodríguez-Rodríguez, 2008).

At least 2 of the 10 PAs assessed can be deemed "paper parks", as no management activities are performed there: NMNI Peña Arcipreste and PPR Soto Henares. Other 2 PAs (FR Laguna San Juan and El Regajal-Mar de Ontígola NR) enjoy only sporadic management activities, including surveillance. All these PAs have, at best, passive management responding reactively to management and conservation needs. Another PA (PP Pinar Abantos) has regular surveillance as the only habitual management activity. The other 5 PAs (4 parks and the NSNI Hayedo Montejo) do enjoy diverse active management regularly.

Figure 7. Fauna observation installation in Sureste Regional Park (left figure) and information panels at the entrance of El Regajal-Mar de Ontígola Nature Reserve (right figure). Poor management and inadequate visitor behaviour result in the deterioration of PAs and their infrastructure

The indicators that scored highest for the ten PAs were the "existence of sufficient management staff", "evolution of investment in the PA", and "existence of environmental education and volunteering activities", all with a mild 1.1 score over a maximum of 2 points.

In contrast, the indicators that scored lowest were the "effectiveness of public participation bodies" (0.0 points) and the "easiness to identify the PA" (0.4 points). Additionally, two of the 12 indicators assessed under "Management" could not be assessed due to lack of rough data: "evolution of the feature(s) for which the PA was designated", and "sanctioning procedures".

A set of ultimate reasons could explain the poor quality of management of the PAs of the Autonomous Region of Madrid, as mentioned by Nolte et al. (2010) and Pomeroy et al. (2005): competence and information dispersal among different administrative units; poor coordination among these units on policies and activities related to PAs; shortage of human and material resources devoted to management; deficient scientific knowledge; and weak institutional support to biodiversity conservation policies.

In contrast to what we had expected, the MAI was not correlated with any other index used in the study, although it might have some degree of relatedness with the Planning Index (Rodríguez-Rodríguez & Martínez-Vega, in press). The non-significant, low correlation between the MAI and the State of Conservation Index and the non-significant, moderate correlation between the MAI and the Sustainability (Effectiveness)[5] - Index suggest that management is not as determinant a factor for the effective conservation of PAs.

Although management might not be as fundamental a factor for the effective long-term conservation of PAs as it had been stated, it remains nevertheless one of the most important components of PA effectiveness. The importance of management increases in contexts of numerous and intensive inner and outer pressures on PAs, as it happens in the Autonomous Region of Madrid. Thus, much more effort will be needed regarding sustainable territorial management at different scales in this urban, densely-populated region to effectively safeguard its valuable natural and cultural resources for future generations (Mata et al., 2009; Rodríguez-Rodríguez, 2012b).

NOTES

[1] - See heading 4

[2] - See heading 7

[3] - See heading 6

[4] - In our preliminary work (Rodríguez-Rodríguez & Martínez-Vega, in press), we identified "environmental" or "hard" "sustainability" as the core concept to be assessed when evaluating PAs, according to the goal of long-term conservation of natural and cultural resources established in the latest definition of PAs by the IUCN (Dudley, 2008) and in the original definition of sustainable development (WCED, 1987). However, some colleagues were reluctant to the use and formalization of the term "sustainability" as a result of the vagueness of its original definition and its different interpretations, so we changed the term to "PA effectiveness", also meaning the PA capacity or potential to conserve its resources in the long-term. Nevertheless, we think "sustainability" is a valid focus for assessing PAs with no more difficulties in its definition and formalization than any other complex integrated concept such as "effectiveness".

[5] - See Chapter 7

REFERENCES

1. D. Alcaraz-Segura, J. Cabello, J. M. Paruelo, M. Delibes, 2009Use of Descriptors of Ecosystem Functioning for Monitoring a National Park Network: A Remote Sensing Approach. Environmental Management 4313848
2. M. B. Araújo, D. Alagador, M. Cabeza, D. Nogués-Bravo, W. Thuiller, 2011Climate change threatens European conservation areas. Ecology Letters 145484492
3. D. A. Barrado, 1999Actividades de ocio y recreativas en el medio natural de la Comunidad de Madrid.

4. La ciudad a la búsqueda de la naturaleza [Nature leisure and recreacional activities in the Autonomous Region of Madrid.The city in search of nature]. Consejería de Medio Ambiente, Comunidad de Madrid, Madrid, Spain.
5. P. Beier, R. F. Noss, 1998Do habitat corridors provide connectivity? Conservation Biology 12612411252
6. M. Bertzky, S. Stoll-Kleemann, 2009Multi-level discrepancies with sharing data on protected areas: What we have and what we need for the global village. Journal of Environmenal Management 901824
7. C. A. Bonham, E. Sacayon, E. Tzi, 2008Protecting imperilled "paper parks"; potential lessons from the Sierra Chinajá, Guatemala. Biodiversity Conservation 1715811593
8. L. Boitani, A. Falcucci, L. Maiorano, C. Rondinini, 2007Ecological Networks as Conceptual Frameworks or Operational Tools in Conservation. Conservation Biology 21614141422
9. G. Borrini-Feyerabend, A. Kothary, G. Oviedo, 2004Indigenous and Local Communnities and Protected Areas: Towards Equity and Enhanced Conservation. IUCN. Gland, Switzerland and Cambridge, UK.
10. D. Brunckhorst, 2000Bioregional Planning. Resource management beyond the new millennium. Routledge, London.
11. F. Burel, J. Baudry, 2003Landscape Ecology. Concepts, Methods and Applications. Science Publishers INC. Enfield, USA.
12. S. H. M. Butchart, M. Walpole, B. Collen, A. van Strien, J. P. W. Scharlemann, R. E. E. Almond, J. E. M. Baillie, B. Bomhard, C. Brown, J. Bruno, K. E. Carpenter, G. M. Carr, J. Chanson, A. M. Chenery, J. Csirke, N. C. Davidson, F. Dentener, M. Foster, A. Galli, J. N. Galloway, P. Genovesi, R. D. Gregory, M. Hockings, V. Kapos, J. Lamarque, F. , F. Leverington, J. Loh, M. A. Mc Geoch, L. Mc Rae, A. Minasyan, M. H. Morcillo, T. E. E. Oldfield, D. Pauly, S. Quader, C. Revenga, J. R. Sauer, B. Skolnik, D. Spear, D. Stanwell-Smith, S. N. Stuart, A. Symes, M. Tierney, T. D. Tyrrell, J. C. Vié, R. Watson, 2010Global biodiversity: indicators of recent declines. Science 328598211641168
13. CBD.Convention on Biological Diversity. (1992Available from: http://www.cbd.int/convention/about.shtml
14. S. Chape, M. Spalding, M. D. Jenkins, 2008The World's Protected Areas: Status, Values and Prospects in the 21st Century. UNEP World Conservation Monitoring Centre, University of California Press, Berkeley, USA.
15. J. B. . Davis, Ed, 2001Paper Parks: Why they happen and what can be done to change them? MPA News. International News and Analyses on Marine Protected Areas 21116Available from: http://depts.washington.edu/mpanews/MPA20.pdf
16. N. . Dudley, Ed, 2008Guidelines for Applying Protected Area Management Categories. IUCN,Gland, Switzerland.
17. N. Dudley, S. Stolton, A. Belokurov, L. Krueger, N. Lopoukhine, Kinnon. K. Mac, T. Sandwith, N. Sekhran, 2010Natural Solutions. Protected areas helping people cope with climate change. IUCN-WCPA, TNC, UNDP, WCS, The World Bank and WWF. Gland, Switzerland, Washington DC and New York, USA.
18. L. Emerton, J. Bishop, L. Thomas, 2006Sustainable Financing of Protected Areas: A global review of challenges and options. IUCN. Gland, Switzerland, and Cambridge, UK.
19. J. Ervin, 2003aRapid Assessment and Prioritization of Protected Area Management (RAPPAM) Methodology. WWF, Gland, Switzerland.
20. J. Ervin, 2003bProtected area assessment in perspective. BioScience 539819822
21. T. A. Farrell, J. L. Marion, 2001Identifying and assessing ecotourism visitor impacts at eight protected areas in Costa Rica and Belize. Environmental Conservation 283215225
22. T. A. Farrell, 2002The Protected Area Visitor Impact Management (PAVIM) Framework: a simplified process for making management decisions. Journal of Sustainable Tourism 103151

23. S. Fernández-Muñoz, 2008Participación pública, gobierno del territorio y paisaje en la Comunidad de Madrid [Public participation, territorial governance and landscape in the Autonomous Region of Madrid]. Boletín de la A.G.E 4697119
24. E. D. G. Fraser, A. J. Dougill, W. E. Mabee, M. Reed, P. Mc Alpine, 2006Bottom up and top down: Analysis of participatory processes for sustainability indicator identification as a pathway to community empowerment and sustainable environmental management. Journal of Environmental Management 782114127
25. C. Gago, M. Serrano, F. J. Antón, 2004Repercusiones de las carreteras orbitales de la Comunidad de Madrid en los cambios de usos del suelo [Land-use impacts of orbital roads of the Autonomous Region of Madrid]. Anales de Geografía 24145167
26. K. J. Gaston, K. Charman, S. F. Jackson, P. R. Armsworth, A. Bonn, R. A. Briers, C. S. Q. Callaghan, R. Catchpole, J. Hopkins, W. E. Kunin, J. Latham, P. Opdam, R. Stoneman, D. A. Stroud, R. Tratt, 2006The ecological effectiveness of protected areas: the United Kingdom. Biological Conservation 13217687
27. M. Hillery, B. Nancarrow, G. Griffin, G. Syme, 2001Tourist perception of environmental impact. Annals of Tourism Research 28853867
28. M. Hockings, S. Stolton, N. Dudley, 2000Evaluating effectiveness: a framework forassessing the management of protected areas. IUCN. Gland, Switzerland, and Cambridge, UK.
29. M. Hockings, S. Stolton, F. Leverington, N. Dudley, J. Courrau, 2006Evaluating effectiveness. A framework for assessing management effectiveness of protected areas. (Second Ed.) IUCN, Gland, Switzerland, and Cambridge, UK.
30. C. I. Ioja, M. Patroescu, L. Rozylowicz, V. D. Popescu, M. Verghelet, M. I. Zotta, M. Felciuc, (2010, The efficacy of Romania's protected areas network in conserving biodiversity. Biological Conservation 1431124682476
31. S. C. Jameson, M. H. Tupper, J. M. Ridley, 2002The three screen doors: Can marine'protected' areas be effective? Marine Pollution Bulletin 4411771183
32. C. N. Jenkins, L. Joppa, (2009, Expansion of the global terrestrial protected area system. Biological Conservation 1421021662174
33. L. Jiménez, F. Prieto, J. Riechmann, A. Gómez-Sal, (2005, Sostenibilidad en España 2005. Informe de Primavera [Sustainability in Spain 2005. Spring Report]. Observatorio de la Sostenibilidad de España, Alcalá de Henares, Spain.
34. R. Jongman, 2002Homogenisation and fragmentation of the European landscape: ecological consequences and solutions. Landscape and Urban Planning 58211221
35. R. Kerry, J. Paavola, P. Cooper, S. Farber, V. Jessamy, S. Georgiu, 2003Valuing nature: lessons learned and future research directions. Ecological Economics 463493510
36. S. L. Koplylova, N. R. . Danilina, Eds., 2011Protected Area Staff Training: Guidelines for Planning and Management. IUCN, Gland, Switzerland.
37. F. Leverington, K. Lemos, J. Courrau, H. Pavese, C. Nolte, M. Marr, L. Coad, N. Burguess, B. Bomhard, M. Hockings, 2010Management effectiveness evaluation in protected areas- a global study. (Second Edition). University of Queensland, Brisbane, Australia.
38. J. M. Mallarach, J. Germain, X. Sabaté, X. Basora, 2008Protegits de fet o de dret? Primera avaluació del sistema d'espais naturals protegits de Catalunya [Actually or legally protected? First assessment of the system of protected areas of Catalonia]. Institució Catalana d'Història Natural. Available from: http://ichn.iec.cat/Avaluaci%C3%B3%20d%27espais.htm
39. R. Mata, L. Galiana, F. Allende, S. Fernández, P. Lacasta, N. López, P. Molina, C. Sanz, 2009Evaluación del paisaje de la Comunidad de Madrid: de la protección a la gestión territorial [Evaluation of the landscape of the Autonomous Region of Madrid: from protection to territorial Management]. Urban 143457
40. R. I. Mc Donald, T. M. Boucher, 2011Global development and the future of the protected area strategy. Biological Conservation 1441383392
41. B. Mc Kercher, K. Weber, 2008Rationalising inappropriate behaviour at contested sites. Journal of Sustainable Tourism 164369385

42. MEA. Millennium Ecosystem Assessment.2005Ecosystems and Human Well-Being: Synthesis. Island Press. Washington DC.
43. E. Metera, T. Sakowsky, K. Sloniewsky, B. Romanowicz, 2010Grazing as a tool to maintain biodiversity in grasslands. A review. Animal Science Papers and Reports 284315334
44. C. Mora, P. F. Sale, 2011Ongoing global biodiversity loss and the need to move beyond protected areas: a review of the technical and practical shortcomings of protected areas on land and sea. Marine Ecology Progress Series 434251266
45. J. Morales, 2001Guía Práctica para la Interpretación del Patrimonio. El arte de acercar el legado natural y cultural al público visitante [Practice Guide for the Interpretation of Patrimomy. The art of approaching natural and cultural legacies to visitors].Junta de Andalucía, Seville, Spain.
46. M. Múgica, J. . Gómez-Limón, Coords., 2002Plan de Acción para los espacios naturales protegidos del Estado español [Action Plan for the protected areas of the Spanish State]. Fundación Francisco González Bernáldez, Madrid, Spain.
47. M. Múgica, J. V. De Lucio, C. Martínez-Alandi, P. Sastre, J. A. Atauri-Mezquida, C. Montes, 2002Integración territorial de espacios naturales protegidos y conectividad ecológica en paisajes mediterráneos [Territorial integration of protected areas and ecological connectivity in Mediterranean landscapes]. Consejería de Medio Ambiente, Junta de Andalucía, Seville, Spain.
48. J. M. Naredo, J. Frías, 2005Desarrollo: la síntesis del "desarrollo sostenible" con especial referencia a la Comunidad de Madrid [Development: the síntesis of "sustainable development" with special referente to the Autonomous Region of Madrid]. In Sánchez-Herrera, F., 738Cuartas Jornadas Científicas del Parque Natural de Peñalara y del Valle de El Paular. Conservación y desarrollo socioeconómico en Espacios Naturales Protegidos [Fourth Scientific Meeting of Peñalara Natural Park and El Paular Valley. Conservation and Socioeconomic Development in Protected Areas]. Consejería de Medio Ambiente y Ordenación del Territorio, Comunidad de Madrid, Madrid, Spain.
49. L. Naughton-Treves, M. Buck, K. Brandon, 2005The Role of Protected Areas in Conserving Biodiversity and Sustaining Local Livelihoods. Annual Review of Environmental Resources 30219252
50. C. Nolte, F. Leverington, A. Kettner, M. Marr, G. Nielsen, B. Bomhard, S. Stolton, S. Stoll-Kleemann, M. Hockings, 2010Protected Area Management Effectiveness Assessments in Europe. A review of application, methods and results. University of Greifswald, Germany.
51. D. M. Olson, E. Dinerstein, E. D. Wikramayanake, N. D. Burgess, G. V. N. Powell, E. C. Underwood, Amico. J. A. D', I. Itoua, H. E. Strand, J. C. Morrison, C. J. Loucks, T. F. Allnutt, T. H. Ricketts, Y. Kura, J. F. Lamoreux, W. W. Wettengel, P. Hedao, K. R. Kassem, 2001Terrestrial Ecoregions of the World: A New Map of Life on Earth. BioScience 5111933938
52. J. D. Parrish, D. P. Braun, R. S. Unnasch, 2003Are we conserving what we say we are? Measuring the ecological integrity within protected areas. BioScience 539851860
53. A. . Phillips, Ed., 2000Financing Protected Areas. Guidelines for Protected Area Managers. IUCN, Gland, Switzerland and Cambridge, UK.
54. R. S. Pomeroy, J. E. Parks, L. M. Watson, 2005How is your MPA doing? A methodology for evaluating the management effectiveness of marine protected areas. IUCN, Gland, Switzerland, and Cambridge, United Kingdom.
55. R. L. Pressey, M. Cabeza, M. E. Watts, R. M. Cowling, K. A. Wilson, 2007Conservation planning in a changing world. Trends in Ecology and Evolution 2211583592
56. A. Pullin, 2002Conservation Biology. Cambridge University Press, Cambridge, UK.
57. V. C. Radeloff, S. I. Stewart, T. J. Hawbaker, U. Gimmi, A. M. Pidgeon, C. H. Flather, R. B. Hammer, D. P. Helmers, 2010Housing growth in and near United States protected areas limits their conservation value. PNAS 1072940945
58. D. Rodríguez-Rodríguez, 2008Los espacios naturales protegidos de la Comunidad de Madrid. Principales amenazas para su conservación [The protected areas of the

Autonomous Region of Madrid. Main threats to their conservation]. Editorial Complutense, Madrid, E-Book, Available from: http://www.ucm.es/BUCM/ecsa/36254.php?id=187

59. D. Rodríguez-Rodríguez, 2009Mitigación de los impactos del turismo en espacios naturales protegidos y mejora de su financiación a través de medidas económicas. El caso de la Comunidad de Madrid [Tourism impact mitigation and enhanced financing of protected areas through economic measures. The case of the Autonomous Region of Madrid]. Boletín de la A.G.E. 50217238
60. D. Rodríguez-Rodríguez, 2012aPerception use and valuation of protected areas by local populations in an economic crisis context. Environmental Conservation 392162171
61. D. Rodríguez-Rodríguez, 2012bIntegrated Networks. A territorial planning proposal for biodiversity conservation in urban, densely populated regions. The case of the Autonomous Region of Madrid, Spain.Journal of Environmental Planning and Management 555667683
62. D. Rodríguez-Rodríguez, J. . Martínez-Vega, press. In, Proposal of a system for the integrated and comparative assessment of protected areas. Ecological Indicators, doi:j.ecolind.2012
63. D. Rodríguez-Rodríguez, J. . Martínez-Vega, review. Under, Results of implementation of the System for the Integrated Assessment of Protected Areas (SIAPA) to the protected areas of the Autonomous Region of Madrid (Spain). Ecological Indicators.
64. E. W. Sanderson, M. Jaiteh, M. A. Levy, K. H. Redford, A. V. Wannebo, G. Woolmer, 2002The Human Footprint and the Last of the Wild. BioScience 5210891904
65. T. Sandwith, C. Shine, L. Hamilton, D. Sheppard, 2001Transboundary Protected Areas for Peace and Co-operation. IUCN, Gland, Switzerland and Cambridge, UK.
66. J. H. Spangenberg, 2002Environmental space and the prism of sustainability: frameworks for indicators measuring sustainable development. Ecological Indicators 23295309
67. L. Thomas, J. Middleton, 2003Guidelines for Management Planning of Protected Areas. IUCN, Gland, Switzerland, and Cambridge, UK.
68. C. Tisdell, 2001Tourism Economics, the Environment and Development. Analysis and Policy. Edward Elgar Publishing, Cheltenham, UK.
69. TNC. The Nature Conservancy.2001Conservation by Design: A Framework for Mission Success. Arlington, Virginia, USA. Available from: http://www.fws.gov/southeast/grants/pdf/cbd_en.pdf
70. K. Tzoulas, K. Korpela, S. Venn, V. Yli-Pelkonen, A. Kazmierczak, J. Niemela, P. James, 2007Promoting ecosystem and human Health in urban areas using Green Infrastructure: A literature review. Landscape and Urban Planning 813167178
71. UNEP-WCMC.United Nations Environment Programme-World Conservation Monitoring Centre. (2008State of the world`s protected areas 2007. An annual review of global conservation progress. UNEP-WCMC, Cambridge, UK.
72. UNESCO.2002Biosphere reserves: Special places for people and nature. UNESCO, Paris, France.
73. V. V. A. A. (2000, La conservación de la naturaleza: aspectos clave y retos de futuro. Seminarios de Conservación de la Naturaleza (1996-1999) [Nature conservation: key aspects and future challenges. Nature Conservation Seminars (1996-1999)] Serie Documentos nº 30, Centro de Investigación "Fernando González Bernáldez", Soto del Real, Madrid, Spain.
74. M. J. Walpole, H. J. Goodwin, K. G. R. Ward, 2001Pricing policy for tourism in protected areas: lessons from Komodo National Park, Indonesia. Conservation Biology 151218227
75. WCED.World Commission on Environment and Development. (1987Our Common Future. Oxford University Press, Oxford, UK.

76. F. Witte, T. Goldschsmidt, P. C. Goudswaard, W. Ligtvoet, M. J. P. Van Oijen, J. H. Wanink, 1991Species Extinction and Concomitant Ecological Changes in Lake Victoria. Netherlands Journal of Zoology 422214232
77. R. Zamora, 2010Las Áreas Protegidas como Observatorios del Cambio Global [Protected Areas as Observatories of Global Change]. Ecosistemas 19214

CHAPTER 7

International Trends in Protected Areas Policy and Management

Joel Heinen

[1] *Department of Earth and Environment, Florida International University, Miami, FL,, USA*

1. INTRODUCTION

Traditional human societies have protected natural areas for various cultural purposes for millennia. Examples include the sacred forests of South Asia and parts of Africa, sacred burial grounds of some native American groups and traditional royal hunting reserves in many parts of Europe, Asia and Africa, which were generally only seasonally opened for hunting by royalty (Borgerhoff Mulder & Coppolillo, 2006). The modern concept of the national ownership and protection of natural areas for the benefit of society at large is a much more recent phenomenon; the United States became the first country to conserve nationally protected areas with the creation of Yellowstone National Park in 1872. This Category II (below) protected area is now close to 9,000 square kilometers in size and was inscribed as a World Heritage Natural Site in 1978.

The management of many of the earliest protected areas would be at odds with modern conservation practices. For example, for several decades after its creation, the US Calvary managed Yellowstone and mounted soldiers regularly hunted bison and elk for food - and wolves as vermin - within its borders. By the 1930s, wolves had been eradicated from the park, and remained absent until the mid-1990s when the US Park Service and US Fish and Wildlife Service jointly reintroduced the species from animals captured in Canada. Within several decades of the creation of Yellowstone National Park, Canada, New Zealand and Australia all had set aside protected areas and had begun developing national legislation to manage them, and the United States began establishing wildlife refuges as a separate category of protected area (Fischman, 2003).

Much has been written about the historical and cultural context of this (then) new phenomenon, and the similarities of its earliest adherents. All were British colonies, spoke English as their national language, and were being quickly populated by immigrating Europeans. All four countries also had *de facto* policies of subduing their native peoples to the point of what many now consider cultural genocide. This had the effect of depopulating large natural areas, within even larger countries with low population densities to begin with, in a rather short time period during the late 19th and early 20thCenturies. Some historians also note that the European Diaspora naturally tended to look to Europe for its cultural inspiration. The countries of the Old World had great universities, museums, artworks, palaces and ruins dating back to ancient Greece and Rome, while the New World had scenery and natural areas unsurpassed by anything in densely populated Europe. This school of thought considers the movement to create national protected areas to be motivated, at least in part, by 'Europe envy' (Zaslowsky and Watkins, 1986). By the early 20th Century, all four of those countries and a few others (*e.g.* Sweden) had set aside multiple natural areas and had created professional management authorities to protect them. Canada was the first country to create a national park management agency (in 1911) followed by the USA (in 1916).

In any case, there is much evidence to suggest to that the earliest parks (and many still; see below) were not set aside with particular reference to conservation in any form, and thus the 'Europe Envy' thesis is generally accepted. Most of the earliest units contained spectacular scenery, but their borders did not relate to the habitat needs of native species, much less the dynamics of entire ecosystems (Norton, 2005). By the 1930s, the American park system received criticism from within with a report by Dixon and Wright, two Federal employees, that received widespread attention. The authors stated that most units were "mountain top parks" and preserved only scenery with no regard for wildlife. Seasonal movements of many species were such that large populations of birds and mammals were left outside park boundaries (and therefore subjected to hunting); the early American 'mountain top parks' were, therefore, ineffective for many conservation purposes (Dombeck & Williams, 2003).

Modern conservation biology has also greatly expanded our ideas of the geometric design and placement of protected areas across landscapes (Primack, 2006), but the problem of 'mountain top parks' still remains. For national governments, it is simply easier to set aside large protected areas in places such as high elevations, deserts, tundra, *etc.*, *i.e.* where there are few competing economic demands, than in areas of high biological productivity. The latter tend to be at low elevations, in temperate or subtropical zones, and receive adequate rainfall. In short, the most productive ecosystems are also those where humans tend to concentrate. Tropical rainforests, with their primary productivity largely found in the canopy and frequently harboring human diseases, are possible exceptions to this generality, but in that cases too, they are at risk worldwide (Wilson, 1999).

Canada and the United States also pioneered several other conservation movements during the Progressive era of the early 20th Century. They developed the world's first international treaties on the protection of migratory wildlife, with separate instruments for wild salmon, fur seals and migratory birds (Dorsey, 1998). The last, The Migratory Bird Treaty Act of 1918, is still in force. Canada and the United States also developed the world's first transboundary protected area in 1932, with the creation of Waterton-Glacier International Peace Park, a large area that conserves habitat on both sides of the international border in the northern Rocky Mountains. With this came the recognition that many species and ecosystems cannot be conserved within the borders of single nations and these legal instruments were watershed events in the history of conservation worldwide (Susskind, 1994). From these humble beginnings, many other bilateral, regional and global conservation conventions have been developed for the protection of both migratory species and natural areas.

The largely Western ideal of protected areas as raw nature devoid of humans (except for tourism) was never really true to begin with; most areas set aside in the nations that began the movement had been occupied by pre-industrial people who were removed. This concept was also largely out of synch with realities on the ground in developing nations During the post World War II period of decolonization, many seminal wildlife studies were conducted in various places in Africa and Asia and the world became much more aware of their unique heritage. The International Union for the Conservation of Nature and Natural Resources (now IUCN – The World Conservation Union; www.iucn.org) was begun in 1948 with a charter to develop world wide standards for conservation and the World Wildlife Fund (www.worldwildlife.org; now the Worldwide Fund for Nature) was established several years later, initially as a fund raising mechanism for IUCN. Having been developed in the West, with essentially all funding coming from West, meant that Western standards of nature conservation were becoming global (Swanson, 1997). IUCN's World Commission on Protected Areas (WCPA) was organized in the 1950s, and developed internationally-recognized categories of protected areas by the 1970s, which were modified in the 1990s (see below).

Post-colonial governments in developing nations began setting aside protected areas by the 1960s, but the 'fences and fines' approach of the West had its limits in this context. Some, such as Kenya, Tanzania and India, already had the semblance of a protected area system as a result of colonial British rule, but these were areas largely set aside for use by British government officials and indigenous elites for hunting reserves, and effectively prohibited rural residents, who were dependent on natural resources, from entry (*e.g.* Gillingham & Lee, 2003; Bruyere *et al.* 2009). In 1962 and 1972, IUCN held its First and Second World Conferences on Protected Areas, respectively. Both were characterized by over-representation of delegates from developed countries and there was little focus on the issues relevant for newly emerged developing countries. This began to change with the Third World Conference on Protected Areas, held in 1982, in Bali, Indonesia. The Conference was renamed "National Parks, Conservation and People" and the theme was the role of protected areas in economic development; a majority of participants came from developing countries. The Fourth and Fifth World Conferences were held in Venezuela (1992) and South Africa (2002) respectively, and the global agenda for protected areas in each decade expanded from the one preceding it.

The relative success of national parks in the United States, Canada, Australia and New Zealand was due at least in part to the fact that population densities were low in those countries to begin with and that indigenous peoples had been largely removed from many ancestral areas as part of national policy as those countries were developing. Such was not the case in the developing world, and there is now near universal agreement that the Western national park model is generally inappropriate for the situation in most developing countries with their large rural populations dependent (at least in part) on extractive activities in natural areas (*e.g.* Campbell & Vainio-Mattila, 2003; De Boer & Baquete, 1998; Groom & Harris, 2008; McShane and Wells, 2004)). The WCPA recognized this with the liberalization of rules regarding national parks and more strictly protected areas, and with the modification of protected area categories recognized worldwide in 1994 (below).

2. IUCN – WCPA CATEGORIES OF PROTECTED AREAS

IUCN defines a protected area as "an area of land and/or sea especially dedicated to the protection and maintenance of biological diversity, and of natural and associated cultural resources, and managed through legal or other effective means." According to the World Database on Protected Areas compiled by the WCPA, there were over 7,000 separate units covering over 17,000,000 square kilometers as of 2007. This includes about 3.3% of Earth's total surface area and nearly 10 % of Earth's land surface, but less than 0.5% of its sea surface, although there has been recent growth in the designation of near-shore marine protected areas as well. The WCPA's mission is to "promote the establishment and effective management of a world-wide representative network of terrestrial and marine protected areas as an integral contribution to IUCN's mission." A general goal is to bring 10% of the Earth's land surface, including 10% of all recognized ecosystem types, under one or another internationally recognized category of protected area. The growth of such areas has been very rapid during the past several decades, but, based on the aforementioned criteria, some ecosystem types are over-represented, while most, and especially the more productive ones, are under-represented (Chape *et al.* 2008).

The WCPA uses a system in place since 1994 to define these areas (Table 1). Here I describe the major management categories, but please note that many nations have additional protected natural areas that do not fit within the IUCN criteria and are thus not included on the United Nations List of Protected Areas. Based on IUCN criteria, national protected areas are those managed by the "highest competent authority" which, in most cases, is the national government. Yet many countries have State, County, Provincial or Urban parks, recreation areas, *etc.*, in additional to those designated at the national level. In some cases, depending on the management plan, size and remoteness of such areas, they are included on the World List, but in many other cases they are not.

Similarly, many countries have private reserves (*e.g.* Nature Conservancy reserves, land trusts, *etc.* in the United States and elsewhere) or reserves managed by other entities (*e.g.* university-owned research reserves), that are generally not included based on IUCN criteria. In many cases, national forests or rangelands, which can be important for habitat for many native species, are also not included because their permitted uses exceed that considered appropriate by IUCN. With these caveats in mind, it is generally true that there is much more natural area set aside (about 17% of the Earth's land area; Chape *et al.* 2008), albeit in small reserves and/or under greater degrees of human uses, than is recognized internationally based on IUCN criteria. IUCN categories, based on a numbering system from most to least strictly protected, are as follows (from www.iucn.org/about/union/commissions/wcpa):

Ia. Strict Nature/Scientific Reserve. The main purposes of Category Ia reserves are scientific research and species conservation, and other human uses are generally banned. Because of this, few nations recognize Category Ia reserves within national law, but quite a few have *de facto* Strict Nature reserves. These may include, for example, very remote regions of much larger protected areas in which inaccessibility precludes tourism or other uses.

1b. Wilderness Areas. Wilderness areas are generally large and remote. Tourism is permitted, but since permanent human dwellings and motorable roads generally are not, tourist numbers are few and generally involve backpacking style camping. They provide for the protection of wilderness and maintenance of ecosystem services. This category was added in the WCPA category revisions of 1994.

Table 1. IUCN -The World Conservation Union Protected Area Management Categories (adapted from IUCN 2003)Key to Management Objectives: SR, scientific research; WP, wilderness protection; SD, species or genetic diversity conservation; ES, environmental services; NF, natural or cultural features; TR, tourism and recreation; ED, education; SU, sustainable use; and CA, cultural attributes.Key to importance of objectives by category: *** designates a primary objective; **, a secondary objective; *, potentially not applicable; and -, not applicable.

	Management Objectives							
Category and Name	SR	WP	SD	ES	NF	TR	ED	SU
1a: Strict Nature Reserve	***	**	***	**	-	-	-	-
1b: Wilderness Area	*	***	**	***	-	**	-	-
II: National Park	**	**	***	***	**	***	**	*
III: National Natural Monument	**	*	***	-	***	***	**	-
IV: Habitat or Species Management Area	*	*	***	***	*	*	**	**
V.: Protected Landscape Or Seascape	**	-	**	**	***	***	**	**
VI: Managed Resource or Extractive Reserve	*	**	***	***	*	*	*	***

Category II. National Parks. This has been the most used protected area category worldwide. National parks are generally large areas that protect more than one important natural feature and/or wildlife population, and in which tourism is generally permitted and promoted. Other important functions include providing environmental services and opportunities for environmental education as well as scientific research. National parks tend to be the best known and most important protected areas economically, and many of the best examples worldwide are also recognized internationally as World Heritage Sites.

Category III. National (Natural) Monuments. This category has, in general, the same aims as Category II, but national monuments are generally smaller than national parks and are set aside to protect one or several important natural features. In some cases, these can be combined with cultural features in a natural setting (*e.g.* Mt. Rushmore in the United States). Because of their generally smaller size, they are usually not important for broader conservation purposes such as ecosystem services, but many contain important wildlife populations.

Category IV. Managed Habitat/Wildlife Reserves. By sheer numbers, this is the second most important protected area category worldwide. In general, these reserves are established to protect one or more important wildlife populations and, for the larger units, they can also be important for providing ecosystem services. Tourism is frequently permitted within them, but not promoted as in the case of II and III, above. In addition, material alteration can take place within Category IV protected areas to enhance habitat for the species of conservation concern. For example, maintaining pastures for ungulate grazing, creating empoundments for waterfowl habitat, *etc.*, may all be permitted within them. These activities are generally not permitted in the previous categories. Sustainable use is frequently a secondary goal of Category IV reserves, and some (limited) hunting of common game species may be permitted within some, or in adjacent areas.

Category V. Protected Landscapes/Seascapes. Category V reserves are perhaps the most interesting for their breadth of permitted activities and management options. These are generally large areas set aside for a combination of their natural and cultural features, and they generally promote tourism. In many places, human habitations are found within them, including small towns with examples of rural working landscapes. As such they are generally designated across landscapes that contain an admixture of public, semi-public and private lands, and may be quite altered from their natural state.

Category VI: Managed Resource/Extractive Reserves. Category VI, like Ia, was added to the list of protected are categories with the 1994 revisions. These are generally large reserves that provide for ecosystems services, but their main purpose is the conservation and sustainable use of important species and their gene pools. Active removal of forest products is permitted and in fact encouraged, and, as such, they tend to be important economically for local communities. The general rule for a protected area to qualify under this category is that no more than one third the area can be subject to intense harvest. Many countries (the United States included) have large areas set aside in which more extensive harvesting is permitted. Such areas may be managed in a semi-natural state for national purposes, but do not qualify as Category VI internationally (*e.g.* National Forests in many countries).

2. SOME CAVEATS OF PROTECTED AREA CATEGORIES

In the modern era (post 1990), greater areas under Category V and VI reserves, both terrestrial and marine, have been created worldwide than other reserve types. This is especially true in developing countries, but is also true in some of the large marine protected areas created in the United States (*e.g.* the Florida Keys National Marine Sanctuary). Given their more lenient management regimes, this is also not surprising due to the dependence in many places that rural residents have on natural resource extraction and use. However, there has been a great deal of concern expressed, especially by natural scientists, about this phenomenon. Since large predators, especially, are generally not tolerated by humans (and vice versa), and yet are keystone species in many ecosystems, Category V and VI reserves are especially problematic from a purely ecological standpoint (*e.g.* Heinen and Mehta, 1999). Yet these reserves can be the most important from a purely economic standpoint (*e.g.* Sherman and Dixon 1990) and from the standpoint of human cultural values.

While debates were ongoing in the western academic literature largely between natural and social sciences on the competing values of different types of protected areas and their uses, with ecologists generally favoring more strict protection and social scientists favoring less strict protection (*e.g.*Redford and Sanderson, 2000), many nations, as well as less philosophically-driven researchers and development workers, were slowly arriving at a different consensus. That is, both sides have valid arguments and large reserves, and the regions in which they are found, can have elements meeting these competing demands via zoning criteria. For example, India and Nepal added less strict regulations to some of their national parks (including some limited extractive uses), while keeping more strict regulations in others, and both also actively supported buffer zone policies in the vicinity of more strictly protected areas beginning in the 1990s (*e.g.* Heinen and Shrestha, 2006). In those cases, many Category II and IV protected areas are surrounded by buffer zones that are managed more like Category V or VI protected areas, whether or not they are recognized as such internationally.

To promote the broad goals of sustainable development as articulated in the 1987 Brundtland Report (Bruntland, 1987), Agenda 21 (Sitarz, 1993) and the 1992 Convention on Biological Diversity (Glowka *et al.,* 1994), rather intensive local development inputs are needed in such areas to reduce demands on the core protected area (*i.e.* the Category IV or lower reserve). This may include rural enterprise development such as farm fisheries, agro- and community forestry and training of local people for tourism related jobs. There is now a large and growing literature on the development and success of community-based conservation (CBC) programs and integrated conservation and development programs (ICDPs) that is generally outside the scope here. Suffice it is to say that, for our purposes, from both socio-economic and ecological standpoints, there is growing evidence that this mixed approach has many advantages and pitfalls (*e.g.* Fiallo & Jacobson, 1995; Lepp & Holland, 2006; Lepp, 2007).

Some ecological factors that may lead to success (or not) include the types of species protected in core areas (*e.g.* large mammals frequently cause much loss to local farmers, including lost human lives on occasion) and the types of natural plant products and other resources (*e.g.* fish), their growth and sustainable harvest rates and local market values, that may be harvested legally from designated extractive zones. Socio-economic and other factors are many and varied. For example, human population density alone, and especially the ethnic heterogeneity and recency of immigration to an area, can determine the degree of difficulty of developing and sustaining CBD programs (Heinen, 1996). Recent research has shown that the creation of protected areas and development inputs into their surroundings can act as attractants for new immigrants, further complicating the issue (Wittemyer *et al.* 2008). In addition, increased wealth (due to tourism and other employment opportunities) of residents around protected areas can also create difficult managerial consequences in their vicinity via increasing demand for many forest products (*e.g.* Fu *et al.,* 2004). But, in general, CBC programs in areas that are more stable demographically and/or especially areas in which they have been in place for longer time periods (and thus institutional trust and social capital has been built), have been shown to be effective over time in many case studies (Baral *et al.* 2007). But this can take many years to a few decades.

The protected area categories used by IUCN's WCPA are broad enough to cover quite a bit of the world's protected natural heritage adequately, but individual countries deviate from international standards frequently. As previously mentioned, they are not inclusive enough to capture many of the world's smaller protected areas (*e.g.* state, provincial and country parks) or important private reserves. Such reserves can be very important for the conservation of local plant and insect species, as stopover areas for birds during migration, as important nesting areas for species such as sea turtles, and for the ever-increasingly important purposes of introducing urban and suburban populations to environmental and science education, which are all very important objectives. For example, the Counties and the State of Florida maintain a system of such reserves in the urban and suburban matrix of Miami-Dade, Broward and Palm Beach in Southeast Florida that are heavily visited by residents and tourists (Alonzo & Heinen, 2011); their combined attendance annually is thought to be greater than for nearby Everglades National Park. As such, small reserves can be disproportionately more important for several simultaneous conservation goals than some internationally recognized large reserves.

Individual countries may also vary quite a bit in terms of management practices and hence in terms of how the WCPA categorizes their protected areas. For example, 'National Parks' under both British and Japanese standards are frequently too materially altered to be considered Category II protected areas by WCPA. Because they may include many important cultural components and have private in-holdings, and, in some cases, entire towns, they are generally classified as Category V by IUCN. Many of the large extractive reserves in the USA (and elsewhere) simply allow too much extraction to be classified as Category VI reserves (*e.g.* US National Forests managed by the Forest Service and Grazing Areas managed by the Bureau of Land Management). Other units within the same system, under less intensive management, do qualify, and thus IUCNs' WCPA must consider each unit on a case by case basis by reviewing individual management plans in deriving the United Nations List of Protected Areas to determine if each qualifies. World wide, the effort required is huge and the list is always in need of updating. In spite of these caveats, the system has proven useful for over 15 years; it is also adaptable and widely recognized, so there is little reason to change it at the present time.

Another issue that is frequently debated and studied is that of 'paper parks.' These can be defined as units that are protected at the national level via appropriate laws, and in some cases, recognized under international conventions as very important protected areas, but in which there is either inadequate or no active enforcement on the ground. This term is now applied to many parks and reserves in developing countries where inadequate budgets and manpower for conservation are the norm. The World Heritage Convention (below) has focused on this issue and has developed the "List of World Heritage in Danger" (whc.unesco.org/en/danger). Inscription of this list should cause national shame, for these are some of the most spectacular parks on Earth, but in fact, the vast majority of protected areas are not World Heritage Sites, and there are many 'paper parks' in which poaching, logging, or other extraction go on regularly in spite of laws. WCPA has no means currently to assess these issues on a case by case basis worldwide, so many listed sites (especially Category II) either should be placed in another less strict category or removed from the List.

3. INTERNATIONAL REGIMES CONCERNING PROTECTED AREAS

There are currently dozens of international instruments related to the conservation of species, natural areas, or both. The vast majority are bilateral or regional; some are quite well studied while others are more obscure (Klemm & Shine, 1993). Regional treaties in this area exist in Europe, the Commonwealth of Independent States (*i.e.* most of the former Soviet Republics), Southeast Asia, sub-Saharan Africa and Central America. Here I briefly discuss the major international regimes that are subject to ratification or acceptance by all United Nations members, but interested readers are referred to Klemm and Shine (1993) for more information on some of the regional agreements.

3.1. The Man and Biosphere Program

In the late 1960s, the Man and Biosphere Program (MAB) was conceived under the auspices of the United Nations Educational, Scientific and Cultural Organization (UNESCO) based in Paris (Batisse, 1982, 1986). By 1971, MAB was implemented with the broad goals studying human relationships with the biosphere, especially for studying long-term human induced impacts and conservation for sustainable development. A major objective of MAB since its beginning was to develop a worldwide network of international biosphere reserves and, and there are currently (2009) 553 MAB-designated Biosphere Reserves in 107 countries (www.portal.unesco.org).

Based on MAB criteria, biosphere reserves are established in representative ecosystems for research purposes, with several secondary goals. These include: preserving traditional forms of land use, disseminating knowledge to manage resources, and promoting cooperation in solving resource related problems. The biosphere reserve model is one in which more strictly protected core reserves are surrounded by nested buffers permitting more human uses with distances from the core. Given the time period, the goals of the program and the concept of reserves with functional buffers, were quite progressive and a number of countries have since followed suit with the zoning implicit in the MAB reserve design (above). Education and training are also promoted under MAB, as is ecosystem level management. A number of countries in Latin America and the former Soviet

Union now use 'biosphere reserve' as a category of nationally-protected areas; many of these sites are listed on UNESCO's international list, while others are not.

The type and scale of reserves listed under MAB vary quite a bit based on national norms and conventions, and national MAB programs are given a great deal of leeway in nominating sites (Heinen & Vande Kopple, 2003). Any nomination is subject to acceptance by the international MAB program. MAB is also quite fluid in maintaining ties with international organizations (*e.g.* the World Wide Fund for Nature, WWF; the World Wildlife Fund in the USA), and United Nations agencies such as the United Nations Environmental and Development Programs, respectively, and with Secretariats of international conventions such as Ramsar and CBD (below). Through international as well as regional programs, MAB fosters exchanges among reserves and facilitates interactions through networks such as AfriMAB and EuroMAB.

Sites listed under MAB range from large and nationally protected areas (*e.g.* Everglades National Park) to much smaller reserves maintained by sub-national entities (*e.g.* The University of Michigan Biological Station). While some may perceive this as a weakness in that such disparity in size and purpose leads to little uniformity in management of these reserves, this can also be considered a strength of MAB. That is, individual national programs can promote the overarching goal with a number of different reserve types, and can take part in various levels of international cooperation, as long as the reserve meets the general criteria of research, education and outreach and includes some semblance of the zoning criteria in which core areas are well protected. As such the program is quite flexible and unique. The fact that it is not legally binding can also be considered both a strength and a weakness as it, again, promotes more flexibility but less uniformity. In any case, the MAB program has existed for four decades and in many ways set additional standards for protected areas management internationally. As such, it is quite important for the movement worldwide.

3.2. The Ramsar Convention

The Convention on Wetlands of International Importance, Especially for Waterfowl Habitat was conceived in Ramsar, Iran in 1971, and is generally known as the Ramsar Convention (Hails, 1996). Ramsar was the first truly international convention promoting the protection of natural areas and, in many ways, it remains the most important. Like MAB, it was also very progressive for its time (below). To date (2009) there are 159 contracting parties and 1,847 sites included on the List of Wetlands of International Importance, which collectively cover about 1.8 million square kilometers of area (www.ramsar.org). The purpose of Ramsar as articulated in the

Preamble is to recognize the interdependence of humans and the environment and to consider the ecological and economic functions of wetlands as fundamentally important. Parties are instructed to develop national wetlands policy with the aim of decreasing wetland loss, and to recognize that waterfowl, by virtue of their annual migrations, are an important international resource. All Parties must nominate at least one Wetland of International Importance from within their borders.

Ramsar defines wetlands very broadly, to include fens, bogs, marshes and swamps as well as near-shore marine areas in which low tide does not exceed 6m in depth. In this way, the Convention was progressive in that near-shore marine areas can be included. At the time of its formulation, very few nations had created marine reserves, but this movement has increased greatly in the decades since, and many coastal areas are now listed as Ramsar sites. Similarly, Ramsar provides a very broad definition of waterfowl to include any species of migratory bird that uses wetlands for any part of its life cycle. As such, waterfowl in the traditional sense are included (*i.e.* ducks, geese and swans), as well as all species of waders and fishing birds, and a number of passerine species that breed in wetland areas.

Various Articles of Ramsar further articulate the responsibilities of Parties to conserve wetland areas. Article 4, for example, encourages Parties to create wetland reserves whether or not they are listed sites, and to train personnel for research and management of wetlands, while Article 5 instructs Parties to consult about implementing the Convention, an important provision for sites that may cross international borders. Ramsar was also historically important in promoting the concept wise use of wetland resources for sustainable development. This also made it very progressive for its time in the sense that the Bruntland Report was released 15 years after Ramsar, and the Convention on Biological Diversity followed Ramsar by 2 decades. It also preceded the changing concepts of the WCPA about protected area management categories and the promotion of sustainable human uses within more categories than had been the case previously (above). Ramsar also maintains a Trust Fund for which Parties that are developing nations can apply for funding for special projects to maintain sites, offer trainings, *etc.*

In another sense, however, Ramsar can be criticized for having relatively little control over Parties as to how they manage wetlands overall. The idea of no net loss of wetlands, inherent to Ramsar, has frequently not been met, even in the United States, and wetland drainage continues in many Party States. Listed Ramsar sites themselves vary quite a bit in terms of their importance. For example, Canada and the United States have relatively few listed sites, but all are large and of obvious importance for the broad goals of the Convention (*e.g.* Everglades National Park). While many of the smaller and much more densely populated European countries list large numbers of small sites, some of which are of dubious importance. Even with these caveats in mind, Ramsar is very important for international conservation for promoting wetland protection and for many broader issues related to protected areas management. It could also be used as a template to form other Conventions focused on single broad ecosystem types (*e.g.* tropical forests or tropical grasslands), although none currently exist.

3.3. The World Heritage Convention

The International Convention for the Protection of World Cultural and Natural Heritage (The World Heritage Convention or WHC) as adopted by UNESCO in Paris in 1972. As of 2009, WHC had 186 Parties; of these, 148 have sites listed on the World Heritage List. Of the 890 sites listed worldwide, the majority are Cultural heritage sites (689) and will not be considered here (whc.unesco.org). Of the remaining, 176 are Natural heritage sites (mostly national parks) while 25 are mixed sites containing both cultural and natural heritage. WHC came into force in 1975 with the purposes of conserving both natural and cultural areas of outstanding universal importance. As such, Parties recognize that many sites are of importance to world's heritage and not just to the heritage of the countries that may contain them. In addition, to maintaining the World Heritage List, WHC's Secretariat also maintains the World Heritage Trust, under which developing countries can apply for project funds to help maintain sites. Lastly, the Secretariat also maintains a list of World Heritage Sites in Danger for previously listed sites that are under improper management or for some other reason at risk. Sadly, the site nearest my own desk, Everglades National Park, is currently on this list due to the lack of progress of the Comprehensive Everglades Restoration Project under the Bush Administration. Compliance is a major issue, and focus of study, regarding many such legal instruments (*e.g.* Faure & Lefevere, 1999).

Many of the Articles of WHC pertain solely or mainly to cultural sites but there are several important provisions that relate to natural sites. Article 2, for example, states that natural heritage consists of "physical and biological formations or groups of such formations, which are of outstanding universal value." The definition is further clarified to include areas that constitute important habitat for endangered species of universal value, outstanding geological formations, or other natural features of outstanding beauty. Parties to the Convention are responsible for proposing sites within their borders for listing, and providing strong evidence based on a set protocol for each place that allegedly constitutes a site of outstanding universal value. Most natural sites on the list were already world famous before they were listed (*e.g.* The Grand Canyon, The Serengeti, Mount Everest, The Great Barrier Reef, The Galapagos Islands), and in fact, most were already protected under national law as National Parks or other types of internally-recognized protected areas. None-the-less, there is national prestige to having sites listed as World Heritage, and the added (although rather meager) incentive for developing nations to garner some funds through the Trust. National governments and private tour operators alike frequently use listing in advertising as an incentive to attract more tourism, and World Heritage Natural Sites are among the most-visited protected areas on earth. In Nepal, for example, of 16 nationally protected areas, the two World Heritage Sites alone (Everest and Chitwan National Parks) typically account for over one third of tourist entries in protected areas in the country (Heinen and Kattel, 1992).

3.4. The Convention on The Conservation of Migratory Species of Wild Animals

The Convention on the Conservation of Migratory Species of Wild Animals, also known as CMS or the Bonn Convention, came into force in 1979; as of 2009, there were 112 Parties (www.cms.int). Throughout its history, CMS has attracted fewer Parties than the other Conventions described here, in part because many nations of the Western Hemisphere were already party to an older regional convention protecting migratory wildlife (the Western Convention of 1940). As such, most Parties to CMS are in the Eastern Hemisphere, but more recently, a number of Latin American countries have ratified it. As the name implies, CMS focuses on migratory wildlife and not with protected areas *per se*. It is thus much more of a species-based than area-based conservation convention, but within its 20 Articles there are some clauses that are germane for the topic at hand.

CMS's Article 1 considers conservation status to be favorable if (among other things) the distribution and abundance of migratory species approach historic coverage, suitable ecosystems for conservation exist, and that there is sufficient protected habitat to maintain migratory species. The Article further describes unfavorable conservation status as being those in which these (above) conditions are not met. CMS's fundamental principles (Article 2) similarly contains an important clause outlining the importance of conserving habitat: The Parties acknowledge the importance of migratory species... "taking individually or in cooperation appropriate and necessary steps to conserve such species and their habitats."

Through its long history, and in conjunction with other Conventions (especially Ramsar) CMS has been indirectly importance in expanding protected area networks, and especially in Europe and Africa due to the large avian migrations between those two continents. A number of small reserves were created along flyways that offer staging and stepping-stone habitats, and many of these also contain significant wetland resources. None-the-less, with its focus on migratory species, its relatively few signatory nations, and its appendices of species under varying degrees of threat, it is not nearly as important for protected areas as the other instruments described here, but is important for species protection. Under its auspices, various important regional agreements have been established and form some rather interesting case studies in species (and area) conservation. Among these are: the Agreements for the Conservation of Cetaceans in the Black and Mediterranean Seas and Contiguous Atlantic Area; the Africa-Asia Migratory Waterbird Agreement, several agreements on sea turtles in the Pacific and Indian Oceans, and the Agreement on Gorillas and their Habitats. Some habitat protection clauses are found within all of these.

3.5. The United Nations Convention on Biological Diversity

The United Nations Convention on Biological Diversity (CBD), formulated prior to and during the 1992 United Nations Conference on Environment and Development in Rio de Janeiro, is far and away the broadest of the international conservation agreements. It came into force on 29 December 1993 and has three main objectives: to conserve biological diversity, to use biological diversity sustainably and to share the benefits of biological diversity equitably (www.cbd.in). The CBD currently (2009) has 156 Parties. Many provisions of the Convention do not deal with protected areas *per se*, so I only highlight important aspects of CBD and focus on those few aspects that do relate to this topic.

Of its 42 Articles and 3 Annexes, Article 8 (*In Situ* Conservation) is the main one dealing with protected areas. Therein, contracting Parties are encouraged, as far as possible and as appropriate, to, establish systems of protected natural areas and to develop guidelines for their selection and management.

Article 8 of CBD further requests Parties to manage important biological diversity appropriately whether it is located within the protected area network or not, and to promote general ecosystem protection. Parties are also requested to promote environmentally sound sustainable development in the vicinity of protected areas, to promote restoration of degraded ecosystems, and to control exotic species that pose a risk to conservation. Parties are further encouraged to use innovative practices in management and to involve local and indigenous communities in protected areas management. The final clauses of Article 8 instruct Parties to develop appropriate regulatory legislation to conserve endangered species, cooperate in financial support for *ex situ* conservation, and regulate processes that may adversely affect biological diversity in accordance with Article 7, which addresses the identification and monitoring components of biological diversity. Annex 1 (referenced in Article 7) defines components to be monitored to include ecosystems and habitats with high diversity, large numbers of endemic or threatened species, wilderness, and/or important habitat for migratory species. It further instructs Parties to identify and monitor communities and species that are threatened, contain wild relatives of important domesticates or other value, or are important for research, conservation and sustainable use.

Much of the rest of CBD deals with issues of domestic biodiversity, genetic complexes, appropriate uses of biological diversity and trade, *ex situ* conservation, equitability and sustainability management of biodiversity, and not with protected areas *per se* (Glowka *et al.,* 1994). None-the-less, CBD is far and away the broadest in scope of any conservation treaty and recognizes explicitly that protected natural areas are essential for biodiversity conservation at all levels of integration (*i.e.* from genes to ecosystems) and such protected area systems are important to effectuating the CBD objectives. It is also useful to note that CBD articulates quite well the more modern view of protected areas as places in which human are an integral part, as opposed to the older view of raw nature and 'fences and fines' management characteristic of the first American model. Shortly after CBD came into force, WCPA modernized its protected area categories (above). Another aspect of CBD that has proven important since its passage is that many Parties have undertaken the task of creating national conservation strategies and action plans, which is promoted by Article 7 and Annex 1. Such plans have allowed for a fuller inventory of important biological diversity, and have generally contained parts that deal with the existing protected areas network of each Party, with recommendations for expansion and for better management of existing units.

4. GENERAL DISCUSSION AND OVERVIEW

From its humble American beginnings in 1872, the international movement to conserve protected areas at the national level has mushroomed in the modern era. Most nations now maintain systems of protected areas, and the majority are now Parties to all of the international conservation conventions discussed above. The intellectual breadth of protected area management categories recognized worldwide, and of the types of both traditional and non-traditional uses permitted within them, has also expanded greatly, as has the use of zoning large areas to permit more or less uses in specific tracts depending both on the need for biological conservation and human enterprises. None-the-less, the stamp of the earlier history of a Western and largely American model still pervades many protected area systems. For example, of the categories recognized worldwide, four were derived largely from American law. These include Category Ib (Wilderness), II (National Parks); III (Natural Monuments) and to a lesser degree, Category IV (Wildlife Reserves, refuges, managed habitat areas, *etc.*). Categories V (Managed Landscapes and Seascapes) also had some precedent in the national protected area system of the United States, with its National Seashores, National Recreation Areas, *etc.* The nation that began the movement is thus still the most dominant player, at least in terms of general categories and many accepted management practices.

However, the internationalization of the protected areas movement created many opportunities and altered many previously accepted practices, which proved important for the continuous expansion of protected area systems. Allowing private in-holdings and some limited extractive uses from National Parks, for example, did not become recognized until the 1980s, and was only recognized by WCPA as a result of several national experiments to remove local people completely from Category II reserves. One such case happened in Nepal, where two local villages were removed from high-altitude Rara National Park, and their residents relocated into the western Terai (lowlands) of the country. Within a generation, over half of the original population had either left or died from malaria or other lowland diseases, and Nepal's Department of National Parks and Wildlife Conservation ended any plans to remove larger (but still rather small) local populations from other Himalayan national parks such as Langtang and Everest (Heinen & Kattel, 1992). The international movement, and the international organization in the form of IUCN's WCPA, was sensitive to these issues and adjusted Category II accordingly by allowing the zoning out of traditional villages (*i.e.* they are not recognized as part of the park, although they are surrounded by it). Similarly, the development of Category VI in 1994, and its subsequent worldwide expansion, was an important acknowledgement of the needs of many people in developing countries by recognizing that traditional local uses, and even some more modern commercial, uses of at least some protected areas should be permitted.

The literature on many facets of protected area management is similarly expanding greatly, and both space and topical content of this volume does not permit a closer look at some of the more scientific aspects. Suffice it is to say that the literature in conservation biology, a field only recognized since the 1980s, includes literally hundreds of well-done studies on issues such as placement of reserves, how to prioritize areas for protection based on scientific criteria, the appropriate size of reserves, the utility of maintaining natural corridors to promote gene flow between reserves, the placement and uses of buffer zones, *etc.* (Primack, 2006). Modern conservation agencies thus have at their disposal a great arsenal to help them plan the most efficient uses of scarce resources in conserving biodiversity. But conservation biology, first defined by Michael Soule (one of its founders) as a "mission oriented, crisis discipline" also recognizes that time is running out for many wild species on earth, and for the places that harbor them.

We are in the midst of a mass extinction, recognized by science as the 6th such event in the history of life on Earth, and the only one to be caused by one species: ourselves (Wilson, 1999). Modern humans threaten to have a commensurate impact, albeit more slowly, of the great asteroid that landed in the Western Caribbean and wiped out the dinosaurs - and about half of all other life forms - some 65 million years ago. While much conservation literature, and the Convention on Biological Diversity, also discusses the importance of *ex situ* conservation in the form of seed banks, zoos, botanical gardens, *etc.* (and doubtlessly they are all important), science and much of society recognizes that *in situ* conservation of species - and complexes where they occur naturally - is a much more cost effective and efficient way to conserve biodiversity. It has the added advantage of keeping ecological phenomena (*e.g.* predation, competition, migratory behavior, pollination) intact (or at least partly so) and allows for the evolutionary game to continue. *Ex situ* conservation provides, at best, a short term buffer.

So the international movement to conserve protected areas will increase over time. As the planet becomes increasingly crowded, it will remain the major way to conserve biodiversity and, ultimately, ourselves. New ideas - and ideals are constantly expanding the field with more recent foci on issues such as landscape-level conservation across wide regions, a recent major increase in the study of invasive species and their removal, and better ways to manage areas already protected. But like the situation with so much else, the legal and policy instruments, both within nations and across nations in the form of international treaties and programs, lag greatly behind the science (*e.g.* Jacobson and Weiss, 2000). To further this field, we will need much more land set aside, which may mean including even broader protected area categories recognized in the future, and we will need to pay much more attention and legal recognition to dynamic processes across landscapes. In short, we need to become much smarter and more adept at living with the functional natural world.

REFERENCES

1. J. Alonzo, J. T. Heinen, 2011Miami-Dade County's Environmentally Endangered Lands Program: Local efforts for a global cause. Natural Areas Journal 312500506
2. N. Baral, M. Stern, J. T. Heinen, 2007Integrated conservation and development project life cycles in the Annapurna Conservation Area, Nepal. Biodiversity and Conservation 161029032917
3. M. Batisse, 1982The biosphere reserve concept: A tool for environmental conservation and management. Environmental Conservation 9101114
4. M. Batisse, 1986Developing and focusing the biosphere reserve concept. Nature and Resources 223111
5. Mulder. M. Borgerhoff, P. Coppolillo, P. , 2006Conservation: Linking Ecology, Economics and Culture. Princeton University Press, Princeton, NJ.
6. G. H. Bruntland, 1987Our Common Future. Oxford University Press, Oxford, UK.
7. B. L. A. Bruyere, Beh, G. Lelngula, 2009Differences in perceptions of communication, tourism benefits and management issues in a protected area in rural Kenya. Environmental Management 4314959
8. L. M. Campbell, A. Vainio-Mattila, 2003Participatory development and community-based conservation: Opportunities missed for lessons learned? Human Ecol. 313417437
9. S. Chape, M. Spalding, M. Jenkins, 2008The World's Protected Areas: Status, Values and Prospects for the 21st Century. Berkeley: University of California Press.
10. W. F. De Boer, D. S. Baquete, 1998Natural resource use, crop damage and attitudes of rural people in the vicinity of Maputo Elephant Reserve, Mozambique. Environmental Conservation 253208218
11. M. P. Dombeck, C. A. Woods, J. E. Williams, 2003From Conquest to Conservation: Our Public Lands Legacy. Island Press, Washington, DC.
12. K. Dorsey, 1998The Dawn of Conservation Diplomacy: U.S.- Canada Wildlife Protection Treaties in the Progressive Era. University of Washington Press, Seattle, WA.
13. M. Faure, J. Lefevere, 1999Compliance with international environmental agreements. 138156in Vig, N. J. and R. S. Axelrod, editors. The global environment: institutions, law, and policy. Congressional Quarterly, Inc., Washington, DC.
14. E. A. Fiallo, S. K. Jacobson, 1995Local communities and protected areas: Attitudes of rural residents towards conservation and Machalilla National Park, Ecuador. Environmental Conservation 223241249
15. R. L. Fischman, 2003The National Wildlife Refuges: Coordinating a Conservation System through Law. Island Press, Washington, DC.
16. B. J. Fu, K. L. Wang, Y. H. Lu, K. M. L. D. Chen, G. H. Liu, 2004Entangling the complexity of protected area management: The case of Wolong Biosphere Reserve, southwestern China. Environmental Management 336788798
17. S. Gillingham, P. C. Lee, 2003People and protected areas: A study of local perceptions of wildlife crop damage conflict in an area bordering the Selous Game Reserve, Tanzania. Oryx 373316325
18. L. Glowka, F. Burhenne-Guilmin, H. Synge, 1994A Guide to the Convention on Biological Diversity. IUCN Publications, Gland, Switzerland.
19. R. Groom, S. Harris, 2008Conservation on community lands: The importance of equitable revenue sharing. Environmental Conservation 353242251
20. A. J. Hails, 1996Wetlands, Biodiversity and the Ramsar Convention: The Role of the Convention on Wetlands in the Wise Use and Conservation of Biodiversity. IOUCN Publications, Gland, Switzterland.
21. J. T. Heinen, 1996Human behavior, incentives and protected area management. Conservation Biology 102681684

22. J. T. Heinen, B. Kattel, 1992Parks, people and conservation: AS review of management issues in Nepal's protected areas. Population and Environment 1414984
23. J. T. Heinen, J. N. Mehta, 1999Conceptual and legal issues in the designation and management of conservation areas in Nepal. Environmental Conservation 2612129
24. J. T. Heinen, S. K. Shrestha, 2006Evolving policies for conservation: an historical profile of the protected area system of Nepal. Journal of Environmental Planning and Management 2006; 4914158
25. J. T. Heinen, R. Vande, Kopple, 2003Profile of a biosphere reserve: The University of Michigan Biological Station and its conformity to the Man and Biosphere Program. Natural Areas Journal 232165173
26. IUCN.2003United Nations List of Protected Areas. Gland, Switzerland, IUCN Publications.
27. H. K. Jacobson, E. B. Weiss, 2000A framework for analysis. 118in Weiss, E. B. and H. K. Jacobson, editors. Engaging countries: strengthening compliance with international environmental accords. MIT Press, Cambridge, MA.
28. C. Klemm, de , C. Shine, 1993Biological diversity: legal mechanisms for conserving species and ecosystems. IUCN- The World Conservation Union, Environmental Law and Policy Paper 29Cambridge, UK.
29. A. Lepp, 2007Residents' attitudes towards tourism in Bigodoi village, Uganda. Tourism Management 283876885
30. A. Lepp, S. Holland, 2006A comparison of attitudes toward state-led conservation and community-based conservation in the village of Bigodi, Uganda. Society & Natural Resources 197609623
31. T. O. Mc Shane, M. P. Wells, 2004Getting Biodiversity Projects to Work: Towards more Effective Conservation and Development. Columbia University Press, New York.
32. B. G. Norton, 2005Sustainability: A Philosophy of Adaptive Ecosystem Management. University of Chicago Press, Chicago, IL.
33. R. B. Primack, 2006Essentials of Conservation Biology (4th edition). Sinaur: Sunderland, Mass.
34. K. H. Redford, S. E. Sanderson, 2000Extracting humans from nature. Conservation Biology 14513621364
35. P. B. Sherman, J. A. Dixon, 1990Economics of Protected Areas: A New Look at Costs and Benefits. Washington, DC: Island Press.
36. H. I. Sitarz, 1993Agenda 21: The Earth Summit Strategy to Save our Planet. EarthPress, Boulder, CO.
37. L. E. Susskind, 1994Environmental diplomacy: Negotiating more effective global agreements. Oxford University Press, Oxford, UK.
38. T. Swanson, 1997Global action for biodiversity. IUCN Publications, Cambridge, UK.
39. E. O. Wilson, 1999The Diversity of Life (New Edition). Harvard University Press, Camridge, MA.
40. G. Wittemyer, P. Elsen, W. T. Bean, A. Coleman, O. Burton, J. S. Brashares, 2008Accelerated human population growth at protected area edges. Science 3215885123126
41. D. Zaslowsky, T. H. Watkins, 1986These American Lands: Parks, Wilderness and Public Lands. The Wilderness Society, Washington, DC.

CHAPTER 8

Woody Species Diversity and Structure of Agroforestry and Adjacent Land Uses in Dallo Mena District, South-East Ethiopia

Bikila Mengistu[1*], Zebene Asfaw[2]

[1]*Sinana Agricultural Research Center, Bale-Robe, Ethiopia*

[2]*Wondo Genet College of Forestry and Natural Resources, Hawassa University, Shashamane, Ethiopia*

ABSTRACT

Sustainable farming practice that utilizes and conserves biodiversity, reducing negative impacts of agriculture on biodiversity and provides wood and energy to local community is a good solution to reduce deforestation and forest degradation. This study was carried out in Dallo Mena district of Bale zone to identify woody species diversity in homegarden agroforestry practices, shade grown coffee agroforestry practices and adjacent natural forest, and to show how land use and management practice determine the plant species diversity. The study site was selected based on spatial analogue approach. In each land use, a plots of 10 m × 10 m size was drown by using systematic sampling method following the transect line. A total of 36 sample plots were sampled along the transect line laid down inside each land uses. In each plot, woody species were counted and the diameter and height of trees and shrubs were measured. Based on this inventory a total of 39 woody plant species which are about 23, 10, 15 woody plant species were recorded from natural forest, shade grown coffee agroforestry and homegarden agroforestry practices respectively. These species were classified belonging to 24 families. The Shannon Wiener diversity index used to estimate species diversity ranged from 0.14 to 2.54 with a mean of 1.47. Also 21% woody species similarities were observed between those three land uses. The result shows that both agroforestry practices (shade grown coffee and home garden agroforestry practices) and adjacent natural forests are conserving several woody species diversity in its system.

Keywords: Agroforestry, Conservation, Diversity, Practice, Woody Species

1. INTRODUCTION

Natural forest is one of the richest ecosystems in plant species diversity and used as a home of many and diversified plant species. Natural forest has wide ecological and environmental values and is a source of biodiversity [1] . Ethiopia is also one of the richest countries in having plant species diversity. The geographical location of Ethiopia covers wide agro-climatic zones and important center of biological diversity. This wide ecological condition of Ethiopia has created diverse and conducive environments for the development of a variety of flora. There are about 7000 different flowering plants grown in Ethiopia, out of which about 12% of them are endemic for the country [2] .

The rapid expansion of agriculture is one of the greatest threats for the losses and degradation of forest biodiversity. This uncontrolled expansion of agriculture has led to deforestation and forest degradation, which in turn led to food insecurity and poverty in many parts of the world including Ethiopia. However, there are evidences that indicate sustainable farming practices, like agroforestry, utilize and conserve biodiversity, improve environmental quality and limit agricultural expansion into natural forests as well as the negative impacts of agriculture on biodiversity [3] .

Agroforestry is a dynamic, ecologically based, natural resources management system that, through the integration of trees on farms and in the agricultural landscape, diversifies and sustains production for increased social, economic and environmental benefit for land users at all levels [4] . Agroforestry is the production of trees and of non-tree crops or animals on the same piece of land and provide diverse output from the same land units. These land use types were conserve different types of plant species diversity in pieces of land and minimizes the impacts of community in the natural forest.

Homegarden agroforestry practices are among the agroforestry systems with the potential to harbor native forest biodiversity. Homegardens are a place where a clearly bounded piece of land immediately surrounding the dwelling house is cultivated with a mixture of perennials and annuals [5] . According to [6] , homegarden agroforestry in the Ethiopian highlands hosts higher woody species diversity than their nearby natural woodlands or forest lands by conservation of diverse native plant species. This high species diversity ecosystem provides important ecological services such as nutrient recycling, soil and water conservation, and reduces environmental deterioration. In Ethiopia homegarden agroforestry is the most common practices which are familiar to small holder farmers [7] .

A shade grown coffee agroforestry practice is one of the land uses that contain various types of plant species diversity. As the name indicates it contains various types' of shade tree species and coffee shrubs as the major component. Due to the high income source, coffee based agroforestry practices are strongly developed and cultivated by many farmers. Because of the reduction of land and the need to increase income from coffee monoculture system, the introduction of multi-purpose timbers species in to this system is a good system to get different benefits simultaneously. As [8] reported many small coffee farms around the world incorporate trees as part of the production system, as wind breaks, for land scaping around the home and to protect the coffee plants from excessive sun and high temperature. In Ethiopia the coffee shade based agroforestry practices also conserve various native woody species [9] .

Agroforestry and other land uses of Dallo Mena districts are providing various environmental as well as economical benefits for the community. However, there was no information in relation to comparing the agroforestry and adjacent natural forest in woody species diversity conservation. Most of the reports which are studied in Dallo Mena districts are focus on the plant species diversity of the natural forest. Also there are very few studies focusing on the plant diversity potential of agroforestry practices [10] . But the status of woody species diversity in natural forest and agroforestry land uses is not well studied. Therefore, this study designed to show the contribution of agroforestry and adjacent natural forest of Dallo Mena district for woody species diversity conservation and to compare the amount of woody species diversity conserved in those land uses.

2. MATERIAL AND METHOD

2.1. Study Area

Dallo Mena district is located in Oromia regional state, Bale zone and covers about 461,665 hectares. It lies between latitudes 5°51'N and 6°45'N, and longitudes 39°35'E and 40°30'E (Figure 1). The altitude ranges from 1314 to 1508 m.a.s.l. The major soil type of the woreda is Nitosol which is the dominant soil in the area [11] . Climatically the area characterized by bimodal rainfall with the main rainy season occurring early March through June and the short rain late September through November. The mean annual rainfall is 986.2 mm and the mean annual temperature is 22.5°C.

According to 2007 Population and Housing Census of Ethiopia, the population number of Dallo Mena was 89,670 [12] . The indigenous inhabitants of the area are primarily Oromo peoples. In addition to this few ethnic groups Gurage and Amhara are also settled in the area. Dallo Mena forest has mixed broad-leaved and coniferous montane forest species. According to [13] classification, this forest is categorized as Dry Evergreen Afromontane forest type. The dominant farming activities in Dallo Mena woreda is mixed farming systems, livestock and subsistence crop production farming.

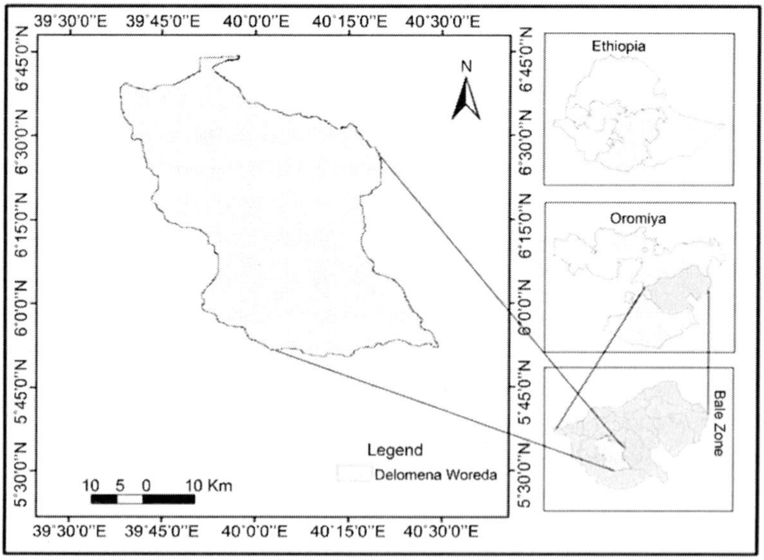

Figure 1. Map of Dallo Mena Woredas of Bale zone, Southeast Ethiopia.

2.2. Method

2.2.1. Sampling Techniques

Before staring the sample collection, reconnaissance survey was carried out to identify the study site, verify the land use type and sampling method. The land use were selected based on its arrangement, vegetation structure and the size of land uses. The three land uses are found adjacent to each other in stratified manner in the order of natural forest, shade grown coffee and homegarden. Due to the similarity of topography, climate and land use history of all land uses which are originated from the forest land, special analog approach was chosen for site selection. The sites differed only in the land use type and soil management practices.

A systematic sampling method was applied to locate the sample plots to assess woody species diversity and composition. For all land uses the data was collected following the transect line. The distance between each transect and plot in each land uses were about 500 m and 400 m respectively. In the process of data collection the first transect line and the first plot were selected purposely inside the land uses. For this study a plot size of 10 m × 10 m was used to encompass all woody seedling, sapling and tree grown in side plot.

2.2.2. Sampling Design

Plant Identification and Biomass Estimation

On each plot woody species seedlings (<2.5 cm diameter and height < 1 m), saplings and shrub (2.5 - 5 cm diameter and height 1 - 2 m) and trees and shrub (≥5 cm diameter and height ≥ 2 m) were recorded by complete count of each [14] [15] . For all woody species having DBH ≥ 5 cm the diameter at breast height (DBH at 1.3 m) were recorded. For the coffee shrubs, the diameter was measured at 15 cm from the ground (D at 15 cm) [16] .

The diameter and height was measured by using Caliper, Diameter tape and Hypsometer. Species identification for common species was done in the field by using different plant identification keys as references [17] [18] [19] [20] . But for the less common species identification was done by taking plant sample specimen.

2.3. Data Analysis

Floristic Composition, Population Structure and Diversity

The floristic composition and structure of the agroforestry systems and adjacent natural forest were calculated in terms of size, relative frequency, relative density, relative dominance, basal area, and importance value index.

Diversity Indices

A diversity index is a mathematical measure of species diversity in a community. Diversity indices provide more information about community composition than simply species richness (i.e., the number of species present); they take the relative abundances of different species into account. Diversity indices provide important information about rarity and commonness of species in a community. For this study the diversity indices of the sampled land use types were calculated by using Shannon diversity which is very widely used index for comparing diversity between various habitats [21] .

The Shannon diversity index is calculated as:

$$H' = -\sum_{i=1}^{S} p_i \ln p_i \qquad (1)$$

where, H' = Shannon-Wiener index of species diversity

s = number of species in community

p_i = proportion of total abundance represented by i^{th} species

Evenness:

Species evenness is a diversity index, a measure of biodiversity which is used to measure the homogeneous distribution of tree species in sample plot [22] . The evenness of a population was calculated by;

$$E = \frac{H'}{H_{max}} = \frac{H'}{\ln S} \text{ with } H_{max} = \ln S \qquad (2)$$

where, E = Evenness

H' = Calculated Shannon-Wiener diversity

H_{max} = ln(s) [species diversity under maximum equitability conditions]

S = the number of species,

P_i = proportion of individuals of the ith species or the proportion of the total

Species

Simpson's diversity Index:

Simpson's diversity index was also used for this study. Simpson's diversity index is the most sensitive to changes in more abundant species and hence places more weight on the most abundant species in the community. The Simpson's diversity index was derived from probability theory and it is the probability of picking two organisms at random which are of different species [23] . Simpson's diversity index can be represented by;

$$\textbf{Simpson's index of diversity} = 1-(D)^2\,; D=\frac{n}{N} \tag{3}$$

where D = Simpson's index

n = The total number of organisms of a particular species

N = The total number of organisms of all species

Similarity Indices (S_s):

Similarity indices measure the degree to which the species composition of different systems is alike. Many measures exist for the assessment of similarity or dissimilarity between vegetation samples or quadrates. The Sorensen similarity coefficient is applied to qualitative data and is widely used because it gives more weight to the species that are common to the samples rather than to those that only occur in either sample [24] . The Sorensen coefficient of similarity (S_s) is given by the formula:

$$S_s = \frac{2a}{2a+b+c} \tag{4}$$

where, S_s = Sorensen similarity coefficient.

a = number of species common to both samples.

b = number of species in sample 1.

c = number of species in sample 2.

Multiple-Site Similarity Indices (S_s):

The similarity of woody species diversity among the three land use types were analyzed by using a multiple-site similarity index through the use of a multiple-site similarity measure. This method was used because similarity measures have restriction that is limited to pair wise comparisons even in a multiple-site study and this method overcomes this problem of covariance between pair wise similarities in a multiple site study [25] . This index was defined as:

$$MSSI = \frac{ab + ac + bc - abc}{a + b + c} \quad (5)$$

where, MSSI = multiple site similarity index

a = number of species found in NF

b = number of species found in SC

c = number of species found in HG

ab = number of species common to NF and SC system

ac = number of species common to NF and HG systems

bc = number of species common to SC and HG systems

abc = the number of species found in the three systems

Basal area/Dominance:

Basal area is the cross-sectional area of woody stems at breast height. It measures the relative dominance (the degree of coverage of a species as an expression of the space it occupies) of a species in a forest. Basal area was calculated for each woody species with diameter ≥ 2.5 cm and calculated as:

$$BA = \frac{\pi\left(DBH^2\right)}{4} \quad (6)$$

where, p = 3.14

BA = basal area (m^2)

DBH = diameter at breast height (cm)

$$\text{Relative dominance} = \frac{\text{Dominance of a species}}{\text{Dominance of all species}} \times 100 \quad (7)$$

Density:

The density of woody species is one of the most important structural parameters to be considered during vegetation data analysis. Density was calculated by summing up all stems across all sample plots and converting into hectare basis.

$$\text{Density} = \frac{\text{Total number of individuals}}{\text{Sample area (ha)}} \quad (8)$$

$$\text{Realtive Density} = \frac{\text{Total number of individuals of a species}}{\text{Total number of individual of all species}} \times 100 \tag{9}$$

Frequency:

Frequency is defined as the probability or chance of finding a species in a given sample area or quadrant [26] . Thus, it shows the presence or absence of a given species within each sample plot. Two frequency values were computed for each woody species encountered within the study plots: the absolute frequency and the relative frequency;

$$\text{Frequency}(\%) = \frac{\text{Frequency of a species}}{\text{Total number of sample plot}} \times 100 \tag{10}$$

$$\text{Realtive Frequency} = \frac{\text{Frequency of a species}}{\text{Frequency of all species}} \times 100 \tag{11}$$

Important Value Index (IVI):

The IVI of a species is used to express the relative ecological significance of the species in the forest ecosystem. It was calculated by summing up the relative dominance, relative density and relative frequency of the species [26] . It indicates the significance of species in the system. It is calculated as follows:

$$\text{IVI} = \text{Relative Frequency} + \text{Relative Density} + \text{Relative Dominance} \tag{12}$$

The plant species diversity data were analyzed by using PAST paleontological statistical software for biodiversity. Mean comparison of the four systems in plant species diversity were tested by least significant difference (LSD) test at $P < 0.05$ by using SAS statistical software version 9.1.3.

3. RESULTS

3.1. Woody Species Diversity

3.1.1. Species Diversity and Frequency

Overall, 39 woody plant species were recorded from natural forest (NF), homegarden agroforestry practices (HG) and shade grown coffee agroforestry practices (SC). Out of this, 23 species were recorded from the natural forest while 25 were from adjacent agroforestry land uses (10 from shade grown coffee and 15 from homegarden) respectively. The entire recorded species were belonging to 24 families. From the overall plant family categorization, Rutaceae and Myrtaceae are families with relatively higher number of species. Coffea arabica, Mangifera indica and Saxifraga hederifolia were the most frequently observed woody species in SC, HG and NF which are belonging to Rubiaceae, Anacardiaceae and Saxifragaceae families respectively. The number of individual woody species recorded at HG, NF and SC were 335, 194 and 1548 respectively. For coffee shrub, the highest numbers of individuals were observed in the shade grown coffee agroforestry practice (Table 1). In the homegarden agroforestry practice, farmers manage both exotic (46.7%) and native trees/shrubs species (53.3%).

To assess species diversity Shannon, Simpson and evenness indices were employed in this study. In terms of Simpson and Shannon diversity indices there is no significance difference between the two agroforestry practice (HG and SC), but both are significantly different from NF (Table 2).

In the natural forest, the overall frequency of occurrence of woody species varied between 1 - 56 in the surveyed plot. Saxifraga hederifolia (56), Rhus ruspolii (23) and Lepidotrichilla volkensii (17) were the most frequently observed woody species. These species accounted for 29%, 11.9% and 8.8% of the plant community. Species like Allophyllus abyssinicus, Flacourtia indica, Carissa spinarum and Commiphora habessinica were less frequently encountered than the other woody species (Figure 2).

Table 1. Number of individual woody species and coffee shrubs recorded in natural forest (NF), shade grown coffee (SC) and homegarden agroforestry (HG) land uses of Dallo Mena Woreda.

System	Woody Species	Coffee Shrubs	Total Abundance
NF	194	–	194
SC	35	1513	1548
HG	233	102	335

Table 2. Mean Simpson, Shannon and Evenness diversity indices among the three land uses (natural forest, shade grown coffee and homegarden) in Dallo Mena Woredas of Bale.

System	Mean		
	Simpson (1-D)	Shannon (H')	Evenness (J)
HG	0.237[b]	0.447[b]	0.537[b]
NF	0.617[a]	1.168[a]	0.721[ab]
SC	0.308[b]	0.466[b]	0.812[a]

Means with the same letters across column are not significantly different (P > 0.05) with respect to the system; NF = Natural Forest, SC = Shade grown coffee and HG = Homegarden.

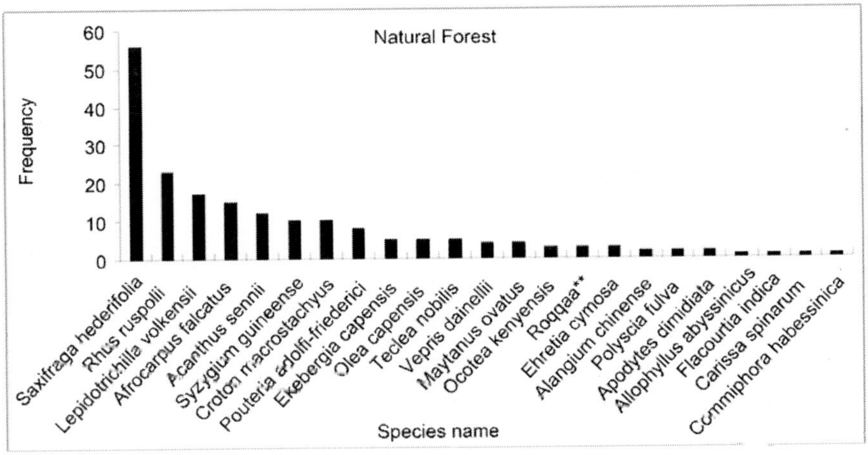

Figure 2. Overall frequency of woody species in the natural forest of Dallo Mena Woredas of Bale Zone.

For shade grown coffee agroforestry practice, the most frequently observed woody species was Coffea arabica (92%) followed by Afrocarpus falcatus (Figure 3).

Woody species frequency in the homegardens was in the order of: Mangifera indica (36.7%), C. arabica (30.5%) and Carica papaya (13.7%) for most frequent category, and Citrus grandis, Ehretia cymosa, Jatropha curcus and Mimusops kummel for less frequent categories (Figure 4). The studied homegardens are dominated by fruit species hence can be classified as fruit tree based homegarden agroforestry practice.

3.1.2. Woody Species Similarity Indices

The similarities in woody species composition were compared among the land uses (Table 3). The highest similarity in woody species compositions (24.2%) was recorded between homegarden and shade grown coffee, while the lowest (9.5%) was recorded between natural forest and homegarden. Syzygium guineense, Afrocarpus falcatus, Croton macrostachyus, Ocotea kenyensis and Polyscia fulva woody species were common to NF and SC with 23.3% similarities. Whereas, Croton macrostachyus, S. guineense and Ehretia cymosa were common for NF and HG with 9.5% similarity. S. guineense, Croton macrostachyus, Persea americana, Psidium guajava and C. arabica are species commonly grown both in the SC and HG land uses having 24.2% similarity. About 21% species were overlapping between the NF, SC and HG.

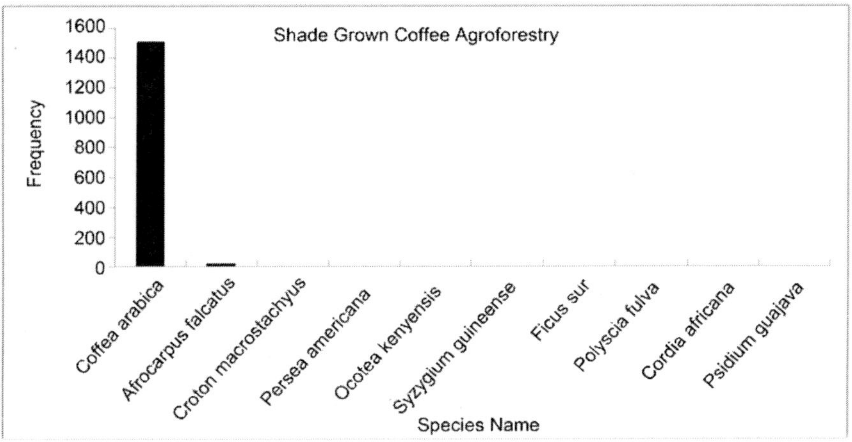

Figure 3. Overall frequency of woody species across the shade grown coffee agroforestry practice of Dallo Mena Woredas of Bale Zone.

Table 3. Similarity index of woody species among the land uses.

Land Use Type	Similarity (%)
Natural Forest vs Shade Grown Coffee	23.3
Natural forest vs Homegarden	9.5
Shade Grown Coffee vs Homegarden	24.2
Natural Forest vs Shade Grown Coffee vs Homegarden	21

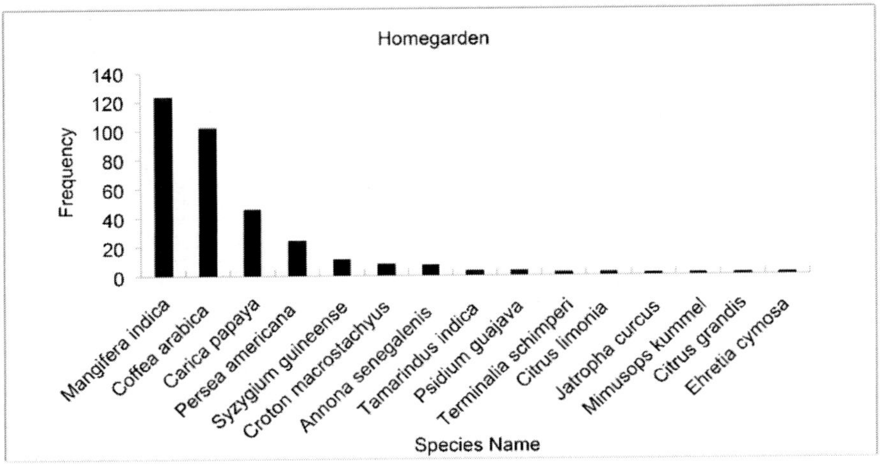

Figure 4. Overall frequencies of woody species across the homegarden based agroforestry practice of Dallo Mena Woredas of Bale.

3.1.3. Importance Value Index

Species with higher IVI among NF, SC and HG are different. The five most important woody species with the highest IVI's is presented in Table 4. Of all species, S. guineense (77.57), C. arabica (142.79) and Mangifera indica (98.95) were the top ranking in natural forest, shade grown coffee and homegarden, respectively. The top five highest IVI species contribute about 60.22%, 86.44% and 80.77% of the total population in NF, SC and HG land uses respectively (Table 4).

3.2. Density and Population Structure of Woody Species

A total of 2077 individual woody species were recorded from all sample plots of natural forest, shade grown coffee agroforestry and homegarden agroforestry land uses. The highest individual woody species were recorded from homegarden agroforestry practices and the lowest were from shade grown agroforestry practice. Saxifraga hederifolia and Rhus ruspolii are the most abundant woody species with the lower diameter class (<5 cm) in natural forest. Where as the major woody species with highest diameter class (>35 cm) were S. guineense, Ocotea kenyensis and Pouteria adolfi-friederici. The graphical descriptions of the frequency distributions of woody species among each diameter class were inverted J-shaped graph for NF and HG, and broken inverted J-shaped graph for SC (Figure 5).

The overall mean value in DBH, height and density were significantly ($P < 0.05$) different among all land uses. The higher mean DBH were observed in the SC (55.55 ± 9.83). Whereas the lowest mean DBH were observed in the HG (24.09 ± 1.64). From the three land uses the highest height were recorded in NF (23.32 ± 2.37). The overall mean value of woody species density of the three land uses were significantly ($p < 0.05$) different from each other. The highest woody densities were observed in HG (164.68 ± 47.35) and the lowest were in SC (37.30 ± 13.21) (Table 5). But for BA there is no significant difference among all land uses.

Table 4. The top five woody species with the highest IVI values in the natural forest, shade grown coffee and homegarden agroforestry practice of Dallo Mena Woredas of Bale.

System	Species Name	IVI
	Syzygium guineense	77.57
	Saxifraga hederifolia	42.39
NF	*Afrocarpus falcatus*	23.28
	Rhus ruspolii	19.02
	Pouteria adolfi-friederici	18.41
	Coffea arabica	142.79
	Syzygium guineense	35.94
SC	*Afrocarpus falcatus*	35.32
	Croton macrostachyus	31.45
	Ficus sur	13.81
	Mangifera indica	98.95
	Persea americana	47.78
HG	*Coffea arabica*	41.31
	Carica papaya	39.15
	Annona senegalenis	15.13

For Coffee shrub woody species among shade grown coffee and homegarden agroforestry practices, there is no significant difference in mean diameter size (d_{15}) and basal area. But for coffee height and density values are significantly ($P < 0.05$) different from each other. The result shows that the largest heights were observed in HG (4.13 ± 0.16). The value of coffee density at SC (1869.14 ± 421.93) is significantly higher than the value for HG (283.33 ± 154.62) (Table 6).

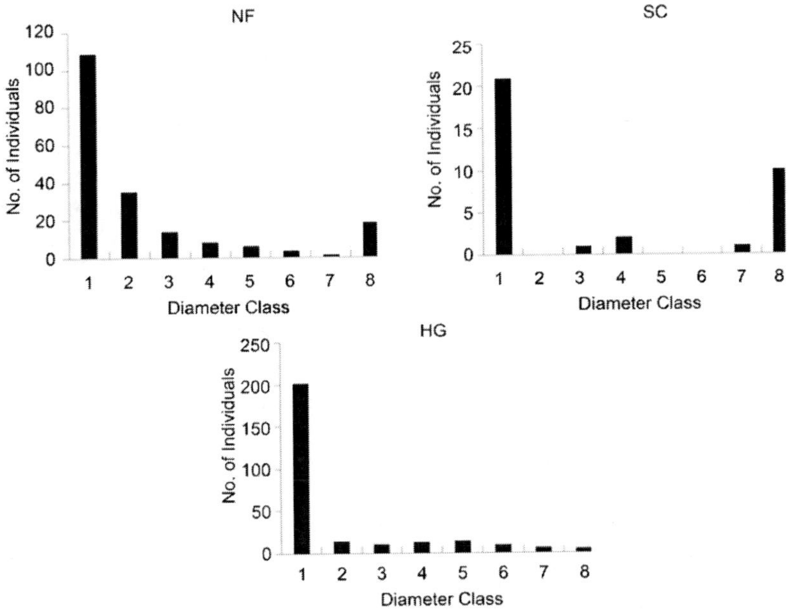

Figure 5. Frequency distributions of woody species by diameter class in Dallo Mena Woredas of Bale. Diameter class in cm: 1, 0 - 5, 2, >5 - 10, 3, >10 - 15, 4, >15 - 20, 5, >20 - 25, 6, >25 - 30, 7, >30 - 35, 8, >35.

Table 5. Stand Structure Mean (±SE) of woody species in three land use system (NF, SC and HG) in Dallo Mena Woreda of Bale.

System	Mean (±SE)			
	Dbh (cm)	Height (m)	Density (ha^{-1})	BA (m^{2} ha^{-1})
HG	24.09 (±1.64)^{b}	8.74 (±0.54)^{b}	164.68 (±47.35)^{a}	2.42 (±0.54)^{a}
NF	41.75 (±11.99)^{ab}	23.32 (±2.37)^{a}	109.63 (±41.85)^{ab}	3.22 (±1.73)^{a}
SC	55.55 (±9.83)^{a}	17.25 (±3.47)^{a}	37.30 (±13.21)^{b}	3.79 (±1.25)^{a}
P-Value	<0.0001	0.0006	<0.0001	0.7457

Means with the same letters across column are not significantly different (P > 0.05) with respect to the system; SE = Standard error, DBH = Diameter at breast height and BA = Basal area.

Table 6. Mean (±SE) of population structure of coffee shrub species in the two agroforestry practices (SC and HG) in Dallo Mena Woreda of Bale.

System	Mean (±SE)			
	Dsh (cm)	Height (m)	Density (ha^{-1})	BA (m^2ha^{-1})
HG	5.30 (±0.28)[a]	4.13 (±0.16)[a]	283.33 (±154.62)[b]	0.29 (±0.03)[a]
SC	5.45 (±0.21)[a]	3.60 (±0.16)[b]	1869.14 (±421.93)[a]	0.30 (±0.02)[a]
P-Value	0.6603	0.0196	0.0028	0.8309

Means with the same letters across column are not significantly different (P > 0.05) with respect to the system; SE = Standard error, DSH = Diameter at stump height and BA = Basal area.

Diameter Distribution of Woody Species with Highest IVI

Natural Forest

In natural forest woody species with the top five highest IVI were S. guineense, S. hederifolia, A. falcatus, R. ruspolii and Pouteria adolfi-friederici (Table 4). In terms of population structure of S. guineense, individuals with the lower and middle diameter classes was absent (Figure 6(a)). On the other hand population structure of S. hederifolia was observed in less than 25cm dbh class (Figure 6(b)).The number of individuals of A. falcatus are present in the lowest two and highest diameter classes. But the medium diameter classes were absent (Figure 6(c)). For Rhus ruspolii the number of individuals were only limited to the smallest diameter class (Figure 6(d)). In the case of Pouteria adolfi-friederici, few lower classes and significant higher diameter class individuals were observed. But the medium diameter class was absent (Figure 6(e)).

Shade Grown Coffee Agroforestry

In shade grown coffee agroforestry practice the top five woody species with highest IVI are C. arabica, S. guineense, A. falcatus, Croton macrostachyus and Ficus sur (Table 4). There is fair population structural distribution for C. arabica bush having highest number of individuals with diameter range of 5.6 - 7.0 cm followed by 1.4 - 2.8 cm diameter class (Figure 7(a)). For S. guineense, A. falcatus and Ficus sur only individuals of higher diameter class were recorded (Figure 7(b), Figure 7(c), Figure 7(e)). In the case of Croton macrostachyus only few individuals of medium and higher diameter class were observed (Figure 7(d)).

Homegarden Agroforestry

In homegarden agroforestry practice the top five woody species with higher IVI are

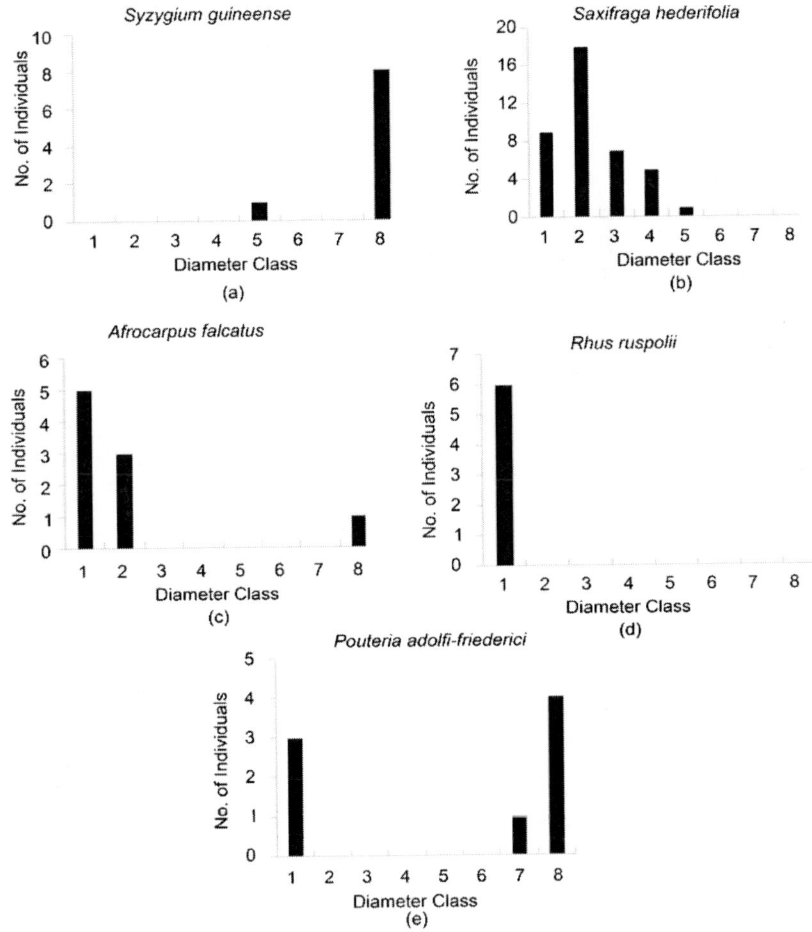

Figure 6. Diameter class distribution of woody species with the highest top five IVI in NF. Diameter for trees in cm: 1, 0 - 5, 2, >5 - 10, 3, >10 - 15, 4, >15 - 20, 5, >20 - 25, 6, >25 - 30, 7, >30 - 35, 8, >35.

indicated in Table 4. Mangifera indica was one of the frequently observed woody species. The overall population structure of this species represented with few individuals in smaller diameter and larger diameter sizes, and with higher larger number of individuals in the medium diameter class (Figure 8(a)). For Persea americana large amount of individuals were observed in the medium and larger diameter class species. For this species the lower diameters class individuals were absent (Figure 8(b)). For C. arabica larger amount of individuals were recorded in the medium diameter classes. But the number of individuals was declined as the diameter increased (Figure 8(c)). In the case of Carica papaya higher amount of small diameter class were observed (Figure 8(d)). For Annona senegalenis higher number of individuals was recorded in lower and medium diameter class (Figure 8(e)).

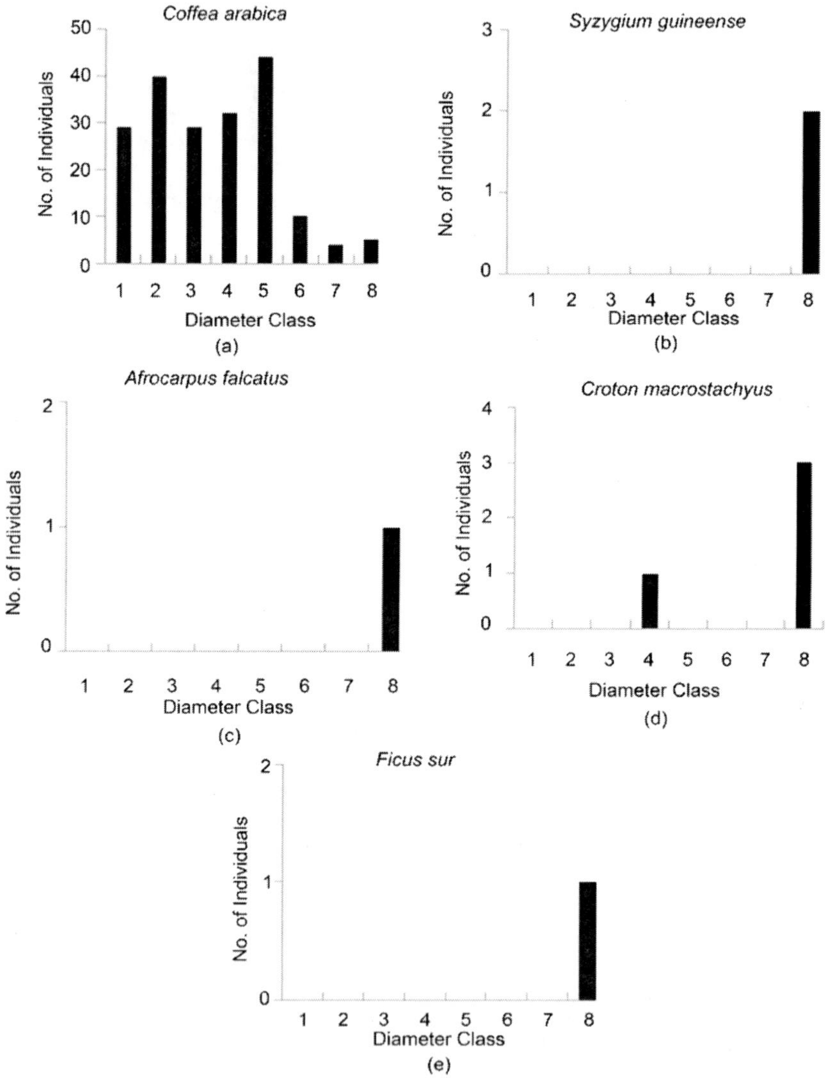

Figure 7. Diameter class distribution of woody species with the highest top five IVI in SC. Diameter for trees in cm: 1, 0 - 5, 2, >5 - 10, 3, >10 - 15, 4, >15 - 20, 5, >20 - 25, 6, >25 - 30, 7, >30 - 35, 8, >35 and diameter class for coffee in cm: 1, ≤1.4, 2, >1.4 - 2.8, 3, >2.8 - 4.2, 4, >4.2 - 5.6, 5, >5.6 - 7.0, 6, >7.0 - 8.4, 7, >8.4 - 9.8, 8, >9.8.

4. DISCUSSION

4.1. Woody Species Diversity and Composition

Different land uses encompasses various types of biological diversity. Among several of them, woody species are one of the dominant types basically grown naturally or manually. From a total 39 woody species diversity recorded during this inventory, about 23, 10, 15 species were recorded from natural forest, shade grown coffee and homegarden agroforestry practices. Among all land uses higher species richness were recorded in the NF forest (H' = 1.168). Due to farmers' activity to increase the land use efficiency and due to intensive thinning of other plant species in order to reduce competition from the coffee lower species richness were recorded from both agroforestry practices. This activity might affect and limit the number of woody species which are grown in system.

The degree of species similarity between natural forest and shade grown coffee agroforestry, natural forest and homegarden agroforestry, and shade grown coffee and homegarden agroforestry is about 23.3%, 9.5% and 24.2%. This implies that the degree of species similarity within the three land use system is low and different woody plant species diversity was observed within each land use. The multiple-site similarity index of natural forest, shade grown coffee and homegarden agroforestry was found to be 21%. This indicates that, the three land uses have less overlapping species with each other. This could be explained by the fact that farmer intensive species selection for different use and coping strategies for constraint facing them. For example, in case of homegarden agroforestry practices due to land size shortage and irrigation facilities farmers preferred high market demanded species which are more interested on fruit species and high market value species than native species. For shade grown coffee agroforestry practice and natural forest due to the nature of the land and its closeness to each other they share some species together. Generally due to the above mentioned reasons the species similarity index of the three land uses of the area is very less. This less spices overlapping of the land uses are shows that the degree the land uses which are supports each other are less. These shows that the levels of agricultural land escape to conserve the native woody species which are grown in the natural forest are relatively less. This implies that still there are great pressure on the natural forest to satisfy wood consumption of the community.

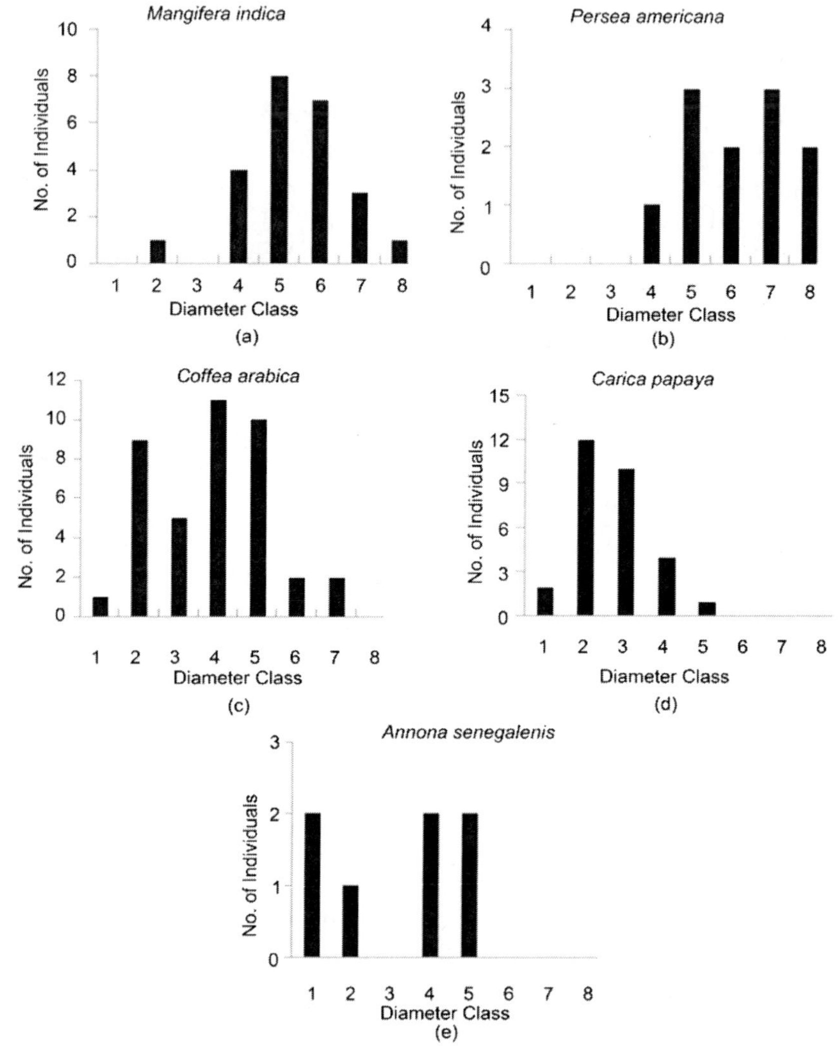

Figure 8. Diameter class distribution of woody species with the highest top five IVI in HG. Diameter for trees in cm: 1, 0 - 5, 2, >5 - 10, 3, >10 - 15, 4, >15 - 20, 5, >20 - 25, 6, >25 - 30, 7, >30 - 35, 8, >35 and diameter class for coffee in cm: 1, ≤1.4, 2, >1.4 - 2.8, 3, >2.8 - 4.2, 4, >4.2 - 5.6, 5, >5.6 - 7.0, 6, >7.0 - 8.4, 7, >8.4 - 9.8, 8, >9.8.

When we see the overall species composition of the system, the natural forests were dominated by few pioneer species which are very dominant in area. But the composition of middle aged and regeneration structure of woody species is that much not promising. It is only dominated by few old aged (matured) trees which cover the upper canopy. Most of the trees which are grown in this system are very huge indigenous species with larger diameter size, height and canopy coverage. In case of shade grown coffee agroforestry practice the system is covered by very few woody species. The type of shade plant species are vey limited and only covered by over populated coffee shrubs. But when we observe the overall status of the woody plant structure most of the plant in this system are higher DBH class shade tree species. This is due to the factor that shade trees are the major component of coffee production of the area and the tree are conserved for the longer period of time. This larger diameter size species takes larger space and create great pressure on coffee plants through resource computation. In addition to this the continuity of this important woody species is under question. Also in order to increase the productivity of the system this matured higher diameter size species must be replaced by other new species. Homegarden agroforestry practice is also conserving a number plant species. But due to high market demand for vegetables the farmers concentrated only on the short rotation vegetables. But fruit trees are one of the dominant woody species which are grown in area. This farmer's direction will change their interest to focus on short rotation exotic species. This will affect the number of native woody species conserved in the farmer's field.

4.2. Population Structure and Importance Value Index

From all system the highest mean DBH was observed in shade grown coffee agroforestry practice (55.55 ± 9.83 cm). This is due to the factor that in the coffee production process the tree species which are important for shade are purposely selected and conserved for the longer period of time. Due to selective thinning very few or limited tree species are grown on the larger area without competition. While the lowest DBH size was observed in the homegarden agroforestry practice (24.09 ± 1.64 cm). This is due to farmers' intensive tree selection, the tree characteristics, and high resource competition within the system. As we know farmers are very sensitive for the land and very much selective for the tree grown on their field. There for due to this factor most of the trees grown in their field are fruit tree species which are not that much larger in diameter. In addition to this, farmers are cut matured and aged tree species which grown in the farm for the construction (timber) purpose and to replace by other new productive tree species.

In the case of height, the highest and lowest size were recorded in the NF and HG land uses (23.32 m and 8.74 m) respectively. This is due to the reason that NF is the free growing area and the tree grows without any human influence which affects the height growth. In the case of SC (17.25

m) due to the shade purpose the height of the trees are not managed. But for HG in order to reduce the shade and minimize light computation from the under growth plants the height of the trees are managed (pruned) repeatedly. In addition to this most of the trees which are grown in the farmers filed are multi-purpose tree species which used as animal fodder. This all factor affects the height growth.

Homegarden agroforestry practice is the highest in terms of woody species density per hectare (164.68 trees ha^{-1}). This is due to farmer's day to day activity in order to maximize the land use efficiency to increase their income. In this study area the average homegarden sizes of the land which are hold by each farmer are very limited (0.31 ha). Due to this limitation farmers incorporate different types of plant species on their farm. In other way the amount of tree density in the shade grown coffee agroforestry practice is relatively low. This is due to farmers' intensive selective tree growing and gives priority for the coffee plant only. Also this lower density of shade tree species is due to the selective removal of other plant species to promote and to produce suitable environment for coffee development and selective management of shade tree for their ecological and economical values. In addition to this the intensive soil management for coffee shrubs is also damages the regeneration and growth of plant species of the system. All this reasons shows that the tree density is depend both on the intensity of management and biophysical conditions. For BA there is no significant difference among all land use.

Due to the variation of main target of the land uses, the coffee density of shade grown coffee agroforestry practices (1869.14 trees ha^{-1}) was significantly different and higher than that of HG (283.33 trees ha^{-1}). That is the main objective of coffee farm is only coffee production but in HG it contains other types of annual and perennial plant species. In addition to this, the mean coffee height of HG (4.13 m) was higher than that of SC (3.60 m). This is due to light competition with other tree species which are grown together in the HG and due to higher computation of over populated coffee shrub in the SC.

For IVI, species with higher IVI are varied and depend on the systems. In the case of agroforestry practice the species importance is depend on the product value of the species. For example in shade grown coffee agroforestry practice woody species with the highest IVI are C. arabica and few other shade tree species. In Homegarden agroforestry practice woody species with highest IVI are fruit tree species and other high market value species. According to this study species with high IVI is associated with the land uses and based on farmer's species preference. When we see the overall IVI of the whole system, in all land uses more than half of IVI is covered by very few limited species. In the case of natural forests, species with higher IVI is linked with those species which are dominant and well adapted in that agroecology. Where as in the case of agroforestry practices species with higher IVI is associated with framers species preference and product value. The interest of farmers for selection of species is linked with species market demand and service value. The highest IVI species of the natural forest land uses were covered by five pioneer species which are highly dominant in the area. For homegarden agroforestry practices the highest IVI value is covered by fruit tree species which provide higher income for the farmers. Where as in the case of shade grown coffee agroforestry practice, about large percent of the practice were covered by coffee shrub and other important shade tree species which are highly familiar and positive interaction with coffee plants.

If we see the status of each individual species separately within the land uses, from the five higher IVI woody species of the NF, S. guineense is the one which are in problem with no seedling and sapling. For P. adolf-friederici even if it is disturbed there is also in seedling stage. S. hederifolia and A. falcatus is relatively species with promising future. Also abnormal distributions of Rhus ruspolii were observed in the system with only in the seedling stage. For SC from all species C. arabica is species with good status. But due to the absence of seedling and sapling stages the futurity of the rest species (S. guineense, A. falcatus, Ficus sur and Croton macrostachyus) is in problem. In HG from all species C. arabica, C. papaya and Annona senegalenis is species with promising future. But abnormal distribution of M. indica and Persea americana with poor lower diameter class species were observed.

When we see the overall population structure of woody species which are grown in the both agroforestry practices, abnormal structure were observed. This is due to human interferences the growth and the densities of undergrowth woody species were affected. This may minimize the number of lower diameter class woody species which are grown in the land uses. In addition to this, in order to produce suitable environment for the growth of major species (e.g. Coffee), the regeneration and density of other species were affected by human activity.

5. CONCLUSION

The overall woody species diversity of the studied land uses was different from each other. Highest woody species diversity was recorded in the natural forest. From the two agroforestry practices, homegarden was conserves higher woody species diversity than shade grown coffee agroforestry practice. In the case of shade grown coffee agroforestry practice due to farmer's species selection for shade purpose, selective tree thinning and intensive human interference or management shade tree species diversity are less in number. Low species similarly in species compositions were observed among the three land uses. Generally, there is significant difference among natural forest, shade grown coffee agroforestry practice and homegarden agroforestry practice in woody species diversity composition.

ACKNOWLEDGEMENTS

The work was financed by Oromia Agricultural Research Institute, Sinana Agricultural Research center.

CITE THIS PAPER

Mengistu, B. and Asfaw, Z. (2016) Woody Species Diversity and Structure of Agroforestry and Adjacent Land Uses in Dallo Mena District, South- East Ethiopia. Natural Resources, 7, 515- 534. http://dx.doi.org/10.4236 /nr.2016.710044

REFERENCES

1. 1. Sayer, J., Chokkalingam, U. and Poulsen, J. (2004) The Restoration of Forest Biodiversity and Ecological Values. Forest Ecology and Management, 201, 3-11.
2. Food and Agriculture Organization (FAO) (2001) Global Forest Resources Assessment 2000. Main Report. FAO Forestry Paper 140, 479 p.
3. Khumalo, S., Chirwa, P.W., Moyo, B.H. and Syampungani, S. (2012) The Status of Agrobiodiversity Management and Conservation in Major Agroecosystems of Southern Africa. Agriculture, Ecosystems and Environment, 157, 17-23.
4. International Center for Research in Agroforestry (ICRAF) (2002) What Is Agroforestry? World Agroforestry Center, Nairobi.
5. Kabir, M.E. and Webb, E. (2008) Can Homegardens Conserve Biodiversity in Bangladesh? Biotropica, 40, 95-110.
6. Tolera, M., Asfaw, Z., Lemenih, M. and Karltun, E. (2008) Woody Species Diversity in a Changing Landscape in the South-Central Highland of Ethiopia. Agriculture, Ecosystems and Environment, 128, 52-58.
7. Yakob, G. (2011) Diversity and Management of Woody Species in Homegarden Agroforesty in Gimbo Woreda, South West Ethiopia. Unpublished MSc Thesis, Hawassa University, Wondo Genet College of Forestry and Natural Resources, Hawassa, 82 p. http://www.worldagroforestrycentre.org
8. Travis, I. and Adel, Y. (2010) Managing Shade Trees for Coffee Can Benefit the Soil.
9. Negash, M., Starr, M. and Kanninen, M. (2013) Allometric Equations for Biomass Estimation of Enset (Ensete ventricosum) Grown in Indigenous Agroforestry Systems in the Rift Valley Escarpment of Southern-Eastern Ethiopia. Agroforestry System, 87, 571-581.

10. Didita, M. and Mengistu, B. (2012) The Role of Home Garden Agroforestry in Plant Diversity Conservation in Bale. In: Tadesse, W., Desalegn, G. and Yirgu, A., Eds., Forestry and Forest Products: Technologies and Issues, Ethiopian Institute of Agricultural Research, Addis Ababa, 1-14.
11. Lulekal, E., Kelbessa, E., Bekele, T. and Yineger, H. (2008) Plant Species Composition and Structure of the Mana Angetu Mosit Montane Forest, South-Eastern Ethiopia. Journal of East African Natural History, 97, 165-185.
12. Central Statistics Authority (CSA) (2012) Population and Housing Census: Administrative Report. Addis Ababa, 117 p.
13. Institute of Biodiversity Conservation (IBC) (2005) National Biodiversity Strategy and Action Plan. Government of the Federal Democratic Republic of Ethiopia, Addis Ababa, 103 p.
14. Lu, X., Xia, J. and Jang, J. (2010) Diversity and Composition of Understory Vegetation in Tropical Seasonal Rainforest of Xishuangbanna, Southwest China. International Journal of Tropical Biology, 59, 455-463.
15. Jiangshan, L., Xiangcheng, M., Haiba, R. and Keping, M. (2009) Species-Habitat Associations Change in Subtropical Forest of China. Journal of Vegetation Science, 20, 415-423.
16. Segura, M., Kanninen, M. and Suarez, D. (2006) Allometric Models for Estimating Aboveground Biomass of Shade Trees and Coffee Bushes Grown Together. Agroforestry System, 68,143-150.
17. Breitenbach, F.V. (1963) The Indigenous Trees of Ethiopia. 2nd Edition, Ethiopian Forestry Association, Addis Ababa, 306 p.
18. Kelecha, W.M. (1980) A Glossary of Ethiopian Plant Names. 3rd Edition, Addis Ababa, 262 p.
19. Mooney, H.F. (1963) A Glossary of Ethiopian Plant Names. Dublin University Press Ltd., Dublin, 79 p.
20. Bekele, A. (2007) Useful Trees and Shrubs for Ethiopia, Identification, Propagation and Management for 17 Agroclimatic Zones. RELMA in ICRAF Project, Nairobi, 552 p.
21. Clarke, K.R. and Warwick, R.M. (2001) Change in Marine Communities: An Approach to Statistical Analyses and Interpretation. 2nd Edition, PRIMER-E, Plymouth, 172 p.
22. Krebs, C.J. (1999) Ecological Methodology. 2nd Edition, Benjamin Cummings, Menlo Park, 620 p.
23. Magurran, A.E. (1988) Ecological Diversity and Measurement. Springer, Berlin, 354 p.
24. Kent, M. and Coker, P. (1992) Vegetation Description and Analysis: A Practical Approach. CRC Press, Boca Raton, 363 p.
25. Diserud, O.H. and Odegaard, F. (2006) A Multiple-Site Similarity Measure. Biology Letters, 3, 20-22
26. Moreno-Casasola, P., Mata, D.I., Madero-Vega, C., Castillo-Campos, G. and Warner, B.G. (2011) Floristic Composition and Soil Characteristics of Tropical Freshwater Forested Wetlands of Veracruz on the Coastal Plain of the Gulf of Mexico. Forest Ecology and Management, 262, 1514-1531.

CHAPTER 9

Long-Term Projections of Global Food Requirements: Why Were We Wrong?

Ira Sohn

Department of Economics and Finance, Montclair State University, Upper Montclair, USA.

ABSTRACT

This article reports on and analyzes long-term projections of world food requirements compared with observed 2000 data reported by the United Nations' Food and Agriculture Organization. The importance of this "post-mortem" is to strengthen the case for carrying out long-term projections of essential resources—food, energy, and non-fuel minerals— because of the long-lead times needed to insure that adequate global output levels of these basic ingredients of living standards will be met. This study should prove useful to those preparing today's long-term projections, with world population projected to rise to over 9bn by mid-century.

Keywords: Long-Term Economic Models; Global Food Projections; Assessment of Long-Term Agricultural Projections

1. INTRODUCTION: GLOBAL AGRICULTURE IN THE 21ST CENTURY AND LONG-TERM PROJECTIONS

The run-up of global food prices since 2004 that reversed a three-decade old pattern of declining or stable prices is causing concerns in developed and developing countries alike, as well as in international organizations such as the World Bank and the United Nations' Food and Agriculture Organization (FAO). Higher prices for food and energy—the two principal components of economic wellbeing—are putting at risk the accessibility, availability, and affordability of these two critical resources for millions of the world's most vulnerable people. Some of the gains made over the last decade or two in reducing the number of people living in poverty worldwide have been lost due to the sharp increases in food and energy prices. Even though the FAO's global food price index at the end of 2012 has receded modestly from its peak in 2010- 11, currently, it is still 100% higher than it was in 2000 [1]. As a consequence, since the 2007-08 civil unrest in some 30 food-short developing countries [2], the issue of "food security" is becoming as important as the issue of "energy security" in the international public policy debate.

Increasing geo-political tensions among the leading economic powers who are trying to lock-in their resource supplies, along with mounting scientific evidence of climate change due to the emissions of greenhouse gases, is making it clear to political leaders and policy-makers that transformational changes in the global energy, transport, manufacturing, and agricultural systems will be required if living standards are to increase over the next 40 years, when the world's population is projected to reach more than 9bn people [3]. Consequently, there is a critical need for long-term projections of these resources—because of the long lead-times to develop additional food and energy capacities—if rising geo-political stress and domestic civil unrest are to be contained.

Before turning to the subject of this article, it is important to provide the context for writing it. In September 1981, a symposium sponsored by the Norwegian Nobel Institute was convened to "explore the long-term perspectives of world demographic and economic growth with particular emphasis on international differentials in demographic and economic characteristics, on resources and supplies, and on implications for emerging patterns of cooperation and conflict" [4].

This article reviews long-term global projections made more than 30 years ago of four agricultural sectors: livestock products (including meat, eggs, and milk products); oil crops (which include oilseeds, peanuts, and soybean products, among other crops); grains (which include maize (corn), rice, wheat, barley, and oats, among others); and root crops (which include potatoes, sweet potatoes, cassava, and yams, but not sugar crops), with the observed data from FAO's detailed Food Balance Sheets [5].

2. LONG-TERM STUDIES ON AGRICULTURE, FOOD, AND NUTRITION

Two studies published within a year of each other that focused on long-term trends provide the facts to support the view that without a successful global agricultural sector that incorporates production, transportation, distribution, and changing consumption patterns, improved material well-being in the future—characterized by higher life expectancy, improved nutrition, higher incomes, and lower poverty rates—will not be possible.

Robert Fogel [6] used statistical data—with particular emphasis on Europe and America—to highlight the remarkable extension of life expectancy in the 20th century and the decline in mortality and morbidity rates which he attributed largely to increasing (and nutritionally-improved) food supplies that provided the higher energy levels needed for a growing and increasingly healthy work force to fend off a long list of, often fatal, infectious diseases. In turn, these improvements in human nutrition contributed to economic growth and development and technological change, and vice versa. Without this synergy, it is doubtful that mortality and morbidity rates would have declined sufficiently to achieve the remarkable average life expectancy levels that we are enjoying today.

Giovanni Federico [7] reviewed the success of agriculture over the last two centuries. Federico argued that the world agricultural system has successfully fed an ever-growing population with an increasing variety of products at falling prices, and, simultaneously over the 200 hundred years since Malthus's hypothesis [8], it did this while releasing a growing number of workers from agricultural work to the rest of the economy. According to Federico, the world was able to feed 6.5bn (in 2005), up from 1bn in 1800, because of steady progress in the "primary forces"—scientific advance, technological and institutional change, and globalization.

However, Federico questioned whether the still growing population in the developing countries can be provided with the nutritional standards of the developed countries that reflects a much higher proportion of meat and dairy products that are much more land-intensive than a diet comprised primarily of cereals.

Another important issue that Federico raised is the front burner issue of the "sustainability of modern agriculture", that is, how do we mitigate the environmental damage caused by the excessive use of fertilizers and pesticides, diminishing bio-diversity, increasing scarcities of water and land, and increasing levels of greenhouse gas emissions, with the continuing improvement in the nutritional value of food in particular, and material wellbeing in general, in Asia and Africa in the 21st century?

The US Department of Agriculture (USDA) publishes 10-year projections every year of commodities, trade, and prices based on projected US and world economic and population growth, agricultural policies, and other assumptions that affect the demand and supply conditions in the global agricultural sectors. The United Nations' Rome-based Food and Agriculture Organization recently published long-term projections of global food and nutrition levels for 2050 [9].

Two other organizations that develop long-term projections of global food and agriculture requirements are the Washington-based International Food Policy Research Institute (IFPRI) that is supported by the Consultative Group on International Agricultural Research (CGIAR), and the International Institute for Applied Systems Analysis (IIASA), in Laxenburg, Austria, that conducts policy-oriented research on problems that are both global and inter-disciplinary in nature. Their recent work on long-term projections of global agricultural requirements to 2050 appears in the FAO report, "Looking Ahead in World Food and Agriculture: Perspectives to 2050" [9].

In addition, a number of recent policy papers by government and academic experts [10,11] and review articles in the international press have appeared [3,12,13] on the challenges and opportunities for farming and food to the middle of the 21st century.

3. THE UNITED NATIONS WORLD INPUT-OUTPUT MODEL

In the early 1970s with the recognition of the increasingly adverse environmental effects caused by worldwide industrialization and the first major oil crisis in the decade, the United Nations voiced its concerns regarding the growing gap in per capita incomes between the less-developed countries of the world and the highly industrialized ones.

As a result, in 1973 the United Nations commissioned the construction of a general-purpose model of the world economy. In hindsight, writing in mid-2012, this modeling effort should still be recognized as an intellectual "tour de force". As far as the resource sectors were concerned, the model—which was developed by Professor Wassily Leontief, who was awarded the 1973

Nobel Memorial Prize in Economic Science—tracked three fossil-fuels, six metallic minerals, and four agricultural resources. Other agriculturaland food-related variables included in the model were irrigation investment, land development, cultivated land area, calories and proteins per day, and fish catch. The model also included 30 manufacturing and service sectors, as well as eight types of major pollutants, and five pollution abatement activities [14].

Economic activity was regionalized—the countries of the world were originally aggregated into 15 regional blocks—but were unified by export and import flows of goods and services, capital flows, aid transfers, and "cross border" payments of interest on borrowed capital. An upgrade of the model around 1980 separated the two countries—Canada and the United States—included in the "North America" region, making a total of sixteen regions.

Once assembled, the model was designed to provide quantitative projections of regional and global resource requirements, pollution levels, cumulative resource use, and required inter-regional financial flows, etc., under varying assumptions regarding future income growth in the developed and developing countries with a view towards narrowing the income gap between the two groups of countries from 12:1 (in 1970) to 7:1 (by the year 2000). Curiously, according to the observed per capita GDP for the year 2000 used in the tables in this article, the income gap between the two groups of countries declined from 12:1 in 1970 to 6.6:1 in 2000.

4. PROJECTED AND OBSERVED POPULATION AND INCOME GROWTH: THE MAIN DRIVERS OF GLOBAL FOOD REQUIREMENTS

This section presents the principal long-term determinants of food consumption—population and income growth. The population projections made in the mid-1970s for the year 2000 by the United Nations Population Division [15] were insignificantly different from the observed global rate for 2000 (**Table 1**).

This table presents the projected and observed annualized regional population, Gross Domestic Product (GDP), and GDP per capita growth rates for the projection interval. (Please see Tables A1 and A2 in the Appendix for the country and region aggregation schemes). The lower observed population growth rates for most of the regions (with the exception of Africa, Asia, one European region, and the United States) as compared to the projections are consistent with long-term demographic trends. The demographic changes in two of the European regions WEH (western Europe, high income) and EEM (eastern Europe, medium income) are the result of the re-unification of Germany in the early 1990s and the migration of many eastern-Europeans to western Europe, in part, as a result of the economic stress in the eastern countries in the wake of the collapse of the Soviet Empire in the late 1980s. As for the US, the higher observed population growth rate compared with the projected rate for the projection interval is most likely the result of increased immigration during the last two decades. The observed growth rate for (non-Japan) Asia exceeded the projected rate by less than 10%, and the only substantial "error" appears to be Africa (AF), where the observed rate exceeded the projected rate by 0.5% per year.

Observed world GDP growth was less than the rate projected by the World Model, despite the formidable growth rates achieved by China (ASC), and India and other Asian countries (ASL), from 1980 to the terminal year, 2000. The observed lower growth rates in GDP relative to the projected rates over this interval in the middle-eastern oil-producing countries (OIL), Central and South America and the Caribbean (LA), South Africa (SAF), the Soviet Union (SU), and Eastern-Europe (EEM), and the lower observed growth in Japan and Western Europe (WEH), can be attributed to an array of well-known political, social, economic and financial problems that these regions confronted during the 1980- 2000 interval. These problems also explain, in part, the lower observed world GDP growth rate relative to the projected rate despite the successes in Asia.

The last two columns of **Table 1** present projected and observed annual growth in GDP per capita. For economists, per capita GDP represents an imperfect, though adequate, metric to measure relative "living standards" and their growth (or decline) over time.

In addition to population and GDP growth, regional and global food requirements are also impacted by other important variables such as urbanization rates, which along with higher living standards, result in changing diets such as a movement away from root crops to grains, and finally, to increased livestock products, characterized by more meat, eggs and cheese, in daily diets.

Despite not having access to the documentation regarding the technical relationships governing these regional changes in food diets from decade to decade over the projection interval when preparing this article, it is my understanding that because the World Model tracked both the changing regional urbanization rates and per capita GDP levels from decade to decade, the projections of the regional food requirements were, in part, determined by these factors that were modified region-byregion and decade-by-decade.

5. GLOBAL AGRICULTURE, 1970-2000: THE PROJECTIONS AND THE OBSERVED DATA

5.1. Production and Domestic Supply

This section provides an overview of the projections and observed data for the 30-year projection interval. Please see Appendix B1 for some caveats about the data, and Appendix B2 for a comment about "consumption" in the World Model. Tables 2 and 3, respectively, present the projected and observed annual growth rates in production and domestic supply (and, in the case of the observed data, food consumption) for the four agricultural sectors.

Table 1. Projected and observed growth in population, gross domestic product, and gross domestic product per capita, 1970-2000 (annual percentage change).

	Population		Gross Domestic Product		GDP per Capita	
Region	Projected	Observed	Projected	Observed	Projected	Observed
AF*	2.3	2.8	2.9	3.7	0.6	0.9
ASC	1.4	1.5	4.3	8.9	2.9	7.3
ASL	2.0	2.2	3.6	5.4	1.6	3.1
CANADA	1.3	1.2	3.6	5.5	2.2	2.3
EEM***	0.6	−0.2	4.4	2.6	3.8	2.8
JAPAN	0.7	0.6	5.3	3.5	4.6	2.9
LA**	2.3	2.0	6.0	3.6	3.6	1.6
OCH	1.3	1.3	3.8	3.4	2.5	2.1
OIL	3.2	2.9	12.0	2.7	8.5	−0.2
SU	0.8	0.6	3.2	0.0	2.4	−0.6
SAF	2.8	2.3	4.6	2.4	1.8	0.1
USA	0.9	1.1	2.8	3.3	2.3	2.2
WEH***	0.2	0.4	3.3	2.5	3.1	2.1
WEM	1.3	1.2	2.4	3.1	1.0	1.9
WORLD	1.69	1.70	3.96	3.30	2.20	1.57

Notes: *Throughout this article the region designated by AF combines the two regions, AAF and TAF, from the regional representation in **Appendix Table A1**. **Throughout this article the region designated by LA combines the two regions, LAL and LAM, from the regional representation in **Appendix Table A1**. ***Throughout this article the data for 2000 are for a unified Germany that is part of the WEH region. Sources: Leontief and Sohn [16]; Leontief *et al.* [17]; World Bank [18].

At the global level the projected annual growth rates are remarkably close to the observed rates, save oil crops, where the observed growth rate is almost four times higher than the projected rate. I comment on this discrepancy in Section 6, below. It would be reasonable to expect greater divergence at the regional level, not only between the projected and observed production values, but also between the values for regional domestic supply because of import, export (and, in the case of the observed data, inventory adjustment) activities. This is indeed the case. Section 6 examines the differences at the regional level between the projected and the observed growth rates employing the central explanatory variables—population, GDP, and GDP per capita growth—as the "primary suspects". At the global level, the observed growth rates of livestock products, oil crops and grains all exceeded the rate of growth in population and per capita GDP, with only the rate of growth in production of root crops lagging population and per capita GDP growth.

5.2. Regional (or Country) Dependence on Food Imports

"Food security" has become an issue for countries who rely on food imports to bridge their shortfall between domestic production and food consumption. The World Model also tracked regional import and export activities. Therefore, it would be of interest to examine the changing position of these regions over the 30-year period with respect to their dependence on imports to supplement their domestic food production. It is important to remind readers that the degree of import dependency of a region—or more specifically of a country—on energy or food is only one factor among others that comprise the more complex terms "energy security" or "food security". Some of these other factors include the sources of those imports, the availability of substitutes, the ease or difficulty of increasing efficiency in their use, etc.

The changing rates of projected and observed dependency on agricultural imports are presented in **Table 4**. While there are relatively large differences between the projected and observed data regarding the degree, i.e., the percent, of import dependency in 2000, generally speaking there is broad agreement between the projected and observed data regarding the regions that are most dependent on imports to supplement their domestic production.

5.3. Per Capita "Consumption" of Livestock Products and Grains: The Projections and the Observed Values in 1970 and 2000

Table 5 presents the projected and observed annual per capita domestic supply (in kilograms) of two of the four agricultural sectors in this study: livestock products and grains. Grains were selected because it was (by far) the largest of the commodity groups in tonnage, and livestock products were selected because increased per capita consumption of livestock products is closely associated with higher living standards. (Readers are reminded that "domestic supply" (see Appendix B2)—domestic production and inventory changes plus net imports—is distributed to uses such as food, feed, seed, processing, etc).

Table 2. Projected growth rates in production and domestic supply in four agricultural sectors, 1970-2000 (annual percentage change).

	Livestock Products		Oil Crops		Grains		Root Crops	
	Production	Domestic Supply	Production	Domestic Supply	Production	Domestic Supply	Production	Domestic Supply
AF	2.1	2.1	0.5	0.9	1.8	2.0	−0.6	−0.6
ASC	2.6	2.6	2.0	2.1	1.7	1.8	1.1	1.1
ASL	0.8	0.8	−1.3	−0.8	1.8	1.8	2.2	2.2
CANADA	2.7	2.6	3.0	3.4	2.9	2.3	2.2	1.7
EEM	2.7	2.8	−0.1	1.0	2.5	1.9	0.9	0.8
JAPAN	4.7	5.2	−1.2	2.0	−0.6	2.1	−3.2	−3.3
LA	3.7	3.5	4.4	3.0	2.5	2.7	−0.3	−0.4
OCH	2.3	3.1	3.7	1.0	2.4	1.7	1.3	1.6
OIL	3.6	4.9	1.0	4.7	4.7	5.4	4.7	4.7
SU	2.9	2.9	−2.1	−1.8	1.4	1.8	−0.1	−0.1
SAF	2.0	2.0	−0.7	3.3	3.4	3.3	0.0	0.8
USA	1.1	0.8	1.2	0.0	2.0	1.1	−0.8	−0.5
WEH	2.3	2.3	2.5	2.3	1.5	1.2	−1.6	−1.7
WEM	0.1	0.3	−2.6	−0.2	1.8	1.9	0.7	0.9
WORLD	**2.3**	**2.3**	**0.7**	**0.7**	**1.9**	**1.9**	**0.5**	**0.5**

Note: Domestic supply equals production plus net imports. Sources: Leontief and Sohn [16]; Leontief *et al.* [17].

Table 3. Observed growth rates in production and domestic supply in four agricultural sectors, 1970-2000 (annual percentage change).

	Livestock Products			Oil Crops			Grains			Root Crops		
Region	Prod.	Dom. Supply	Food Con.	Prod.	Dom. Supply	Food Con.	Prod.	Dom. Supply	Food Con.	Prod.	Dom. Supply	Food Con.
AF	3.1	2.8	3.1	1.4	1.5	1.2	2.1	2.9	2.6	2.3	2.8	2.7
ASC	7.0	7.4	7.4	3.6	3.0	2.5	2.5	2.5	2.3	1.0	1.0	0.0
ASL	4.3	3.3	4.2	2.8	3.3	2.6	2.6	2.5	2.2	2.8	2.7	2.1
CAN	1.2	1.1	1.2	4.1	5.2	2.3	2.0	1.0	2.1	2.0	1.1	1.1
EEM	−1.2	−1.3	−1.1	−0.8	−1.1	0.6	−1.7	−1.6	−2.0	−6.5	−3.3	−2.2
JAPAN	1.9	2.9	3.0	0.0	1.5	0.8	−1.0	0.9	0.0	−1.5	−0.7	0.8
LA	3.4	3.5	3.6	6.2	5.8	4.0	2.2	2.9	2.3	0.0	0.0	0.0
OCH	1.1	0.6	1.1	10.0	4.7	1.4	3.2	2.5	0.7	1.6	0.2	1.4
OIL	4.7	5.1	5.1	2.4	3.1	10.9	2.4	4.3	3.8	3.7	3.6	3.5
SU	−0.8	−0.7	−0.7	0.0	−0.2	0.0	−1.1	−0.9	−0.9	−0.2	−1.1	−1.1
SAF	1.2	1.8	2.2	2.8	4.3	3.7	1.8	2.1	2.4	3.0	3.3	3.3
USA	1.6	1.4	1.4	2.9	2.4	1.4	2.0	1.5	2.3	1.4	1.2	1.5
WEH	1.0	0.7	−0.2	5.0	2.8	2.2	2.0	0.6	0.4	−1.1	−0.8	−0.2
WEM	2.0	1.9	2.9	2.9	3.2	5.0	1.8	1.5	0.9	0.0	1.5	1.0
WORLD	2.3	2.1	2.4	3.3	3.3	2.5	1.8	1.7	1.9	0.7	0.7	1.0

Note: Domestic supply equals production plus changes in inventories plus net imports. Source: FAOSTATS [5].

At the global level the 2000 projection of per capita domestic supply was 17% above the observed FAO level, though most of the difference, 11%, is due to the higher base year value. For livestock products, the projected global level agrees well with the 2000 observed level, even though FAO's base year value was 5% above the World Model 1970 level.

Table 4. Import dependence in agriculture, 1970 and 2000: projected (WM) and observed (FAO) ratios (in percent).

	Livestock Products				Oil Crops				Grains				Root Crops			
	1970		2000		1970		2000		1970		2000		1970		2000	
Region	WM	FAO	WM	FAO	WM	FAO	WM	FAO	WM	FAO	WM	FAO	WM	FAO	WM	FAO
AF	3.8	0.0	4.2	6.0	0.0	0.0	7.6	1.0	5.9	18.0	0.0	28.0	0.0	0.0	0.0	0.0
ASC	0.0	1.0	0.0	0.0	0.0	18.0	2.2	1.0	0.0	5.0	5.1	2.0	0.0	0.0	0.0	0.0
ASL	2.9	2.0	4.0	2.0	0.0	7.0	10.5	7.0	4.4	1.0	2.4	1.0	0.0	0.0	0.2	0.0
CAN	0.0	0.0	0.0	0.0	0.0	0.0	0.0	0.0	0.0	0.0	0.0	0.0	9.7	0.0	0.0	0.0
EEM	0.0	0.0	0.0	0.0	34.0	2.0	52.4	0.0	21.9	1.0	6.5	0.0	0.0	0.0	0.0	0.0
JAPAN	7.5	12.0	18.8	36.0	81.5	98.0	92.9	98.0	32.7	56.0	69.8	77.0	4.5	31.0	0.0	24.0
LA	0.0	12.0	0.0	3.0	1.1	0.0	15.2	0.0	8.8	0.0	5.9	13.0	1.4	0.0	0.5	0.0
OCH	0.0	0.0	0.0	0.0	66.7	0.0	25.0	0.0	0.0	0.0	0.0	0.0	10.0	0.0	0.0	0.0
OIL	6.5	25.0	34.9	25.0	0.0	0.0	51.2	0.0	10.1	55.0	25.6	55.0	1.0	13.0	0.5	1.0
SU	0.0	3.0	0.9	0.0	0.0	0.0	10.6	0.0	0.0	0.0	10.3	0.0	0.0	0.0	0.1	1.0
SAF	0.0	1.0	0.0	7.0	0.0	0.0	50.0	13.0	0.0	0.0	0.0	7.0	0.0	0.0	20.0	4.0
USA	5.5	2.0	0.0	0.0	0.0	0.0	0.0	0.0	0.0	0.0	0.0	0.0	10.9	2.0	0.0	1.0
WEH	0.0	0.0	0.0	0.0	91.6	76.0	90.3	48.0	11.8	17.0	3.3	0.0	3.3	8.0	0.0	9.0
WEM	0.7	0.0	6.7	5.0	6.7	18.0	54.8	27.0	6.2	0.0	9.2	13.0	5.1	0.0	0.0	35.0

Note: An "import dependence" ratio is defined as net imports as a percentage of domestic supply. Sources: Leontief and Sohn [16]; Leontief *et al.* [17]; FAOSTATS [5].

Table 5. Projected (WM) and observed (FAO) annual per capita domestic supply of livestock products and grains, 1970 and 2000 (in kilograms).

	1970				2000			
	Livestock		Grains		Livestock		Grains	
Region	WM	FAO	WM	FAO	WM	FAO	WM	FAO
AF	28.9	25.2	180.9	163.5	23.3	25.5	164.5	173.4
ASC	21.7	11.3	262.5	207.2	30.9	61.0	298.2	274.3
ASL	13.5	16.5	233.9	174.9	9.5	27.8	218.7	195.4
CANADA	158.9	188.5	785.0	1041.9	231.7	187.0	1050.8	1002.0
EEM	122.7	199.0	687.0	958.6	233.2	145.3	1010.4	639.1
JAPAN	38.4	38.8	260.8	259.5	140.0	76.2	389.8	281.9
LA	65.9	64.8	256.5	224.9	93.7	100.7	282.0	286.7
OCH	233.8	256.4	474.0	384.6	393.0	206.5	532.8	543.5
OIL	24.9	20.2	190.9	178.9	40.1	38.1	355.0	270.3
SU	110.4	150.5	728.6	673.0	205.4	102.0	971.6	425.1
SAF	90.9	78.1	355.0	342.1	71.4	66.6	409.8	313.5
USA	221.9	192.1	805.1	797.1	215.3	204.8	868.5	887.1
WEH	138.3	181.2	418.4	431.2	261.2	198.2	574.4	467.1
WEM	124.9	88.3	431.1	463.3	92.4	116.4	503.7	516.2
WORLD	57.7	60.4	339.7	305.8	68.8	69.0	361.9	308.7

Notes: Domestic supply equals production plus net imports plus inventory changes, and 1 kilogram equals approximately 2.2 lbs. Sources: Leontief and Sohn [16]; Leontief *et al.* [17]; FAOSTATS [5].

With a few exceptions, per capita domestic supply of livestock products and grains increased in those regions where living standards advanced and declined in those regions where living standards stagnated or fell. (Please see Section 6). These trends are more sharply discerned in **Table 6**, which presents only the observed changes in annual per capita consumption of livestock products and grains as food consumption.

Table 6. Observed (FAO) changes in annual per capita food consumption of livestock products and grains, 1970 and 2000 (in kilograms).

	Livestock Products		Grains		Change in Consumption from 1970	
Region	1970	2000	1970	2000	Livestock Products	Grains
AF	23.4	24.6	137.4	130.2	1.2	−7.3
ASC	11.1	60.7	133.1	166.7	49.7	33.5
ASL	13.9	25.0	153.1	154.7	11.1	1.6
CANADA	165.9	167.4	87.6	117.3	1.5	29.7
EEM	162.9	124.2	281.7	154.0	−38.7	−127.7
JAPAN	37.3	75.0	143.5	121.5	37.7	−22.0
LA	60.7	96.0	113.7	122.9	35.3	9.2
OCH	179.5	169.6	100.6	85.2	−9.9	−15.4
OIL	18.2	34.5	128.5	166.6	16.2	38.2
SU	108.7	81.8	190.7	154.6	−27.0	−36.1
SAF	61.8	59.3	184.5	184.8	−2.5	0.2
USA	186.9	199.8	81.8	115.4	12.9	33.6
WEH	140.1	116.6	113.2	115.6	−23.5	2.4
WEM	72.3	119.3	173.7	161.3	47.0	−12.3
WORLD	51.0	63.3	139.4	147.9	12.3	8.5

Note: 1 kilogram equals approximately 2.2lbs. Source: FAOSTATS [5].

5.4. Calories and Protein: The "Stuff" of Life

The global increase in daily calorie and protein levels and the decline in poverty rates, particularly in East Asia over the last two decades and more recently in Sub-Saharan Africa, is well documented in Perry [19]. According to Fogel [6], two critical "ingredients"—the total daily calorie level and the amount of daily protein intake—play a crucial role in improving labor productivity, both physical and mental, which, for economists, is the single most important factor responsible for rising living standards.

Table 7 presents the projections and the observed levels of daily per capita calorie and protein levels for 1970 and 2000. For the world, the World Model projected (or targeted) only a very modest increase in daily calories, from 2400 in 1970 to 2500 in 2000, an increase of only 4.2%, along with an increase of 12% in daily per capita protein levels over their 1970 levels. On the other hand, globally, the observed data indicate more impressive gains in both calorie and protein levels over the 30-year interval: a 13.6% increase in global daily per capita calories, along with a 15% increase in daily per capita protein levels.

It is of interest to note the observed changes recorded over the projection interval in three key regions: ASC, which is dominated by China; North America, incorporates both Canada and the United States; and the former Soviet Union. The observed changes in the first region attest to the well-documented improvement in material wellbeing in China over the projection interval, the result of the reform program that began in the late 1970s, including China's re-engagement with the world economy. The changes in the second region provide some additional proof of the ongoing "obesity crisis" in North America with the large observed—though arguably unnecessary—increase in daily calorie levels in both Canada and the US from an already relatively elevated 1970 base year level. The changes in the former Soviet Union portray a region in crisis, with both daily per capita calorie and protein levels falling over the projection interval. In particular, readers will recall that the 1990s was a decade of enormous economic and political stress in the former Soviet Union, accompanied by falling living standards that is borne out in these data and the data on population and income growth (**Table 1**).

6. ANALYZING THE DIFFERENCES

This section addresses the source of the differences in the projected and observed growth rates in domestic supply of livestock products and grains at the regional level. The one sector where there is significant divergence between the projected and observed growth in consumption over the 30-year interval is in oil crops. Pending a follow-up study, part of the discrepancy is likely to be definitional. In addition, part of the difference might be explained by the very large observed increase in global soybean production and consumption over the 30-year projection interval that was unanticipated when the projections were made in the late 1970s [5].

Table 7. Projected (WM) and observed (FAO) levels of daily per capita calories and proteins, 1970 and 2000.

	CALORIES				PROTEINS (grams)			
	1970		2000		1970		2000	
Region	Projected	Observed	Projected	Observed	Projected	Observed	Projected	Observed
AF	2356	2122	2310	2222	67.7	55.5	67.9	56.7
ASC	2025	1900	2220	2834	57.9	47.4	68.3	83.4
ASL	2048	2240	2158	2349	52.7	50.7	56.9	55.5
CANADA	3138	2920	3153	3515	91.8	93.3	97.7	106.7
EEM	3129	3278	3168	3218	93.6	96.5	107.0	94.2
JAPAN	2427	2723	3268	2879	73.0	82.1	125.3	95.5
LA	2345	2440	2987	2809	58.1	64.0	82.3	77.2
OCH	3167	3158	3214	3027	92.3	104.8	100.1	98.6
OIL	1985	1968	3314	2733	53.2	49.0	116.1	80.8
SU	3154	3354	3087	2759	91.2	102.4	100.0	80.0
SAF	2776	2799	2925	2862	79.2	73.3	90.5	76.1
USA	3211	3035	3175	3732	95.8	98.5	98.9	113.5
WEH	3046	3254	3219	3496	92.0	94.5	106.8	105.7
WEM	2746	3012	3003	3351	79.2	89.5	94.8	101.7
World	2400	2398	2500	2725	66.0	64.8	74.0	74.6

Note: A "pint" (14 oz or 414 ml) of Häagen-Dazs ice cream contains 1050 calories. Sources: Leontief and Sohn [16]; Leontief *et al.* [17]; FAOSTATS [5].

Figures 1 and 2, below, provide an overview of the cumulative changes in projected and observed domestic supplies of livestock products and grains, respectively. Differences in projected and observed regional population, GDP, and per capita GDP growth rates over the 30-year projection interval (please see **Table 1**), not unexpectedly, serve as good explanatory variables in accounting for these differences in consumption rates in these two food sectors.

The most obvious differences are in the region that incorporates China—Asia Centrally Planned (ASC). The very large difference in consumption of livestock products is most likely explained by the difference between the projected and observed growth in the region's per capita GDP (**Table 1**). As living standards improve, higher per capita meat and dairy products consumption and increased daily per capita protein levels follow. This is confirmed in the data for China (the ASC region) in Tables 5-7, above. In the ASC region, according to the FAO, annual per capita consumption of livestock products increased from 11.1 kg to 60.7 kg, and annual per capita consumption of grains increased from 133 kg to 167 kg from 1970-2000 (**Table 6**). This improvement in living standards is also confirmed by a 50% increase in daily per capita calorie levels, and a 75% increase in daily per capita protein levels (**Table 7**).

Japan is another region where there is a large discrepancy between the projected and observed changes in the consumption of livestock products and grains (Figures 1 and 2). These discrepancies can be explained by the much slower GDP growth rate that was observed relative to Japan's projected GDP growth over the interval (please see **Table 1**).

Another large regional difference between the projected and observed cumulative change in the consumption of both livestock products and grains over the period was in the region SU, incorporating Russia. According to the two figures, the cumulative difference between the World Model and the FAO data is approximately 120% for livestock products and about 70% for grains.

These differences can be explained by a combination of slower observed population growth and the very poor performance of the regional economy that resulted in a decline in the observed standard of living in the region over the projection period (**Table 1**). The break-up of the Soviet Union and the painful restructuring in its wake exacted a heavy price on the region's standard of living. The region experienced a decline of 27% in per capita consumption of livestock products and a 36% reduction in per capita grain consumption (please see **Table 6**, above). The decline in regional living standards as a result of the momentous political and economic events that culminated in the dissolution of the Soviet Union in 1991 and the major restructuring program that followed are also manifested by 20% reductions in the levels of daily per capita calories and proteins (please see **Table 7**).

Finally, the eastern European countries (EEM) registered large differences between the projected and observed growth rates in food consumption over the 30-year projection interval (Tables 2 and 3). The reasons are similar to those in the case of the Soviet Union.

When the projections were made population growth was assumed to be positive (though modest) and GDP growth was projected at 4.4% p.a., with the growth in food consumption ranging from 0.8% - 2.8% p.a. in the four food categories tracked by the World Model. With the collapse of the Soviet Union—and the Warsaw Pact—eastern Europe was thrown into political turmoil that had profound demographic and economic implications.

Observed annual growth in GDP averaged little more than half the projected rate, and the region's population actually declined over the period (**Table 1**). As a result, living standards fell in the region, including sharply lower observed food consumption in 2000 than projected for that year (Tables 2 and 3), and lower observed consumption in all four food categories in 2000 than observed in 1970 (**Table 3**)! According to the FAO, per capita consumption of livestock products declined by 24%, and per capita consumption of grains declined by 45% over the 1970-2000 interval (**Table 6**). In addition, eastern Europe (EEM) recorded lower levels of daily calories and proteins per capita in 2000 than in 1970 (**Table 7**).

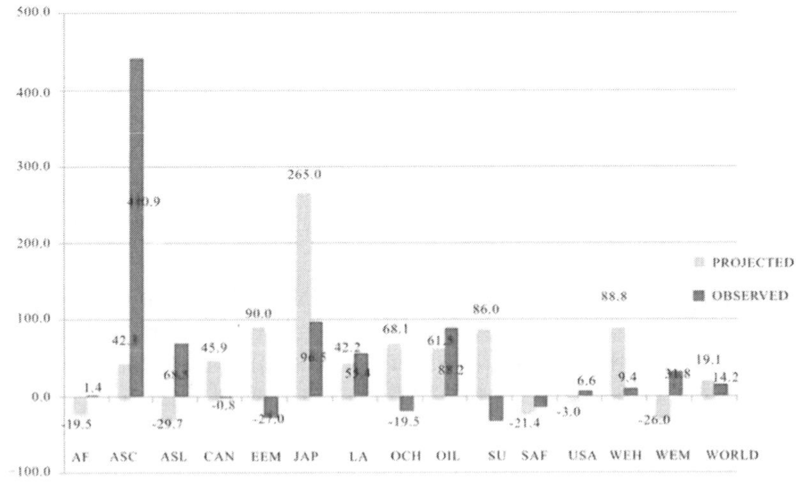

Figure 1. Projected and observed cumulative change in domestic supply of livestock products, 1970-2000 (in percent). Sources: Leontief and Sohn [16]; Leontief et al. [17]; FAOSTATS [5].

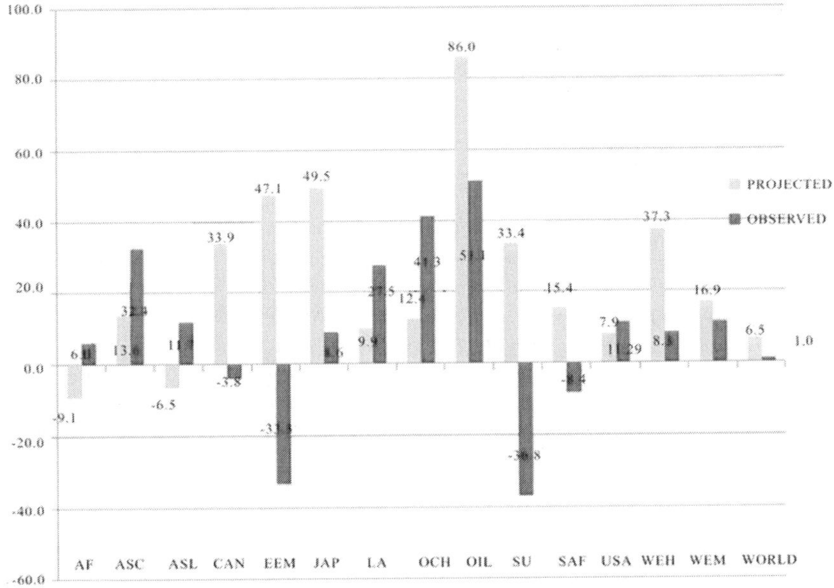

Figure 2. Projected and observed cumulative change in domestic supply of grains, 1970-2000 (in percent). Source: Leontief and Sohn [16]; Leontief et al. [17]; FAOSTATS [5].

Root Crops

Global production and consumption of root crops were projected to increase by only 0.5% p.a. over the 1970- 2000 interval, considerably below projected global population growth, not to mention income growth. According to the FAO, the observed growth rate in production and domestic supply of root crops during this 30-year period was only 0.7% p.a., the slowest growing food category in this study (please see **Table 3**).

A closer examination of the observed regional data on root crops in **Table 3** suggests that this food category exhibits the characteristics of an "inferior good", that is, as income increases, demand increases very modestly (and in some cases declines absolutely). This appears to be the case for most of the high-income developed regions such as Canada, Japan, Oceania, Western Europe, and the United States.

The relatively large decline in root crops consumption in Eastern Europe (EEM) and the Soviet Union (SU) can be attributed to the poor economic performance of these regions during the projection period, as was discussed above. The growth in demand for root crops appears to be highest in the regions experiencing low or negative growth in per capita income over the projection interval—Africa (AF), the Middle-East oil producing countries (OIL), and South Africa (SAF)—or in countries with very low per capita incomes, many of whom are embedded in ASL, the Asia low income region (please see Table A1 in the Appendix).

7. CONCLUSIONS

The "conventional wisdom" regarding minimum required daily calorie levels suggests levels at approximately 2100 calories per capita [3]. According to the observed data reported in **Table 7**, above, at the global level this was easily met even in 1970, not to mention the 14% increase by 2000. It is important to note, that over the 30-year projection interval of this study, the world's population also increased by 65% over its 1970 level.

With a view towards the future, and, in particular, to those engaged in today's long-term modeling efforts, i.e., to 2050 and beyond, there is an indispensable need to look ahead 30 - 40 years using plausible assumptions regarding the forces that will drive future global food requirements. They include the following: a projected 30% increase in the world's current population by 2050; continued economic growth, especially in Asia and Africa; a growing scarcity of agricultural inputs (land and water) and outputs (maize and sugar cane) due to their diversion to the production of bio-fuels and driven by higher population levels and living standards, along with increased urbanization rates, particularly in Asia and Africa; the expected environmental impacts from global warming on agricultural output on the one hand, and, on the other, the projected elevated levels of methane and nitrous oxide emissions as livestock output and fertilizer use increase along with population levels and living standards, particularly in China, and other high-growth countries in Asia and Africa.

On the production side, increasing food output by a projected 70% by 2050 [3] will require the development and adoption of new technologies such as improved irrigation methods, more widespread planting of genetically modified crops that are resistant to herbicides and pesticides, and the introduction of plants that are drought, saline, and/or flood tolerant, reducing the "yield gap" between the "best", "worst", and "average" yielding land, and reducing the amount of food in both developed and developing countries that is lost to waste, currently estimated to be 30% - 40% of the total [11]. The introduction of aquaculture four decades ago, and its increasing share of the world's supply of consumed fish, serves as a shining example of the ability of the global food system to find solutions to critical global challenges. It will be interesting to follow the progress of the development of "cultured meat", that is, meat grown from stem cells, over the next quarter century.

Finally, even though current daily calorie levels on a per capita basis in all 15 regions in this study are above the minimum required 2100 calories (**Table 7**), according to the FAO, 925 m people were classified as "hungry" [20], and, according to Save the Children, the international NGO that promotes children's rights, one in four children in the world is malnourished [21]. Therefore, one of the major challenges in the 21st century is the implementation of food "safety nets" for these vulnerable people through policy and institutional initiatives to the end of reducing, if not eliminating, the daily struggle for food of more than 13% of the world's population.

8. ACKNOWLEDGEMENTS

Ira Sohn I am grateful for the Sabbatical leave provided to me by Montclair State University that enabled me to carry out this study. I am indebted to Claudia Binaghi for building the neat tables and bar graphs that makes the paper more readable. I thank Paul Westcott, at the USDA's Economic Research Service, for taking the time to read through an earlier version of the manuscript, and for bringing to my attention some errors that have been corrected.

REFERENCES

1. L. Lucas and J. Fontanella-Khan, "Food Security: Dampened Prospects," Financial Times, 2012.
2. Financial Times, "Feeding the 9 bn," Editorial, Financial Times, 2012.
3. J. Parker, "The 9 Billion People Question: A Special Report on Feeding the World," The Economist, 2011.
4. J. Faaland, "Population and the World Economy in the 21st Century," Basil Blackwell, Oxford, 1982, p. vii.
5. FAOSTATS, Food and Agriculture Organization, United Nations, Rome, 2012. http://www.faostat.fao.org/site/354/default.aspx#ancor
6. R. Fogel, "The Escape from Hunger and Premature Death, 1700-2100," Cambridge University Press, Cambridge, 2004.
7. G. Federico, "Feeding the World: An Economic History of Agriculture, 1800-2000," Princeton University Press, Princeton, 2005.
8. T. Malthus, "An Essay the Principle of Population," 1798. http://www.econlib.org/library/Malthus/malPop3.html
9. P. Conforti, "Looking Ahead in World Food and Agriculture: Perspectives to 2050," Food and Agriculture Organization, Rome, 2011.
10. United Kingdom Government Office for Science, "Foresight: The Future of Food and Farming," (Executive Summary), The Government Office for Science, London, 2011.

11. H. C. Godfray, et al., "Food Security: The Challenge of Feeding 9 Billion People," Science, Vol. 327, No. 5967, 2010, pp. 797-834.
12. Financial Times, "World Food: A Special Report," Financial Times, 2011.
13. J. Foley, "Can We Feed the World and Sustain the Planet?" Scientific American, Vol. 305, No. 5, 2011, pp. 60-65.
14. W. Leontief, A. Carter and P. Petri, "The Future of the World Economy," Oxford University Press, Oxford, 1977.
15. United Nations, "United Nations Population Projections," UN Population Division, United Nations, New York, 2003. http://www.un.org/esa/population/unpop
16. W. Leontief and I. Sohn, "Population Food and Energy and the Prospects for Worldwide Economic Growth to the Year 2030," In: J. Faaland, Ed., Population and the World Economy in the 21st Century, Basil Blackwell, Oxford, 1982, pp. 96-127.
17. W. Leontief, J. Mariscal and I. Sohn, "Prospects for the Soviet Economy to the Year 2000," The Journal of Policy Modeling, Vol. 5, No. 1, 1983, pp. 1-18.
18. World Bank, "World Development Report, 2002," Oxford University Press, New York, 2002.
19. G. Perry, "Poverty Reduction and Growth: Virtuous and Vicious Circles," The World Bank, Washington DC, 2006.
20. World Hunger, 2012. http://www.worldhunger.org/articles/Learn/world%20hunger%20facts%202002.htm
21. Associated Press, 2012. http://hosted2.ap.org/COGRA /APWorld News/Article_2012-02-15-AF-Africa-Child-Hunger/id-0b95f90587674c1187f3b8b933bb7452

CHAPTER 10

Can Changes in Soil Properties in Organic Banana Production Suppress Fusarium Wilt?

Paul Geense[1], Anthony B. Pattison[2*], Tegan L. Kukulies[2], Johannes M. S. Scholberg[1], Agustin B. Molina[3]

[1]*Farming Systems Ecology, Wageningen University, Wageningen, The Netherlands*

[2]*Agri-Science Queensland, Department of Agriculture, Fisheries and Forestry, South Johnstone, Australia*

[3]*Bioversity International, Los Baños, Philippines*

ABSTRACT

Organic agriculture promotes disease suppression through healthy soils by increasing biological activity and diversity through the application of organic fertilizers and increasing organic inputs. Fusarium wilt of bananas (Fusarium oxysporum f. sp. cubense) (Foc), also known as Panama disease, has been a devastating disease throughout the world. So far, no fungicides or cultural measures have been found that control Foc sufficiently. The aim of this research was to assess whether organic-based farming systems were more resilient than inorganic farming systems to soil borne diseases, in particular Fusarium wilts. A survey was conducted comparing five organic and five conventional banana plantations at paired sites in north Queensland, Australia. Soil samples were collected and analysed for chemical, physical and biological soil health indicators. Disease development of F. oxysporum f. sp. lycopersici in tomatoes and Foc in bananas were studied in pot trials to pursue clues for identifying Fusarium suppressive soil traits. Organic soils from the survey showed higher microbial activity and lower disease symptom expression (both with tomatoes and bananas) than conventional soils. In the survey, nematode diversity and soil sulphate content were recurring indicators in all experiments showing close correlations to pathogen growth, disease expression and plant health. Organic soils were lower in plant-parasitic nematodes and sulphate sulphur levels and higher in nematode diversity, labile soil C and microbial indicators. Soil conduciveness or suppression of Foc appeared to be largely governed by competition for carbon. Measurement of soil microbial enzyme activity, nematode community structure and diversity and possibly sulphate sulphur seem to provide a relatively reliable indicator for general disease suppression. Differences between organic and conventional agriculture cannot be related to single management practices, but may be linked to synergies among system components.

Keywords: Disease Suppression, Fusarium oxysporum f. sp. cubense, Musa spp., Organic Banana Systems, Microbial Activity, Soil Nematode Community

1. INTRODUCTION

The Australian banana industry produced 285,535 tonnes, valued at AU $565 million, from 13,496 ha in 2012 (FAOSTAT, 2014). In 2012, Australia was the 37th largest producing nation of bananas in terms of production, but the 41st largest producer in terms of area and the 50th nation in terms of yield (FAOSTAT, 2014). Due to its favourable climate most of the Australian banana production occurs in north Queensland (93%), which produces about AU $400 million annually, with smaller production areas located in the sub-tropics of Western Australia, northern New South Wales and south-east Queensland and in the tropics around Darwin in the Northern Territory. Banana production in Australia is primarily for domestic consumption.

The banana farms of north Queensland are mainly family-based businesses with a typical farm size of 15 to 20 ha, although some plantations are as large as 500 ha [1] . Production in north Queensland is about 30 - 40 $t \cdot ha \cdot yr^{-1}$, which is characterised by a high level of technology and mechanisation relative to other banana producing countries due to high labour costs [1] . Agrochemicals are commonly used against weeds, bunch pests, foliar diseases and soil-borne pests and diseases. The use of a spraying calendar is still widespread, but there is in- creasing attention, by governmental regulations and policies, to promote integrated pest management (IPM) [2] . In recent years there has been increasing concern about erosion and runoff of agrochemicals, nutrients and sediments, which may pollute rivers and other ecosystems [1] [3] - [5] . The tropical banana industry in Queensland was identified as a contributor of nutrients and sediments to waterways and the Great Barrier Reef lagoon [3] [6] [7] . Sustainability and soil health are now key focal points of the Australian banana industry to improve soil- based resources [4] [5] .

Fusarium wilt of bananas (Fusarium oxysporum f. sp. cubense (Foc)), also known as Panama disease, has been one of the most devastating crop diseases known [8] - [10] . Foc survives in the soil saprophytically, by chlamydospores and micro- and macro-conidia. The chlamydospores can survive in the soil for several years [11] . Currently, there are four races of Foc; Race 1 affects bananas of the Gros Michel, Sugar and Pome varieties, Race 2 affects Bluggoe and related cooking bananas, Race 3 affects Heliconia spp., but is not pathogenic to bananas and Race 4 affects varieties from the Cavendish group, with a distinction between sub-tropical and tropical Race 4 (TR4) [12] - [14] . Fusarium wilt of bananas has been reported in Australia as early as 1874 [15] . However, not all races of the pathogen are present in all production areas, with TR4 being confined to the Darwin region in the Northern Territory [16] [17] . In north Queensland Cavendish cultivars are most commonly grown, which are resistant to Foc Race 1, but there is also production of niche cultivars, Ladyfinger (a Pome variety, Musa AAB) and Ducasse (synonym PlsangAwak, Musa ABB), which are susceptible to Race 1 [8] .

There is growing concern that monoculture of bananas, using the same clone renders commercial banana production systems more susceptible to pests and diseases due to lack of diversity [18] [19] . Moreover, there is a need to address problems in the banana industry such as soil erosion, ecosystem contamination and incidence of pests and diseases, which may be accomplished by using integrated approaches. Organic agriculture is a way of farming that aims to integrate and manage these aspects in an ecological sound system [20] [21] . Organic agriculture aims to achieve disease control through promoting healthy soils featuring enhanced biological activity and thus improved soil suppression of soil-borne pests and diseases. This may be realised by application of organic amendments instead of synthetic fertilizers, thereby stimulating soil microbial biomass and activity [22] - [24] . Moreover, organic agriculture aims to maintain or increase above- and below-ground biodiversity by more effectively employing crop rotation and intercropping techniques compared to conventional systems. However, very little literature is available on organic banana production in north Queensland or in Australia in general. One reason for this might be that there is no systematic collection of data on organic agriculture at the national level [25] . Although, the organic banana sector is still very small compared to the conventional sector, current knowledge gaps tend to hamper any future expansion.

For the banana industry to improve soil management there is a need to develop indicators that are able to quantify changes in soil properties and which may be used to design and promote improved land management practices. Such indicators should encompass a holistic view of soil management techniques and account for changes in physical, chemical and biological soil properties [26] . A better understanding of the soil ecosystem may result in development of a stable, resilient soil system that is able to recover from stress, and demonstrates enhanced biological diversity and more efficient internal nutrient cycling [27] . A selection of soil health indicators for use in the banana industry was developed by Pattison et al. [4] , using physical and chemical indicators, plus soil nematode community structure. However, there needs to be more extensive verification that changes in crop management such as organic production can result in improved soil health characteristics including the suppression of soil borne diseases such as Foc.

Considering the need for a durable management tool for controlling Foc that does not negatively impact on control of other diseases or the environment, this study looks at organic agriculture as a sustainable option to reduce Foc disease incidence. According to Alabouvette [28] and Alabouvette et al. [29] , soil suppression to Fu- sarium wilt is general to Fusarium oxysporum and not specific to certain formae speciales of the fungus. This study aimed to investigate the potential benefits of organic banana production systems through a combination of both a farm survey along with glasshouse experiments. The hypothesis for the research was that organic agriculture altered soil health indicators under bananas by improving microbial activity and diversity relative to con- ventional banana production, resulting in more resilient soil properties that could suppress biotic stresses such as Foc. To test the hypothesis organic and conventional banana production systems were surveyed to compare physical, chemical and biological soil health indicators, followed by different screening methods for comparing soil borne disease suppression.

2. MATERIALS AND METHODS

2.1. Site Selection and Soil Sampling

The area under survey covered tropical north Queensland, banana production area south of Cairns and north of Cardwell, and the Atherton Tablelands. Five organic and five conventional banana farms were selected as paired sites. Selection of organic sites was based on willingness of the farmers to cooperate, while the potential number of available organic farms was rather limited. The conventional sites were selected based on proximity to the organic sites in order to reduce pedo-climatic variation among sites [4] [30] . Observations were made on management practices and soil descriptions from each site were based on the Australian soil classification system [31] , with soil characteristics for drainage and erosion hazard based on SafeGauge [32] . From each site, three banana fields with plantings generally older than five years were sampled and soil samples analysed separately for each field.

Soil samples were collected using a shovel to remove a square block of soil approximately 15 cm × 15 cm to a depth of 15 cm within 30 cm from the base of the banana plant, in front of the following sucker to form a composite soil sample (n = 15) from each field. Stones and large pieces of organic matter were avoided during sampling. Soil was placed in a bucket and thoroughly mixed with a trowel.

2.2. Soil Analysis

Collected bulked soil samples were analysed for physical, chemical and biological soil health indicators and samples were sent to a commercial laboratory, Incitec Pivot Ltd., Weribbee, Victoria, Australia, for further chemical analysis. The following soil health indicators were measured at the Department of Agriculture, Fisheries and Forestry Laboratory at South Johnstone: pH, electrical conductivity (EC), nitrate nitrogen (NO_3-N), labile C, fluorescein diacetate, β-glucosidase, bulk density, water stable aggregates (WSA), soil particle size and nematode community structure. Soil pH, EC and NO_3-N were determined by mixing the soil in water (1 part soil to 5 parts water) and determining values of the suspended solution using a pH and EC multi probe (TPS WP-90) and NO_3-N test strips using the Merck Reflectoquant system with RQflex reflectometer (Merck KGaA, Darmstadt, Germany). Labile carbon contents were determined by measuring the amount of C oxidised by 33 mM $KMnO_4$ in duplicate 5 g sub-samples using the method described by Moody and Cong [33] . Similarly, fluorescein diacetate (FDA) hydrolysis rate was determined from duplicate 5 g sub-samples using a modified version of the method initially proposed by Schnürer and Rosswall [34] . β-glucosidase was determined with the procedure published by Eivazi and Tabatabai [35] except the toluene was substituted with 0.1% Tween solution and the modified universal buffer was replaced with a McIlvaine buffer (pH 6.0).

Soil nematodes were extracted using a modified Baermann funnel technique [36] . A 200 g sample of field soil maintained at the soil moisture capacity at the time of sampling was weighed onto a mesh sieve with a single ply of tissue and then placed into a tray with 250 mL of water and left for 48 hours. The nematodes were collected on a 25 μm sieve and backwashed into a vial. The total number of nematodes in the 200 g soil sample was determined on a counting slide and expressed as the number per 100 g of soil. A 50 μL aliquot was placed on a glass slide, with a minimum of 100 individual nematodes identified to genus level for plant-parasites and to the family level for free-living nematodes. Soil nematode community analysis was made on soil nematode trophic groups (parasites, fungivores, bacterivores, omnivores and predators).

Indices of the nematode community composition were calculated from the number of nematode taxa extracted from each plot. Nematode diversity was determined using the Shannon-Weiner index [37] . The bacterivore to fungivore ratio was calculated from the total abundance of bacterivores (B) and fungivores (F) (B/(B + F)) [37] . Additionally, the weighted functional guilds concept was applied, without plant-parasites. The nematode functional guilds were used to calculate the basal, enrichment index (EI), structure index (SI) and channel index (CI) of the soil food web [38] . Plant-parasitic nematodes were identified to species level where possible and the abundance of each individual classification of plant-parasitic nematode was kept separate.

2.3. Bioassays for Assessing Soil Suppression of Fusarium spp.

Hyphal extension was determined by growing Foc R1 on collected soil samples on laboratory plates in an incubator [29] . Petri dishes were filled with 30 g of 2 mm sieved air dried soil which was spread evenly to obtain a smooth surface (approximate bulk density = 1.0 g cm^3). Sterile water (12 mL) was added to each Petri dish. Wetted soils were left to equilibrate for 7 days in the incubator. Using Foc R1 (VCG 0124) cultures on 1/4 strength Potato Dextrose Agar (PDA) a 1 cm diameter, 4 mm thick plug from the growing margin of the fungal colony was inverted and placed centrally on top of the soil in the Petri dishes. After one week the number of plugs that had hyphae growing on the soil surface was recorded. Each week for 4 weeks of incubation at 25°C, the extension of the mycelium was determined using a binocular microscope and the area of hyphal extension was estimated for a 4-week period. The Petri dishes were maintained at 25°C during this period. A rating system was used to give a one number score for extent of hyphal growth, such that: 1 = less that 5 hyphae and hyphae < 1 cm, 2 = more than 5 hyphae and hyphae < 1 cm, 3 = less than 5 hyphae and hyphae > 1 cm, 4 = more than 5 hyphae and hyphae > 1 cm and 5 = extensive hyphal growth. The test was replicated 5 times per soil sample. The area under the (growth) curve (AUC) for each plate was estimated in Genstat (Edition 11). The AUC values were then statistically analysed using a one-way-ANOVA and tests of multiple comparison of means (Fisher's protected LSD test with $p = 0.05$) to assess differences in suppression of Foc hyphal extension between collected samples.

A glasshouse bioassays using tomato (Solanum lycopersicum cv. Tasty Tiny Tim) with its associated Fusarium pathogen, Fusarium oxysporum f. sp. lycopersici R3 (Fol R3) was used for the assays. Polystyrene trays (72 × 20 × 7 cm) with 5 rows of 18 cells were used for the bioassay. Each cell was 4 cm^2 at the top and narrowing down to 5 mm^2 at the bottom allowing for good drainage but preventing soil from dropping out. To each cell 45 cm^3 of soil was added, equivalent to 50 g of soil. Two inoculation methods were used; inoculation with spore solution in liquid malt extract and inoculation with colonised milled seed.

One litre of liquid malt extract was inoculated with one plug of agar from the edge of a Fol R3 colony and placed on a rotary shaker (150 rpm) for seven days. After four days the solution was placed under near UV light for five hours to stimulate sporulation and placed back on the shaker. The solution was sieved through a sterile funnel (40 µm) and diluted to 10^3, 10^4 and 10^5concentrations. Soil samples of organic and conventional soils were infested with 4 ml of malt-based inoculum suspension at concentrations of 1×10^3, 10^4, 10^5 propagules ml^{-1} soil. Controls received 4 ml of tap water and each treatment was repeated five times. Non-inoculated controls were included for both organic and conventional systems.

Millet seed (500 g) was placed in Erlenmeyer flasks, soaked in water for 24 hours and sterilised by autoclaving. The flasks were inoculated with agar plugs of Fol R3 from the edge of a colony. The flasks were incubated at room temperature for 14 days and were mixed daily by hand shaking. Colonised millet seed was added to 45 ml of soil at rates of 0.5, 1.0 and 1.5 grams of inoculum and thoroughly mixed with approximately 50 g of soil. Each treatment was replicated five times. Non-inoculated controls were included for both organic and conventional systems. One-week-old susceptible tomato seedlings were planted into each cell. Plants were kept in a greenhouse under controlled conditions with air temperatures maintained at 27°C - 32°C, 70% - 80% relative humidity, and 12 h sunlight per day and watered once a day throughout the experiment.

Disease progress was recorded twice a week from 21 to 58 days after sowing. The following rating system was used: 1 = healthy plant, 2 = wilting of lower leaves, 3 = whole plant affected and 4 = plant dead/terminal wilting. Since the tests were done with unsterilized soil from various farms, PDA-plate isolation was done from the roots of each dead plant to determine whether death was caused by Fusarium spp. or by another pathogen. At the termination of the experiments diseased plants were used for pathogen isolation. Randomly selected healthy plants were also used in isolations to determine the presence of Fol R3. The AUC for each plant was estimated in Genstat (Edition 11). The AUC values were statically analysed using a one-way-ANOVA and multiple means comparison (Fisher's protected LSD test with $p = 0.05$). Disease incidence, growth and soil health indicators were tested on two-sided correlations.

Isolation of organisms from dead, diseased and healthy plants were conducted by taking plants from the trays and carefully washing the roots and lower stems. Roots and stems were dipped for 30 sec in 1% bleach solution to remove fungi and bacteria growing on the surface of the plant tissue before being rinsed in sterilised water. The roots and stems were then cut into small (<0.5 cm) sections and a placed on 1/4 strength potato dextrose agar (PDA) in an incubator at 27°C. When the colonies were approximately 1cm in diameter the plates were taken from the incubator and placed under UV light at 23°C to allow morphological characteristics to be identified.

A pot experiment with bananas was conducted to determine whether the previous bioassays for assessing disease suppression provided a good indication of suppression to Foc R1. For this experiment soil from paired organic and conventional sites from the farm survey, which were determined to have the greatest potential for Foc suppression and the least potential for Foc disease suppression were used. Therefore, four farms were selected using soil collected from the three fields from each farm, with a Foc inoculated treatment and uninoculated control from each field. Each treatment had two replicates, making 48 pots in total, which were randomised using a split plot design.

Soils were inoculated with Foc R1 (VCG 0124) colonised millet seed. Millet seed (500 g) was placed in Er- lenmeyer flasks, soaked in water for 24 hours and sterilised by autoclaving. The flasks were left at room temperature to incubate for 14 days and were mixed daily by hand shaking. Approximately 1.5 kg of soil was weighed and placed in plastic bags and 10 g inoculum was added to the soil in each bag. After thorough mixing, the soil was placed in 2 L pots. Control pots were filled with 1.5 kg non-inoculated soil.

Tissue cultured Ducasse (Musa AAB) banana transplants were placed in the pots. The pots were placed on trays and soil moisture content was kept constant by filling the bottom of the trays with water. At five weeks post planting, the trays were left dry for 24 hours to create more favourable circumstances for the disease symptom expression. Each pot received 5 ml of organic liquid blood and bone meal (dilution 1/100) once a week.

Disease progress was recorded once a week from day 21 to day 58 post inoculation. Plant wilting was recorded following a rating system of 1 - 5 developed by INIBAP [39] , where: 1 = plant healthy, 2 = slight yellowing or wilting of lower leaves, 3 = extensive yellowing or wilting of lower leaves, 4 = yellowing or wilting of most or all of the leaves and 5 = plant death. Leaf emergence and growth were measured every two weeks. At the end of the trial the leaf area of the last fully emerged leaf was estimated as in described by Turner [40] , and the corm was dissected to assess vascular discoloration on a rating of 1-6 according to the INIBAP guidelines [39] , where: 1 = rhizome completely clean, 2 = isolated points of discoloration, 3 = vascular discoloration of up to 1/3 of rhizome, 4 = discoloration affecting between 1/3 and 2/3 of rhizome, 5 = greater than 2/3 of rhizome discoloured and 6 = total discoloration rhizome vascular tissue. At termination of the trial, the number of discoloured roots was counted from a random sample of five roots. Soil was washed from roots and corm. Roots, corm and leaves were separately dried in a drying oven for seven days at 75°C.

2.4. Statistics

Soil parameters from banana survey sites were analysed using a one-way analysis of variance (ANOVA). Each pair of sites formed a replicate, with two treatments either organic or conventional. Fields within each site formed sub-samples for each treatment. Where significant F values ($p < 0.05$) were obtained between treatments, means were separated using Fisher's protected LSD method. The area under the curve (AUC) for hyphal growth of Foc and wilting symptoms of tomato and banana was estimated in Genstat (Edition 11). The AUC values were then statistically analysed using a one-way-ANOVA and tests of multiple comparison of means (Fisher's protected LSD test with $p = 0.05$) to assess differences in suppression of between sites. Disease, growth and soil health indicators were tested using a two-sided t-test for correlation and a stepwise multiple regression calculated for soil parameters related to the AUC for Foc hyphal growth and tomato wilt symptoms. The dimensionality of the original data set was reduced by using principal component analysis (PCA), which was performed on those disease indicators associated with Foc of bananas. The PCA was based on the correlation matrix and a varimax factor rotation and results were used to increase the interpretability of the principal components and the construction of a bioplot. All statistical analyses were performed in GenStat for Windows 11th Edition [41] .

3. RESULTS

3.1. Banana Site Survey

Three of five pairs of farms were situated in the wet tropical coastal zone of north Queensland (Table 1). One pair of sites was slightly more inland in the southern zone, which received lower rainfall on an alluvial Dermosol (Table 1). The fifth pair was situated in the Atherton tablelands where both temperatures and rainfall are considerably lower (Table 1). There were visible differences observed among organic and conventional farms in terms of weed and ground cover management. One organic farm used paper and cardboard as mulch but still showed a high incidence of weed growth, while another farm employed a very thick layer of leaf mulch resulting in little weed growth. A third farm had a dense ground cover of pinto peanut (Arachis pintoi), whereas another farm used weed mats. Lastly, one organic farm used a rotation system of three-year production and two- year fallow. Some organic farms put the banana crop residue in the rows in line with banana plants and others in the inter-rows. All organic farms had densely vegetated inter-rows, mainly grass and, in one instance, pinto pea- nut. The conventional farms tended not to use ground cover, with only one farm having densely vegetated inter- rows. Banana crop residue was used as mulch in the rows of one farm, while at most farms banana crop residue was left in the field where it dropped.

There were no significant differences ($p > 0.05$) in physical soil health indicators between organic and conventional soils (Table 2). Based on the similarity of the sand, silt and clay fractions between organic and conventional soils it appeared there were no major difference in soil types between the two management systems for paired sites. Bulk density and WSA did not show difference between the two systems. It could be inferred that soil type and climate have more impact on bulk density and WSA than the management system. Most chemical soil properties did not show significant difference between management systems. However, labile soil C was significantly higher ($p < 0.05$) in organic systems and values were almost twice as high compared to conven tional systems. Sulphate levels were significantly higher ($p < 0.05$) in conventional systems with values about four times higher compared to those measured in organic soils. There was a significant difference ($p < 0.05$) in microbial activity between management systems (Table 3), with both FDA and β-glucosidase levels being higher in organic soils. Moreover, nematode community structure was considerably different between organic and conventional systems (Table 3). The percentage of bacterial-feeding and fungal-feeding nematodes were a factor two and five times greater in organic systems, respectively, while the percentage of predatory and omnivorous nematodes was also more than twice as high in organic systems compared to conventional systems. The percentage of plant-parasitic nematodes in organic systems, on the other hand, was only half of that for conventional soils. The organic systems also demonstrated significantly greater ($p < 0.05$) nematode diversity (Table 3).

Table 1. Location and soil description of surveyed banana farms in North Queensland, Australia [31] [32] .

Site	Banana system	Latitude	Longitude	Rainfall (mm)	Soil	Description	Drainage	Permeability	Erosion hazard
Ric-O	Organic	17°37'33"	145°56'49"	3482	Ferrosol	Friable clay	Poor[1]	High[4]	High[6]
Fra-C	Conventional	17°37'55"	145°57'02"	3482	Ferrosol	Friable clay	Poor[1]	High[4]	Moderate[7]
Frn-O	Organic	17°51'29"	146°05'48"	3149	Ferrosol	Friable clay	Poor[1]	High[4]	High[6]
Dol-C	Conventional	17°51'58"	146°05'57"	3149	Ferrosol	Friable clay	Poor[1]	High[4]	Moderate
Wat-O	Organic	18°37'33"	145°48'11"	2426	Dermosol	Friable clay	Imperfect[2]	Moderate[5]	Slight[8]
Tod-C	Conventional	18°03'50"	145°49'27"	2674	Dermosol	Friable clay	Poor[1]	Moderate[5]	Slight[8]
Grim-O	Organic	17°34'42"	146°00'03"	3390	Ferrosol	Friable clay	Poor[1]	High[4]	High[6]
Dar-C	Conventional	17°35'11"	146°00'14"	3390	Ferrosol	Friable clay	Poor[1]	High[4]	Moderate[7]
Dol-O	Organic	17°07'39"	145°22'57"	967	Ferrosol	Friable clay	Imperfect[2]	Moderate[5]	Moderate[7]
How-C	Conventional	17°06'19"	145°24'47"	924	Ferrosol	Friable clay	Well[3]	High[4]	Slight[8]

[1]Water is removed very slowly; [2]Water is removed only slowly in relation to supply; [3]Water is removed from the soil readily but not rapidly. [4]Vertical transmission of water though the soil profile would take 1 - 12 hours after thorough wetting to reach field capacity, [5]Vertical transmission of water though the soil profile would take 1 - 5 days after thorough wetting to reach field capacity; [6]Significant erosion can occur; [7]Significant erosion can occur during development of a particular land use; [8]No appreciable erosion damage is likely to occur.

3.2. Disease Suppression

The hyphal extension tests did not show a significant difference in Foc hyphal growth between organic and conventional soils although the average hyphal growth of the conventional soils was numerically higher (Table 3). In terms of the incidence of Fol, based on the results from the isolations of dead, diseased and healthy tomato plant tissue suggested that Fol was not the only pathogen present in the soil that potentially affected tomato plant health. Fol was isolated from dead plants of all soils. Other plant pathogens isolated from dead or diseased plants included Chalara spp., Curvularia lunata, Gliocladium spp., Phoma spp., Phomopsis spp., Pythium spp., Rhizoctonia spp., Sclerotinia spp. and Thielaviopsis spp. The bacterial wilt Ralstonia spp. was widespread and presents a serious limitation to tomato cultivation in the region. However, taking all these limitations into account, it appeared that after being exposed to a range of plant pathogens, organic soils demonstrated significantly better ($p < 0.05$) overall tomato plant health than conventional soils (Table 3).

Table 2. Soil physical and chemical properties for organic and conventional soils sampled at selected banana farms (n = 5 for both farm type) in north Queensland.

Group	Indicator	Unit	Organic ± SE		Conventional ± SE	
Physical indicators	Bulk density	kg/l	1.14 ± 0		1.10 ± 0	
	Clay	%	19 ± 1		24 ± 2.1	
	Sand	%	50 ± 1.3		43 ± 1.7	
	Silt	%	31 ± 0.9		33 ± 0.8	
	WSA	%	90 ± 0.7		85 ± 0.9	
Chemical indicators	NH_3-Nitrogen	mg/kg	2.64 ± 0.3		2.08 ± 0.1	
	Available potassium	mg/kg	167 ± 25.3		431 ± 131.5	
	Boron	mg/kg	1.3 ± 0.1		1.7 ± 0.3	
	Calcium	meq/100g	9.1 ± 0.7		10.1 ± 0.8	
	Cation exchange capacity	meq/100g	12.2 ± 1.2		13.6 ± 1	
	Chloride	mg/kg	12.1 ± 1.8		33.5 ± 16.6	
	Copper	mg/kg	5.6 ± 1.3		4.5 ± 1	
	Electrical conductivity	µS	51 ± 2.6		106 ± 6.8	
	Iron	mg/kg	51 ± 10.1		25 ± 5.3	
	Labile C	g/kg	0.62 ± 0	a	0.37 ± 0	b
	Magnesium	meq/100g	2.3 ± 0.2		2.4 ± 0.2	
	Manganese	mg/kg	35.8 ± 5.8		47.5 ± 6.9	
	NO_3-Nitrogen	mg/L	7.4 ± 0.7		11.6 ± 1.2	
	Organic carbon	%	2.4 ± 0.2		2.1 ± 0.1	
	pH		6.65 ± 0.1		6.75 ± 0.1	
	Phosphorouos (Colwell)	mg/kg	129 ± 24.5		241 ± 42.3	
	Phosphorous buffer index		247 ± 48.6		325 ± 40.2	
	Potassium	meq/100g	0.42 ± 0.1		1.25 ± 0.3	
	Sodium	meq/100g	0.067 ± 0		0.047 ± 0	
	Sulphate sulfur	mg/kg	15.5 ± 2.2	b	63.5 ± 11	a
	Zinc (DTPA)	mg/kg	12 ± 3.5		12 ± 1.6	

Mean values are followed by standard errors and means with different letters following were significantly different from one another ($p = 0.05$) determined using Fisher's protected LSD.

Table 3. Soil microbial and nematode properties for organic and conventional soils sampled at selected banana farms (n = 5 for both farm type) in north Queensland.

Group	Indicator	Unit	Organic ± SE				Conventional ± SE			
Microbial indicators	β-glucosidase	µg pNP released g/soil/h	548	±	48.5	a	345	±	34.5	b
	Fluorescein diacetate	mg hydrolysed FDA/kg soil/hr	1.31	±	0.1	a	0.85	±	0	b
Nematode indicators	Bacterial feeding	%	32	±	1.6	a	17	±	1.3	b
	Bacterivore: fungivore ratio		0.71	±	0		0.83	±	0	
	Channel index (CI)		24	±	1.5		19	±	1.1	
	Diversity index (H)		2.05	±	0	a	1.47	±	0.1	b
	Endophytes	%	27	±	3.8		28	±	4.3	
	Enrichment index (EI)		74	±	1		74	±	1.1	
	Fungal feeding	%	14	±	1.5	a	3	±	0.2	b
	Plant parasites	%	31	±	2.2	b	70	±	2.1	a
	Predatory and omnivorous	%	20	±	1.3	a	10	±	1.2	b
	Structure index (SI)		68	±	1.8		69	±	1.6	
Disease indicators	Fusarium plates	AUC	8.2	±	0.4		11.5	±	0.3	
	Tomato trials	AUC	16.2	±	0.8	b	19.6	±	1.0	a

The pot experiment did not demonstrate any significant difference ($p > 0.05$) in terms of plant growth or disease development between the management systems (Table 4). However, the average numeric value for most growth indicators was slightly higher and disease indicators were slightly lower in organic soils than in conventional soils. A principal component analysis (PCA) provided a more effective statistical tool for analysing and visualizing results (Figure 1). The PCA had a varimax rotation applied to latent vectors to determine the variables with the greatest loadings and an overview of the coefficients for specific variable is presented in Table 5. The first principle component (the horizontal axis), explained 72.5% of the variation in the data, where the variables leaf emergence, rhizome discoloration, leaf area and plant growth had the greatest vector loadings. Based on the emergence of variables within PC1 it may be concluded that the component was mainly based on the overall "growth" of the banana plants. The second principle component (the vertical axis), which explained 19.6% of the variation in the data, had wilting and root ratings showing the highest vector loadings. Therefore, the second principle component explains the level of "Fusarium" infection with the plants. The PCA bi-plot separated the site Grim-O from other sampling sites (Dar-C, Dol-O and How-C) included in the experiment. Based on the directions of the vectors it appeared that Grim-O had greater plant growth, leaf area and leaf emergence and was negatively correlated with Foc root ratings. Furthermore, Grim-O was neutral for rhizome discoloration and wilting AUC values (Figure 1). The How-C plants showed high wilting at low rhizome discoloration, whereas, the Dol-O showed low wilting at high rhizome discoloration values (Figure 1).

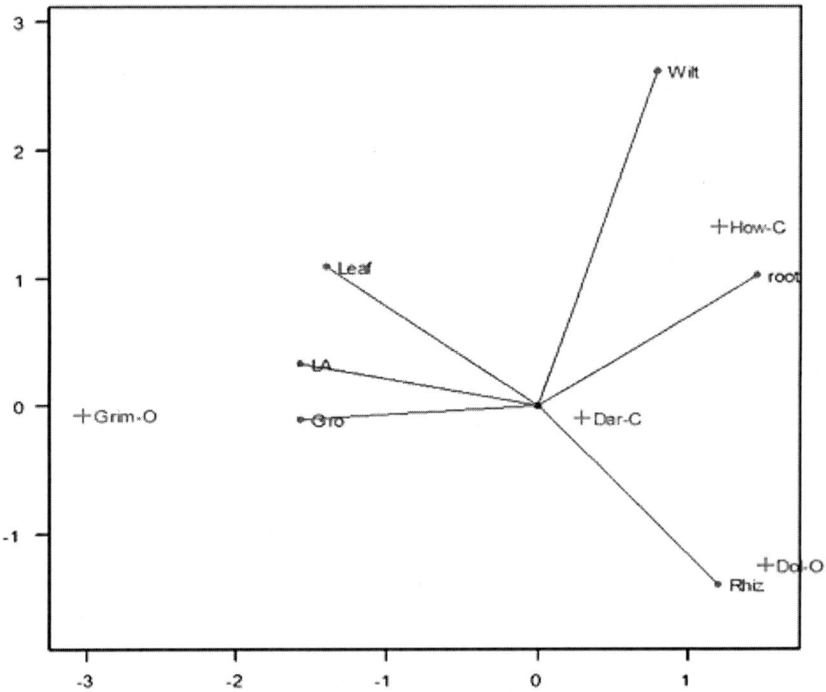

Figure 1. Principal component biplot of farm soils as function of disease and growth indicators. The red crosses represent farms, where: O refers to organic farms and C to conventional farm. Both Dol-O and How-C and Grim-O and Dar-C are farm pairs. Lines represent vectors which for proportion banana growth and disease variables, where: Gro = plant growth cm wk^{-1}, Leaf = leaf emergence leaves wk^{-1}, LA = leaf area cm^2, Rhiz = discoloration of rhizome, Root = discoloration of roots and Wilt = wilting area under disease progress curve.

Table 4. Averages of growth and Fusarium wilt parameters of potted banana plants grown in a selection of organic and conventional soils identified from organic and conventional banana production in north Queensland.

Farm	System	Discoloration		Wilting	Growth (cm week^{-1})	Leaf emergence (leaves week^{-1})	Leaf area (cm^2)
		Rhizome	Root				
Dar-C	Conventional	5.0	3.4	6.67	2.63	2.06	56.7
Dol-O	Organic	6.0	3.2	6.50	1.79	2.00	50.2
Grim-O	Organic	4.8	2.2	6.33	3.71	2.67	85.9
How-C	Conventional	5.3	3.7	8.00	1.88	2.29	56.6

Table 5. Percentage variation based on principal components analysis with discriminate weighting functions for the separate growth and Fusarium wilt parameters of potted banana plants grown in a selection of organic and conventional soils identified from organic and conventional banana production in north Queensland.

	PC1	PC2
Percentage variation of rotated factors	72.5%	19.6%
Plant growth	0.36	−0.31
Leaf area	0.44	−0.20
Leaf emergence	0.53	0.01
Rhizome discoloration	−0.54	−0.12
Wilting	0.28	0.77
Root discoloration	−0.17	0.51

where: PC1 = Principal component 1, PC2 = Principal component.

3.3. Correlation between Soil Health Indicators and Plant Health Indicators

Hyphal growth of Foc on soil in petri-dishes showed a negative correlation with the soil parameters bulk density, nematode diversity, and fungal feeding nematodes. A positive correlation was found between Foc hyphal extension, plant-parasitic nematodes numbers, soil organic carbon, copper, electrical conductivity and sulphate sulphur (data not shown). Area under disease curve (AUC) for tomatoes showed a negative correlation with nematode diversity, nematode community structure index and soil magnesium. Disease in tomatoes was positively correlated with soil potassium and sulphate levels. The height of potted banana plants was positively correlated with nematode diversity, organic carbon and phosphorus content of the soil, while it was negatively correlated with soil sulphate content. The leaf area of banana plants was positively correlated with soil phosphorus values while the reverse was true for rhizome discoloration. Wilting of potted banana plants was negatively correlated with nematode diversity (data not shown for more details see Geense [42]).

Multiple regression models were used to explain disease suppression. Fusarium hyphal growth AUC on soil plates appeared to be a function of nematode diversity, soil organic carbon and copper content (Equation (1)). This model could explain a majority of the variation and resulted in an r^2value of 0.64 ($p < 0.05$), with the corresponding model equation:

$$\text{AUC } \textit{Foc} \text{ hyphal growth} = 29.4 - 17.1 \times \text{H} - 5.96 \times \text{OC} + 1.24 \times \text{Cu} \quad (1)$$

where H = nematode diversity,

OC = organic carbon (%),

Cu = copper (mg/kg).

In terms of tomato plant health, this appeared to be primarily a function of nematode diversity, soil potassium and magnesium content (Equation (2)). Including these parameters into a multiple regression model could account for greater than half the variation in the health of tomato plants with a combined r^2 value of 0.61 ($p < 0.05$) and the multiple regression model:

$$\text{AUC Tomato disease} = 23.69 - 2.1 \times \text{H} - 1.7 \times \text{Mg} + 2.4 \times \text{K} \quad (2)$$

where H = nematode diversity,

Mg = magnesium (meq/100g),

K = potassium (meq/100g).

4. DISCUSSION

This study included farms using both organic and conventional banana systems in the main agro-climatic zones of the north Queensland banana industry, allowing conclusions relevant to the whole north Queensland banana production area. Organic banana farm practices in Australia were found to change soil biological properties, but had little impact on soil chemical properties while there was no change in soil physical properties. Physical soil properties in conventional and organic soils were very similar, especially within farm pairs. It appeared that WSA and bulk density are mainly influenced by climate and soil type, rather than management. Alternatively, it may also be that the soil management components including soil tillage that influence bulk density and WSA are implemented in rather similar manner in both management systems and are more indicative of the type of crop. As bananas in Australia are grown on a semi-perennial system being replanted every six to seven years, tillage was not expected to be a factor contributing to differences between the two systems.

Chemical properties of organic and conventional soils were similar, indicating similar application of nutrients, albeit in different forms. This may also be a reflection of the crops nutritional needs, rather than the form of fertiliser applied. The higher sulphate levels in the conventional soils may be explained by application of sulphur-based pesticides and fertilisers [43] . Additional studies are required for confirmation of this observation, but it seems likely that sulphate levels in soil may be used as an indicator of agrochemical and synthetic fertiliser use. The higher levels of labile soil C in the organic systems can be explained by increased emphasis on application of organic soil amendments, especially thoroughly composted, soluble or liquid organic fertilisers. Observations during the survey showed that organic banana farmers apply appreciable amounts of these types of fertiliser. Furthermore, organic farms typically foster soil organic matter additions by frequently including cover crops in their rotation, while the retention of crop residues was also more common compared to conventional farms.

Soil microbial activity indicators FDA and β-glucosidase were both higher in organic soils than in conventional soils. The majority of nutrients in conventional production are applied in mineral forms, which tend to be readily available to plants. Microorganisms involved in breakdown of organic materials thus lose a food source and their populations decline. Microorganisms engaged in the mineralisation process often have symbiotic interactions with plants, exchanging soil minerals for carbon. However, as plants already receive readily available nutrients in mineral form they no longer benefit from investing carbon in sustaining symbiotic relationships with microorganisms, and these organisms may lose their functionality. Furthermore, several studies have shown a clear relation between decline of soil microbial activity and the use of agrochemicals [44] - [46] . Moreover, all biocides potentially have a negative impact on soil biology. The underlying mechanisms are not entirely clear, but further investigation is required on the combined effects of reduced application of carbon and the antibiotic effects of agrochemicals creating an unfavourable environment for soil microbial activity in conventional agriculture.

The soil nematode community was closely related to soil microbial activity. In this survey it was observed that the percentage of bacterial-feeding, fungal-feeding, predatory and omnivorous nematodes in the organic soils were often a factor two or more times greater than for conventional soils.

Higher levels of bacterial and fungal-feeding nematodes are indicative of higher levels of microbes, which translate to more intense cycling of nutrients, thereby creating a more dynamic, better balanced and healthier soil. The higher level of predatory nematodes indicated a higher degree of self-regulation of the soil nematode community [38] . This was also reflected in the diversity index that was significantly higher in organic systems. Increasing both above-ground and below-ground diversity in an ecosystem, increased dynamics, self-regulation and resilience of that system [24] [47] - [49] . At the conventional banana production sites the percentage of plant-parasitic nematodes was more than double that of the organic systems. In conventional soils, the balancing effect of diversity was weaker and there were less predatory nematodes to regulate the numbers of parasites, leaving the species that can cause economic damage to crops to proliferate.

It is of interest that the environment for beneficial microbes and nematodes was unfavourable in the conventional soils; however, the parasitic nematodes appeared to tolerate that environment. This is possibly because pathogens and parasites have adapted to these circumstances during their association with commercial crops obtaining carbon from root systems. For example, Abang et al. [50] found that pathogen evolution can develop very rapidly in monocultures. Furthermore, pathogens and parasites are less reliant on decomposing organic substrates than other soil micro- and macro-organisms. Co-evolution of host-pathogen systems may lead to local adaptation to environmental conditions of potential pathogens [51] . This effect becomes more pronounced if the pathogen has a greater adaptability and host prevalence is increased, as found in mono-cropped perennial systems [51] .

Yardim and Edwards [52] found under field conditions that frequent use of fungicides, herbicides and insecticides caused a decrease in bacterial-feeding, fungal-feeding, predatory and omnivorous nematodes and an increase in plant-parasitic nematodes, which supports the results of the two-sided correlation tests in this study. Sulphate levels were also positively correlated with disease progress in tomatoes and to Foc hyphal growth. Plant-parasitic nematodes were also positively correlated with Foc hyphal growth. These correlations may provide additional evidence, from which may be concluded that the use of agrochemicals and synthetic fertilizers can foster plant-pathogens and parasites.

It is questionable if the Foc hyphal extension test described by Alabouvette et al. [29] provided a useful insight in soil Fusarium suppression. Firstly, hyphal growth of Fusarium does not guarantee successful infection and subsequent disease expression of the host plant [53] . Secondly, it was difficult to distinguish between different fungi on the plate, especially between Foc and other species of Fusarium. Especially in the second half of the test it was unclear if the Foc on the plug was colonizing the soil or that the fungi in the soil were colonizing the plug. Regardless, the test showed that Foc hyphal growth, probably as well as any other fungi, was favoured by soil organic carbon. From a management point of view this implies that to reduce Foc colonization, organic matter should only be applied if the material is well colonized by other saprophytic microorganisms or if there is sufficient microbial activity in the soil to colonize the material quickly, thereby effectively out-competing the pathogen.

The presence of other fungi and possibly bacteria that are pathogenic to tomatoes made it impossible to make a conclusive statement regarding soil Fusarium suppression. In this context, tomato may not be the most suitable crop to test soil Fusarium suppression because of its susceptibility to a wide range of other plant pathogens. Also, plant health is an important factor in Fusarium infection. Growing tomatoes in a banana soil therefore, may not give a clear picture of potential Foc suppression of a soil. Therefore, it would be preferable to use the specific host and pathogen of interest, or if impossible, to use a wider range of hosts and their associated (Fusarium) pathogens. Another issue was that the Fol R3 strain used had lost much of its pathogenicity, making it difficult to determine how suppressive soils were to the pathogen. However, the tomato bioassay did show that organic management could increase general soil suppression and overall plant health for at least this specific crop.

The principal component analysis of the banana pot trial showed that plant growth and Foc infection are to some extent determined by location and soil type. More specifically, one paired site clearly had less growth and more disease than the others (Figure 1). From the observation that one organic soil demonstrated better growth and reduced disease symptoms suggested that organic management can reduce the effects of Foc. The organic soil from the pair Dol-O and How-C had less wilting than its conventional counterpart, despite higher rhizome discoloration. It appeared that the soil from the organic system allowed plants to tolerate plant infection slowing disease symptom progress.

In all three experiments, hyphal extension test, tomato Fusarium suppression assay test and banana pot trial, nematode diversity and sulphate sulphur were recurring indicators with close correlation to respectively pathogen growth, disease expression and plant health. Foc hyphal extension and tomato disease development were negatively correlated with nematode diversity and positively correlated with sulphate sulphur. Banana plant growth was positively correlated with nematode diversity and negatively correlated with sulphate sulphur. Nematode diversity also played an important role in explaining variation of disease indicators in the multiple regression models. Based on current findings it appeared that nematode diversity, as an indicator of general soil biological diversity and soil sulphate, as an indicator of synthetic inputs may be suitable soil health indicators of suppression of Fusarium and Foc and in general soil health. However, further research is required to confirm these findings for other cropping scenarios.

5. CONCLUSION

It may be argued that stimulating soil microbial diversity and activity, thus increasing competition, can reduce soil pathogen conduciveness. Careful management of organic matter application and reduction or preferably omission of agrochemical use can achieve greater soil biodiversity and microbial activity. Agrochemicals, synthetic fertiliser use, lack of plant diversity and lack of regular organic matter inputs appeared to create soil conditions which were more conducive to plant-pathogens in bananas, specifically plant-parasitic nematodes and Foc. Conducive or suppressive soil conditions to Foc seem to be largely directed by competition for carbon. Soil suppression or promotion to Fusarium wilt of bananas is a complex issue that cannot be assessed by a single indicator. However, measurement of soil microbial enzyme activity, nematode community structure, diversity and possibly sulphate sulphur appeared to provide a reliable indication of general disease suppression for banana- based cropping systems.

ACKNOWLEDGEMENTS

Paul Geense, Anthony B. Pattison, Tegan L. Kukulies, Johannes M. S. Scholberg, Agustin B. Molina This work was jointly funded by the Australian Centre for International Agricultural Research through project Hort/2008/040, with additional support from the Queensland Government through the Department of Agriculture, Fisheries and Forestry and the Biological Farming Systems Group, Wageningen University code number BFS-80436. The banana grower's in north Queensland who contributed to the survey are gratefully acknowledged for allowing this work to be conducted on their properties and the contribution they have made to ensure the experiment reached a conclusion. Thanks are also given to colleagues Dr. Leanne Forsyth, Mr. Mike Male, Mr. Lynton Vawdrey and Mr. David Astridge for their assistance and expert knowledge.

REFERENCES

1. Stirling, G.R. and Pattison, A.B. (2008) Beyond Chemical Dependency for Managing Plant-Parasitic Nematodes: Examples from the Banana, Pineapple and Vegetable Industries of Tropical and Subtropical Australia. Australasian Plant Pathology, 37, 254-267.
2. Kernot, I., Lindsay, S., Campagnolo, D., Daniells, J., Evans, D., Goebel, R., Gunther, M., Lemin, C., Pattison, T., Peterson, R., Pinese, B. and Wharton, D. (1998) Tropical Banana Information Kit. Department of Primary Industries, Nambour.
3. Anonymous (2001) Great Barrier Reef Water Quality Action Plan. Great Barrier Reef Marine Park Authority, Townsville, 64. http://www.gbrmpa.gov.au/ corp_site/ key_issues/water_quality/action_plan/
4. Pattison, A.B., Moody, P.W., Badcock, K.A., Smith, L.J., Armour, J.A., Rasiah, V., Cobon, J.A., Gulino, L.-M., Mayer, R. and Lindsay, S. (2008) Development of Key Soil Health Indicators for the Australian Banana Industry. Applied Soil Ecology, 40, 155-164.
5. Rasiah, R.V., Armour, J.D., Moody, P.W., Pattison, A.B., Lindsay, S. and Florentine, S. (2009) Characterising and Improving the Deteriorating Trends in Soil Physical Quality under Bananas. Australian Journal of Soil Research, 47, 574-584.
6. Kleiese, Y., Prove, B., McShane, T., Moody, P. and Reghenzani, J. (1997) Nutrient Loss Study in the Johnstone River Catchment. Queensland Department of Primary Industries, Brisbane, 8.
7. Moody, P.W. and Aitken, R.L. (1997) Soil Acidification under Some Tropical Agricultural Systems. 1. Rates of Acidification and Contributing Factors. Australian Journal of Soil Research, 35, 163-173.
8. Pegg, K.G., Moore, N.Y. and Bentley, S. (1996) Fusarium Wilt of Banana in Australia: A Review. Australian Journal of Agricultural Research, 47, 637-650.
9. Thusrston, H.D. (1998) Tropical Plant Diseases. 2nd Edition, American Phytopathological Society, New York.
10. Singh, R.S. (2000) Diseases of Fruit Crops. Science Publishers Inc., USA.
11. Marois, J.J. and Ploetz, R.C. (1990) Biological Control of Diseases Caused by Fusarium oxysporum. In: Ploetz, R.C., Ed., Fusarium Wilt of Banana, APS Press, 77-01.
12. Dita, M.A., Waalwijk, C., Buddenhagen, I.W., Souza, M.T. and Kema, G.H.J. (2010) A Molecular Diagnostic for Tropical Race 4 of the Banana Fusarium Wilt Pathogen. Plant Pathology, 59, 348-357.
13. Ploetz, R.C. (1994) Panama Disease: Return of the First Banana Menace. International Journal of Pest Management, 40, 326-336.
14. Ploetz, R.C., Thomas, J.E. and Slabaugh, W.R. (2003) Diseases of Banana and Plantain. In: Ploetz, R.C., Ed., Diseases of Tropical Fruit Crops, CABI Publishing, Wallingford, UK, 73-134.
15. Bancroft, J. (1876) Report of the Board Appointed to Enquire into the Cause of Disease Affecting Livestock and Plants. Queensland, 1876. Votes and Proceedings, 3, 1011-1038.
16. Hennessy, C., Walduck, G., Daly, A. and Padovan, A. (2005) Weed Hosts of Fusarium oxysporum f. sp.cubense Tropical Race 4 in Northern Australia. Australasian Plant Pathology, 34, 115-117.
17. Meldrum, R.A., Fraser-Smith, S., Tran-Nguyen, L.T.T., Daly, A.M. and Aitken, E.A.B. (2012) Presence of Putative Pathogenicity Genes in Isolates of Fusarium oxysporum f. sp.cubense from Australia. Australasian Plant Pathology, 41, 551-557.
18. Pattison, A.B. and Lindsay, S. (2006) Banana Root and Soil Health User's Manual. Department of Primary Industries and Fisheries, Brisbane.
19. Smith, M.K., Whiley, A.W., Searle, C., Langdon, P.W., Schaffer, B. and Pegg, K.G. (1998) Micropropagated Bananas Are More Susceptible to Fusarium Wilt than Plants Grown from Conventional Material. Australian Journal of Agricultural Research, 49, 1133-1139.

20. De Wit, J. and Verhoog, H. (2007) Organic Values and the Conventionalization of Organic Agriculture. NJAS-Wage- ningen Journal of Life Sciences, 54, 449-462.
21. Kristiansen, P. and Merfield, C. (2006) Overview of Organic Agriculture. In: Kristiansen, P., Taij, A. and Reganold, J., Eds., Organic Agriculture, a Global Perspective, CAB International Publishing, Wallingford, 1-23.
22. Fliessbach, A. and Mäder, P. (2000) Microbial Biomass and Size-Density Fractions Differ between Soils of Organic and Conventional Agricultural Systems. Soil Biology and Biochemistry, 32, 757-768.
23. Lundquist, E.J., Scow, K.M., Jackson, L.E., Uesugi, S.L. and Johnson, C.R. (1999) Rapid Response of Soil Microbial Communities from Conventional, Low Input, and Organic Farming Systems to a Wet/Dry Cycle. Soil Biology and Biochemistry, 31, 1661-1675.
24. Mäder, P., Fliessbach, A., Dubois, D., Gunst, L., Fried, P. and Niggli, U. (2002) Soil Fertility and Biodiversity in Organic Farming. Science, 296, 1694-1697.
25. Lockie, S. and Halpin, D. (2005) The "Conventionalization" Thesis Reconsidered: Structural and Ideological Transformation of Australian Organic Agriculture. Sociologica Ruralis, 45, 284-307.
26. Karlen, D.L., Ditzler, C.A. and Andrews, S.S. (2003) Soil Quality: Why and How? Geoderma, 114, 145-156.
27. van Bruggen, A.H.C. and Semenov, A.V. (2000) In Search of Biological Indicators for Soil Health and Disease Suppression. Applied Soil Ecology, 15, 13-24.
28. Alabouvette, C. (1990) Biological Control of Fusarium Wilts in Suppressive Soils. In: Hornby, D., Ed., Biological Control of Soilborne Plant Pathogens, CAB International Publishing, Wallingford, 27-43.
29. Alabouvette, C., Raaijmakers, J.M., de Boer, W., Notz, R., Défago, G., Steinberg, C. and Lemanceau, P. (2006) Concepts and Methods to Assess the Phytosanitary Quality of Soils. In: Bloem, J., Hopkins, D.W. and Benedetti, A., Eds., Microbiological Methods for Assessing Soil Quality, CABI Publishing, Wallingford, 257-269.
30. García-Ruiz, R., Ochoa, V., Viñegla, B., Hinojosa, M.B., Peña-Santiago, R., Liébanas, G., Linares, J.C. and Carreirra, J.A. (2009) Soil Enzymes, Nematode Community and Selected Physio-Chemical Properties as Soil Quality Indicators in Organic and Conventional Olive Oil Farming: Influence of Seasonality and Site Features. Applied Soil Ecology, 41, 305-314.
31. Isbell, R.F. (1996) The Australian Soil Classification. CSIRO Publishing, Melbourne.
32. Simpson, B., Blogg, D., Haydon, G., Ruddle, L., Beswick, A., Rayner, D., Grundy, M. and Smith, D. (2004) SafeGuage 1.1.4. Department of Natural Resources and Mines, Brisbane.
33. Moody, P.W. and Cong, P.T. (2008) Soil Constraints and Management Package (SCAMP): Guidelines for Management in Tropical Upland Soils. Australian Centre for International Agricultural Research, Canberra, 86.
34. Schnürer, J. and Rosswall, T. (1982) Fluorescein Diacetate Hydrolysis as a Measure of Total Microbial Activity in Soil and Litter. Applied and Environmental Microbiology, 43, 1256-1261.
35. Eivazi, F. and Tabatabai, M.A. (1988) Glucosidases and Galacosidases in Soils. Soil Biology and Biochemistry, 20, 601-606.
36. Whitehead, A.G. and Hemming, J.R. (1965) A Comparison of Some Quantitative Methods Extracting Small Vermiform Nematodes from the Soil. Annals of Applied Biology, 55, 25-38.
37. Yeates, G.W. and Bongers, T. (1999) Nematode Diversity in Agroecosystems. Agriculture, Ecosystems and Environment, 74, 113-135.
38. Ferris, H., Bongers, T. and de Goede, R.G.M. (2001) A Framework for Soil Food Web Diagnostics: Extension of the Nematode Faunal Analysis Concept. Applied Soil Ecology, 18, 13-29.
39. Orjeda, G. (1998) Evaluation of Musa Germplasm to Sigatoka Diseases and Fusarium Wilt. INIBAP Technical Guidelines. International Network for Improvement of Banana and Plantain, Montpellier, 60.

40. Turner, D.W. (1972) Dry Matter Production, Leaf Area and Growth Analysis of Bananas. Australian Journal of Experimental Agricultural and Animal Husbandry, 12, 216-224.
41. Payne, R.W., Murray, D.A., Harding, S.A., Baird, D.B. and Soutar, D.M. (2008) GenStat for windows (11th Edition) Introduction. VSN International, Hemel Hempstead.
42. Geense, P. (2010) Soil Health in Organic and Conventional Banana Production and Its Effect on Fusarium Wilt of Bananas. Master's Thesis, Wageningen University, Wageningen.
43. Balik, J., Kulhanek, M., Cerny, J., Szakova, J., Pavlikova, D. and Cermak, P. (2009) Differences in Soil Sulfur Fractions Due to Limitation of Atmospheric Deposition. Plant Soil Environment, 55, 344-352.
44. Duah-Yentumi, S. and Johnson, D.B. (1986) Changes in Soil Microflora in Response to Repeated Applications of Some Pesticides. Soil Biology and Biochemistry, 18, 629-635.
45. Moeskops, B., Sukristiyonubowo, Buchan, D., Sleutel, S., Herawaty, L., Husen, E., Saraswati, R., Setyorini, D. and De Neve, S. (2010) Soil Microbial Communities and Activities under Intensive Organic and Conventional Vegetable Farming in West Java, Indonesia. Applied Soil Ecology, 45, 112-120.
46. Singh, J. and Singh, D.K. (2005) Bacterial, Azotobacter, Actinomycetes and Fungal Population in Soil after Diazinon, Imidacloprid and Lindane Treatments in Groundnut (Arachis hypogaea L.). Journal of Environmental Science and Health, Part B: Pesticides, Food Contaminants, and Agricultural Wastes, 40, 785-800.
47. Elmqvist, T., Folke, C., Nyström, N., Peterson, G., Bengtsson, J., Walker, B. and Norberg, J. (2003) Response Diversity, Ecosystem Change and Resilience. Front Ecological Environment, 1, 488-494.
48. Kennedy, A.C. and Smith, K.L. (1995) Soil Microbial Diversity and the Sustainability of Agricultural Soils. Plant and Soil, 170, 75-86.
49. Torsvik, V. and Øvreås, L. (2002) Microbial Diversity and Function in Soil: From Genes to Ecosystems. Current Opinion in Microbiology, 5, 240-245.
50. Abang, M.M., Baum, M., Ceccarelli, S., Grando, S., Linde, C.C., Yahyaoui, A., Zhan, J. and McDonald, B.A. (2006) Differential Selection on Rhynchosporium secalis during Parasitic and Saprophytic Phases in the Barley Scald Disease Cycle. Phytopathology, 96, 1214-1222.
51. Gandon, S. and Michalakis, Y. (2002) Local Adaptation, Evolutionary Potential and Host Parasite Coevolution: Interactions between Migration, Mutation, Population Size and Generation Time. Journal for Evolutionary Biology, 15, 451-462.
52. Yardim, E.N. and Edwards, C.A. (1998) The Effects of Chemical Pest, Disease and Weed Management Practices on the Trophic Structure of Nematode Populations in Tomato Agroecosystems. Applied Soil Ecology, 7, 137-147.
53. Cook, J.R. and Baker, K.F. (1983) The Nature and Practice of Biological Control of Plant Pathogens. American Phytopathological Society, St. Paul.

CHAPTER 11

Modeling of Soil Erosion and Its Implication to Forest Management

Nuray Misir[1] and Mehmet Misir[2]

[1] *Karadeniz Technical University, Faculty of Forestry, Turkey*

[2] *Department of Forest Management, Turkey*

1. INTRODUCTION

Conservation of natural forest ecosystems will require a land ethic as prelude to understanding the functioning of forest ecosystems, ecological and physiological impacts of disturbances on ecosystems, and the processes involved in recovery of disturbed ecosystems. Many of the harmful effects of pollution, fire, flooding, and soil compaction can be abated by judicious planning measurements to create and perpetuate the critical components of forest stand structure and species composition. Strategies for continuous production of the products and services that can be supplied by forest ecosystems will need to be reinforced by expanded long-term research and close cooperation among various disciplines such as forest biologists, social scientists, economists, and regulatory government agencies [17].

Nowadays, multi-objective planning is necessary in forestry because of increased and varied demand for forest products and services. Management objective such as production of quality potable water, carbon stocking, aesthetic, recreation and community health in forest especially adjacent to big cities are of great importance. Forests have managed to produce wood products at various diameters and quality classes as the society demanded overtime [24]. Afterwards, the importance of these objectives has gradually diminished and overwhelmed by other management objectives such as conservation of water resources, prevention of soil erosion, carbon stocking, creation of landscape aesthetic, camouflaging military facilities and allocation of land for recreation [2]. The forest values on be grouped as static and dynamic forest values (Figure 1).

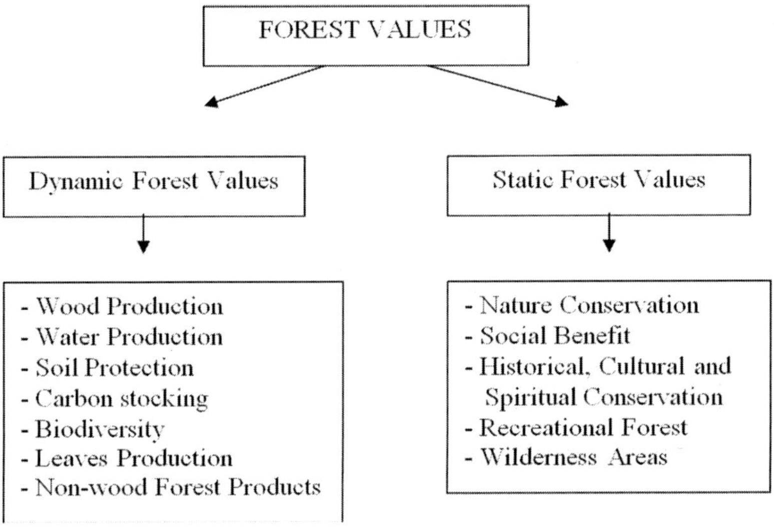

Figure 1. Classification of Forest values [23]

2. SOIL EROSION

Erosion, the detachment of soil particles, occurs by the action of water, wind, or glacial ice. Such 'background' soil erosion has been occurring for some 450 million years, since the first land plants formed the first soil. Only erosion caused by water will be considered here. Water related erosion occurs when raindrops, spring runoff, or floodwaters wear away and transport soil particles. Erosion is a complex natural process that has often been accelerated by human activities such as land clearance, agriculture, construction, surface mining, and urbanization.

Soil erosion by water and wind affects both agriculture and the natural environment, and is one of the most important of today's environmental problems. It isn't easy to find comprehensive information about erosion, as the subobject is multidisciplinary involving geomorphologists, agricultural engineers, soil scientists, hydrologists and others; and is of interest to policy-makers, farmers, environmentalists and many other groups.

Schumm and Harvey [32] believe some of the terms used to describe erosion are misleading. Normal erosion and geologic erosion are often meant to imply pre-agricultural conditions of low erosion rates, whereas accelerated erosion and historic erosion imply greatly increased erosion rates caused by man. Because of the great variability in natural erosion both spatially and temporally at present and throughout geologic time, neither concept is correct. They prefer the term natural erosion for normal and geologic erosion, and the term man-induced erosion for accelerated and historic erosion.

3. FOREST MANAGEMENT AND SOIL EROSION

All forest management activities affect soils, with effects ranging over a continuum from nearly none where the activity is minimal to large. To foster communication, a threshold should be established above which effects merit attention and below which further consideration is not justified. The magnitude of that threshold varies with the state of knowledge, about forest dynamics and must include recognition of uncertainty. Failure to identify thresholds inhibits communication to a wider audience and even among ourselves [12].

There are two kinds of effects of forest management on soils. The first, direct effect is an alteration of soil properties such as an increase in bulk density following passage of heavy equipment. Soil scientists generally agree on those direct effects; recognition of those alterations is literally axiomatic. The second effect of management on soils is indirect; a change in site productivity due to alteration of soil properties. Some of those secondary effects are obvious enough that can be considered corollaries. Specific studies and personal and vicarious experience have led to this worldview. Conversely, some of the indirect effects of management on soils are not as clear, and can be considered postulates. The distinction between axioms, corollaries, and postulates is often in the eye of the beholder, and depends on interpretation of both published reports and personal observations. Papers that support a position are evaluated differently than those in opposition. I offer no excuses for bias; "For every expert, there is an equal and opposite expert" [6].

Erosion is a natural process, but one whose rate and extent is exacerbated by forest management [36]. Most emphasis on erosion has been directed towards its effects on water quality and fish habitat, but because it involves displacement of soil, the growing medium, erosion also can affect site productivity [21]. However, forest management activities are necessary parts of forestry, and there may be minimal control over the circumstances under which they are carried out. Alterations of soil physical properties are extensive, immediate, and their effects in reducing productivity are well-documented. Chemical and biological properties of soils are also changed by management activities, but the effects on productivity are less well-documented and of longer term; their influence is not clear. Historical evidence shows that forest ecosystems are dynamic and resilient. Assessment of the consequences of changes in properties must recognize that shifts in preferred species should not be equated with changes in productivity, and that short-term effects, measured by the length of most experiments or observations, may not be indicative of long-term effects [12]. Accurate assessment of the effects of its change, however, is likely to continue to be obscured by the influence of the many other elements that also affect forest productivity [40]. At our current state of ignorance, a reasonable approach may be a simple sensitivity analysis that uses spatially based techniques (geographic information systems) and reasonable estimates of effects of the many factors that affect forest productivity to develop an impression of changes in soil productivity [12].

Use of more sophisticated simulation models implies greater knowledge than we currently possess. Both ethical and economic considerations demand good stewardship with professional accountability for our natural resources. Extensive forest management, if carried out with both wisdom and prudence, is not antithetical to good stewardship. "All of us have vested interests in making forest management a wise and efficient use of resources. Soil information can immeasurably help us be good stewards of the land" [12].

4. SOIL LOSS CHARACTERIZATION

Characterization of soil loss is very important for environment and natural resources. In erosion control planning, soil loss estimates for a particular site are determined using a prediction model and compared with a T-value for that site [31]. The Universal Soil Loss Equation (USLE) is an example of a model used extensively to predict erosion from croplands and rangelands. More recently, the Agricultural Research Service, Forest Service, and the Bureau of Land Management have joined in a cooperative effort, the Water Erosion Prediction Project (WEPP). WEPP has been implemented to develop an improved model based on modern technology for estimating soil erosion by water. WEPP technology is based on fundamental hydrologic and soil erosion processes and is designed to replace the widely used USLE [8].

Until recently, prediction of soil loss rates on National Forest lands involved using the USLE [8, 22]. Soil losses were evaluated in the context of potential soil losses, natural soil losses, current soil losses and tolerable soil losses. Potential losses were those that would occur after complete removal of the vegetation and litter. Natural losses were associated with the potential natural vegetation community. Current losses were those occurring with current management. Tolerable loss was assumed to be the rate that can occur while sustaining inherent site productivity [8].

The Universal Soil Loss Equation (USLE) is a widely used method for calculating annual soil losses, based on rainfall, runoff, slope, runoff length, soil type and landuse parameters. The equation originally developed on small agricultural plots, but has been adopted for evaluating erosion from large watersheds under a wide range of land uses. [41]

$$A = R \times K \times L \times S \times C \times P \tag{1}$$

where *A* represents the soil loss, commonly expressed in tonnes ha^{-1} $year^{-1}$. *R* refers to the rainfall erosivity factor, calculated by the summation of the erosion index EI30 over the period of evaluation. EI30 is a compound function of the kinetic energy of a storm and its 30-min maximum intensity. The latter factor is defined as the greatest average rainfall intensity experiences in any 30-min period during a storm. *K* is the soil erodibility factor reflecting the susceptibility of a soil type to erosion. It is expressed as the average soil loss per unit of the R factor. *L* is an index of slope length, *S* is a slope gradient index, *C* is an index for the protective coverage of canopy and organic material in direct contact with the ground. It is measured as the ratio of soil loss from land cropped under specific conditions to the corresponding loss from tilled land under clean-tilled continuous fallow conditions. Finally, the protective factor P represents the soil conservation operations or other measures that control the erosion, such as contour farming, terraces, and strip cropping. It is expressed as the ratio of soil loss with a specific support practice to the corresponding loss with up-and-down slope culture [41].

Soil loss rates have been generally estimated in agricultural areas up to now. Various USLE and GIS combinations have been used to estimate soil loss in forest land [25]. But in this kind of studies, soil loss was determined by quantitatively. For example; in study realized in Taiwan estimating watershed erosion using GIS coupled with the USLE in agricultural areas.

Furthermore a WinGrid system was developed to calculate slope length factor (L) in USLE [4].

Samar [30] developed three soil loss prediction models (WEPP, EPIC, ANSWERS) and used them for simulating soil loss and testing their capability in predicting soil losses for three tillage systems (rigde-till, chise-plow, and no-till). In other study (leave a space after point), USLE and GIS combination were used to predict long-term soil erosion and sediment transportation from hillslopes to stream networks under different climate conditions and forest management scenarios. Soil erosion was predicted by the USLE watershed level. The GIS utilities are employed to calculate total mass of sediment moving from each cell to nearest stream network [35]. Mısır et al. [25] developed a soil loss model applicable for forest management scenarios for forested areas in northern Turkey.

Forest values including soil protection function need to be determined quantitatively in multi-objective forest management planning. Relationships between soil loss and stand structure on a particular must be determined before incorporation of soil protection values into multi-objective forest management plans.

5. SOIL LOSS ESTIMATION

The soil loss expressed as ton ha^{-1} year^{-1} is determined using the Universal Soil Loss Equation (USLE). Soil samples are collected from sample plots and analyzed in a laboratory for soil properties including; silt %, sand %, clay %, organic matter %, and classes for structure and permeability. The soil erodibility factor K values of soil samples are calculated using the following equation [41]:

$$K = \frac{2.1 \times M^{1.14} \times 10^{-4} \times (12 - OM) + 3.25 \times (S\text{-}2) + 2.5 \times (P\text{-}3)}{100} \tag{2}$$

where OM is soil organic matter content, M is (%silt + %very fine sand)x(100-%clay), S is soil structure code and P is permeability class. If soil organic matter content was greater or equal to 4%, OM was considered constant at 4%. Moreover, the influence of rock fragments on soil loss was accounted for by a subsurface component in the soil erodibility K factor [29]. The rainfall erosivity was differently obtained from average annual rainfall erosivity map for countries or locations.

The slope length factor L, accounts for increases in runoff volume as downslope runoff lengths increase. The slope stepness factor S accounts for increased runoff velocity as slope stepness increases. These factors were obtained from digitized topographic maps of scale 1:25 000.

For direct application of the USLE a combined slope length and slope stepness (LS) factor was evaluated for each sample plots as [1]:

$$LS = l^{0.5} \times (0.0138 + 0.00965 \times S + 0.00138 \times S^2) \quad (3)$$

where l is runoff length (meter), S is slope (percent).

Crop and management factor is the soil loss from an area with specified cover. C is a function of landuse conditions such as vegetation type, before and after harvesting, crop residues, and crop sequence. Forest management practices create a variety of conditions that influence sheet and rill erosion. The USLE has been used with varying degrees of success to predict these forms of erosion on forest land. Assigning a proper value to cover-management factor (C) in the USLE is a problem, however. An undisturbed, totally covered forest soil usually yields no surface runoff. What erosion does occur on undisturbed forest land comes from stream channels, soil creep, landslide, gullies, and pipes, none of which are evaluated by the USLE. Logging, road building, site preparation, and similar activities that disturb and destroy cover expose the soil to the erosivity of rainfall and runoff [41].

Tree categories of woodland are considered separately:

1. undisturbed forest land,
2. woodland that is grazed, burned, or selectively harvested, and
3. forest lands which have had site preparation treatments for re-establishment after harvest.

Factor C for undisturbed forest land may be obtained from Table 1 [9].

Table 1. Factor *C* for undisturbed forest land

Percent of area covered by canopy of trees	Factor C
100 – 75	0.0001 – 0.001
70 – 45	0.002 – 0.004
40 – 20	0.003 – 0.009

The conservation practice factor *P*, is determined by the extend of conservation practices such as strip, cropping, contouring, and terracing practices, which tend to decrease the erosive capabilities of rainfall and runoff. Values of *P* range from zero to one.

6. DATA ANALYSIS AND MODELING

The candidate variables modeling are numerous and diverse. Hartanto et al. [14] classified such variables in four groups: Soil characteristics, physiographic properties, climatic properties and stand characteristics. The candidate variables of soil loss models can be divided in to two groups:

1. measures of physiographic structure and
2. measures of the stand level of structure and density.

Altitude, exposition, aspect, slope and exposure length have been used as measures of physiographic structure. Mean height, mean diameter, crown closure and stand density may have been used as measures of the stand level of structure.

Several possibilities exist to describe stand density. Hamilton [13], Ojansuu et al. [26], Vanclay [38], Thus [37], all of whom used *BA*, and [3], who used *N*, have provided examples of models with stand density parameters as explicatory variables in modeling. Since *N* and *BA* were directly determined, and did not rely on functional relationships, as opposed to volume (*V*), different stand density indexes [7, 28, 10, 5] may be tested.

The soil loss model should be applicable to different stand structures. Therefore, all variables must be tested. Based on the discussion above, the following soil loss models have been generally hypothesized:

$$\hat{A} = \beta_0 + \beta_1 S_1 + \beta_2 S_2 + \beta_3 S_3 \tag{4}$$

where *S 1* is the physiographic structure (altitude, exposition, aspect, slope and exposure length), *S 2* is the stand structure (d¯q,h¯qd¯q,h¯q and crown closure) and S_3 is the stand density.

Relationship between magnitude of soil loss obtained from sample plots and stand characteristics have been used to model soil protection value one of the forest values for quantifying soil loss by using linear, nonlinear, mixed linear and mixed nonlinear procedures in Regression Analysis Method The significance of parameter estimates was tested by means of *t*=b/ASE, where *b* is the parameter estimate and *ASE* is the asymptotic standard error. The parameters of the model for data have been determined using a software package (e.g. SPPS, SAS). Only were variables which are significant ($P<0.05$) included in the equation. A soil loss model is constructed based on some site and stand characteristics as a predictor and possible insignificant predictor are excluded. The predicted variable in the soil loss model is annual soil loss amount, which resulted in a linear or nonlinear relationship between the dependent and independent variables. The predictors of a soil loss model were chosen from stand level characteristics as well as their transformations. Some of them had to be significant at the 0.05 level without any systematic errors in residuals. The assumption of homoscedasticity has been tested using the Durbin-Watson test.

7. MODEL VALIDATION

The soil loss model was evaluated quantitatively by examining the magnitude and distribution of residuals to detect any obvious patterns and systematic discrepancies, and by testing for bias and precision to determine the accuracy at model predictions [39, 33, 11, 20]. Relative bias and root mean square error have been calculated as follows:

$$Bias = \frac{\sum_{i=1}^{n}\left(A_i - \hat{A}_i\right)}{n} \tag{5}$$

$$RMSE = \sqrt{\frac{\sum_{i=1}^{n}\left(A_i - \hat{A}_i\right)^2}{n-p}} \tag{6}$$

where n is the number of observations, p is the number of parameters in the model, A_i and A^iA^I are observed and predicted soil loss values, respectively.

In addition, the models were further validated by an independent control data set. The validation of a model should involve independant data. Data were partitioned in two independent groups, one for model development of soil loss estimation and the other set for validation. The data set used for model development of soil loss eestimation comprised approximately 80% of the plots, while the remaining 20% of plots were used for validation. Although the number of sample plots determined for development of soil loss estimation was made relatively large in order to provide sufficient data for model development phase, the number of sample plots in the test data still should be large enough for validation and appropriate statistical test. The deviations between predicted and observed values were tested by Student's Paired-t test or Wilcoxon test.

8. SAMPLE SIZE

The size of sample plot for sampling can be an advantage or disadvantage to model soil loss. A plot size of 800 m^2 means that a relatively large number of the trees are not affected by the forest conditions outside the plot. In other words, a relatively number of trees is affected by the forest conditions inside the plot. In this kind of studies plots that might have been subjected to any harvesting operation between the measurements were excluded from the data material because of insufficient information about treatments. If the harvest on these plots was a result of "regular" management practices, there were no problems related to the exclusion [37]. However, if the harvest was a result of an extraordinary situation (i.e. floods), exclusion of the plots may have lead to an underestimated soil loss amount.

9. UNCERTAINTY

There are many sources of uncertainties related to large scale forestry analyses in general, e.g. related to the inventory of input variables used as basis for the analyses [e.g. 16], to model errors of the numerous functions used for predictions [e.g. 15], to the stochasticity of future condition [e.g. 18, 27] and to the stochasticity of future prices and costs [e.g. 34, 19]. Thus [37], as long as the soil loss models are unbiased, they will not introduce any substantial change with respect to the final uncertainty of large scale forestry analyses.

10. CONCLUSIONS

Soil loss is an important variable which is used for multiple forest management planning.

Measuring soil loss is costly; however, foresters usually welcome an opportunity to estimate this function (forest value) with an acceptable accuracy. Missing soil losses may be estimated using a suitable soil loss equation. Based on a comprehensive data set which includes very different stands, such soil loss equation should be fitted for a major tree species in complex.

The stand position and stand density measures used in this kind of studies and variables entered to the soil loss model are easily obtained and are available in forest inventories. In summary, the suggested or developed soil loss models improve the accuracy of soil loss prediction, ensure compatibility among the various estimates in a forest management scenario, and maintain projections with reasonable biological limits.

Linear, nonlinear or mixed models for prediction of soil loss for stand level, designed for use in large scale forestry scenario models and analyses, may been developed. Although soil loss as a phenomenon is complicated to model, and in spite of several uncertain topics revealed from the work, the model fit and the validation tests may be turned out satisfactory.

Provided the many uncertainties of large scale forestry scenario analyses in general, soil loss models seem to hold an appropriate level of reliability, and we feel that it can be applied in such analyses. This does not mean that the model cannot be enhanced, however. With new rotations of permanent sample plots measurements, the models should be evaluated and, if necessary, revised or calibrated.

REFERENCES

1. H. M. J. Arnoldus, 1977 Predicting soil losses due to sheet and rill erosion. FAO conservation guide no.1,Guidelines for watershed management, Rome, Italy
2. Ü. Asan, 1992 letme sınıfı ayrımında fonksiyonel yaklaşım. Orman Mühendisliği Dergisi 5 30 31
3. M. Burgman, W. Incoll, P. Ades, I. Ferguson, T. Fletcher, A. Wholers, 1994 Mortality models for mountain and Alpine Ash Forest Ecology and Management 67 319 327
4. L. Chao-Yuan, L. Wen-Tzu, C. Wen-Chieh, 2002 Soil erosion prediction and sediment yield estimation: the Taiwan experience Soil & Tillage Research 68 143 152
5. H. H. Chisman, A. F. Schumacher, 1940 On the tree-area ratio and certain of its applications. Journal of Forestry 38 311 317
6. A. C. Clarke, 1998 President, Experts and Asteroids Science June 5 1532 1533
7. R. O. Curtis, G. W. Clendenan, D. J. Demars, 1981 A new stand simulator for coast douglas-fir: DFSIM users guide: U.S. forest service general technical report PNW-128
8. L. F. De Bano, M. K. Wood, 1990 Soil loss tolerance as related to rangeland productivity. Proceedings of the Soil Quality Standards Symposium. U.S. Department of Agriculture, Forest, Service Publication WO-WSA-2. 15 27
9. G. E. Dismeyer, G. R. Foster, 1984 Estimating the cover-management factor (C) in the Universal Soil Loss Equation for forest conditions Journal of Soil and Water Conservation 36 235 240

10. T. J. Drew, J. W. Flewelling, 1977 Stand density management: an alternative approach and its application to Douglas-Fir plantations. Forest Science 25 518 532
11. K. Gadow, G. Hui, 1998 Modelling forest development Faculty of forest and woodland ecology, University of Göttingen
12. D. F. Grigal, 2000 Effects of Extensive Forest Management on Soil Productivity Forest Ecology and Management 138 1-3 167 185
13. D. A. Hamilton, 1986 A Logistic model of mortality in thinned and unthinned mixed conifer stands of Northern Idaho. Forest Science 32 989 1000
14. H. Hartanto, R. Prabhu, A. S. E. Widayat, C. Adsak, 2003 Factors affecting runoff and soil erosion: Plot-level soil loss monitoring for assessing sustainability of forest management Forest Ecology and Management 141 1 14
15. A. Kangas, 1996 On the bias and variance in tree volume predictions due to model and measurement errors Scandinavian Journal of Forest Research 11 281 290
16. A. Kangas, J. Kangas, 1999 Optimization bias in forest management planning solutions due to errors in forest variables Silva Fennica 33 303 315
17. T. T. Kozlowski, 2000 Responses of woody plants to human-induced environmental stresses : Issues, problems, and strategies for alleviating stress Critical Reviews in Plant Sciences 19 2 91 170
18. M. . Larsson, 1994 The significance of data quality in compartmental forest registers in estimating growth and non-optimal losses-a study of final fellling Compartments in Northern Sweden. Report 26. Swedish University of Agricultural Sciences Umea, Sweden
19. P. Leskinen, J. Kangas, 1998 Analysing uncertainties of interval judgment data in multiple criteria evaluation of forest plans Silva Fennica 32 363 372
20. D. Mabvurira, J. Miina, 2002 Individual-tree growth and mortality models for Eucalyptus grandis (Hill) maiden plantations in Zimbabwe Forest Ecology and Management 161 1-3 231 245
21. W. F. Megahan, 1990 Erosion and site productivity in Western-Montane forestecosystems. Symposium on Management and Productivity of Western-Montane. Boise, ID, USDA Forest Service General Technical Report INT-280 146 150
22. W. F. Megahan, 1992 Logging erosion sedimentation... Are they dirty words? Journal of Forestry 70 7 403 407
23. M. Mısır, 2001 Developing a Multi objective model forest management plan using GIS and Goal Programming (A case study of Ormanüstü Planning Unit) Karadeniz Technical University, PhD Thesis, Trabzon.
24. M. Mısır, E. Z. Başkent, 2002 The role of GIS in Multi objective forest planning International Symposium on GIS, İstanbul, Tukey, Proceedings 449 465
25. N. Mısır, M. Mısır, U. Karahalil, H. Yavuz, 2007 Characterization of soil erosion and its implication to forest management Journal of Environmental Biology 28 2 185 191
26. R. Ojansuu, J. Hynynene, J. Koivunen, P Luoma, 1991 Luonnonprosessit metsalaskelmassa METSA 2000 385 1 59
27. K. Pasanen, 1998 Integrating variation in tree growth in to Forest Planning Silva Fennica 32 11 25
28. L. H. Reineke, 1933 Perfecting a stand density index for even-aged forests. Journal of Agricultural Research 46 7 627 638
29. K. G. Renard, G. R. Foster, G. A. Weesies, D. K. Mc Cool, D. C. Yoder, 1997 Predicting soil erosion by water: A guide to conservation planning with the Revised Universal Soil Loss Equation (RUSLE). USDA Agricultural Research Service Handbook 703
30. J. Samar, Prasanta. K. Bhuyan, Keith. Kalita, A. Janssen, 2002 Soil loss predictions with three erosion simulation models Environmental Modeling 17 2 137 146
31. B. L. Schmidt, R. R. Allmaras, J. V. Mannering, 1982 Preface in: Determinants of soil loss tolerance. ASA Special Publication,No. 45, Am. Soc. Agr., Madison, Wiscon
32. S. A. Schumm, M. D. Harvey, 1982 Natural Erosion in USA. Schmidt, B.L., ed. Determinants of Soil Loss Tolerance. Madison, WI: American Society of Agronomy, Special Publication 45 23 29

33. P. Soares, M. Tomé, J. P. Skovsgaard, J. K. Vanclay, 1995 Evaluating a growth model for forest management using continuous forest inventory data Forest Ecology and Management 71 251 265
34. G. Stähl, 1994 Optimal stand level inventory intensities under deterministic and stochastic stumpage value assumptions. Scandinavian Journal of Forest Research 9 405 412
35. G. Sun, S. D. Mc Nulty, 1998 Modeling soil erosion and transport on forest landscape Conference 29, NV. Steamboat Springs, Co: International Erosion Control Association 187 198
36. F. J. Swanson, J. L. Clayton, W. F. Megahan, G. Bush, 1989 Erosional processes and long-term site productivity. Perry, D. A., Meurisse, R.; Thomas, B., Miller, R., Boyle, J., Means, J., Perry, C. R., Powers, R. F., eds. Maintaining The Long-Term Productivity of Pacific Northwest Forest Ecosystems. Portland, OR: Timber Press 67 81
37. E. Thus, 1997 Naturlig avgang av traer (Natural mortality of trees). Raport Fra Skogfrosk 6
38. J. K. Vanclay, 1991 Mortality functions for North Queensland rain forests Journal of Tropical Forest Science 4 1 15 36
39. J. K. Vanclay, 1994 Modelling forest growth and yield, applications to mixed tropical forests, CAB International, Wallingford, UK
40. G. F. Weetman, 1998 A forest management perspective on sustained site productivity Forestry Chronical 74 75 77
41. W. H. Wischmeier, D. D. Smith, 1978 Predicting rainfall erosion losses-a guide to conservation planning USDA agricultural research service handbook, no. 537, Washington, D.C.

CHAPTER 12

Optimization of Soil Erosion and Flood Control Systems in the Process of Land Consolidation

Miroslav Dumbrovsky[1] and Svatopluk Korsuň[1]

[1] Brno University of Technology, Faculty of Civil Engineering, Department of Landscape Water Management, Czech Republic

1. INTRODUCTION

Extreme hydrological phenomena of recent years have highlighted a well-known fact that it is necessary to pay greater attention to the problems of flood-prevention and soil erosion control on a large part of the Czech Republic. Case study areas are the most endangered territories. The case study area was selected as a case study mainly for its natural conditions and high risk of soil degradation and occurrence of flash floods. Relief, geomorphology, the present state of the complex system of soil properties, the types of agricultural farming practices and land use, are all contributing to accelerated soil erosion and runoff with all its negative impacts on the built-up areas.

The main soil degradation problems in the case study area are soil erosion caused by water, soil compaction and decline in organic matter. Soil erosion is fostered by i) soil degrading (intensive) farming practices such as up and down hill conventional tillage and other conventional agricultural operations on arable land, ii) frequent extreme hydrological events, and iii) a decreasing ability of soils for water retention (decline in organic matter and land conversion). Soil compaction is a problem due to intensive conventional farming on arable land (using heavy machinery). The decline in organic matter results from the constant soil erosion process. Main causes of decline in organic matter are conventional farming practices without using manure and other organic matter. Decline in organic matter causes a decrease of natural crop productivity of soil and decreases yield.

Great runoffs occur on these areas and transform into flood waves in watercourses. Forest grounds are also affected, especially in case of unsuitable transport, wood cut and growth make-up. Solving of the problems of territory protection from unfavourable and damaging effects of overland water flow must therefore begin in catchments areas and particularly during any interference with landscape. Appropriate conservation measures are required to prevent and reduce runoff and soil degradation resulting from intensive agriculture. The adoption of the most appropriate practices and optimisation the farming conservation system it is necessary to carry out analyses and evaluations of the erosion rate and the basic characteristics of runoff in given sub-catchments. This system of evaluation provides information about erosion and runoff risks plots and serves for decision making regarding soil conservation and flood prevention measures. The success of the system of soil conservation depends on suitable technical assistance and support from responsible state organisations (Ministry of Agriculture and Ministry of Environment), sufficient sources of information as well as the ability and willingness of land users to adopt soil conservation measures. The main motivation for farmers to apply soil conservation measures is the economic motivation through financial subsidies along with penalties for farmers if they fail to comply with the rules of the funding program.

Nevertheless, when introducing a soil erosion control and flood prevention measures in a certain watershed, best management practices are mostly to be able decrease of erosion rate but unable to restrict a surface runoff substantially. For that reason it is necessary to apply a whole system of soil conservation measures. In places with long slopes technical and biotechnical soil erosion control practices (primarily of linear character) are necessary. These technical measures are broad base terraces and channels in case study area. These biotechnical measures together with the implementation of grassed courses of concentrated surface runoff (grassed waterways) create an appropriate network of new hydrolines in the watershed. Biotechnical line elements of soil erosion control serve as permanent barriers or obstacles for water runoff and are designed in order to determine, by their location, the ways of land management. Some technical and biotechnical measures could be suitable regarding their technical feasibility, economic efficiency and environmental effectiveness.

The spatially and functionally limited soil-conservation system in a given teritorry offers spaces and lines in which it would be possible to locate territorial systems of ecological stability under certain conditions. Soil conservation and flood prevention practices, connected with territorial systems of ecological stability can be characterized as desirable anthropogenic landscape-forming elements. These would form the appearance of the landscape and significantly enhance natural processes in the region. They create suitable biological conditions in spite of the fact that they mostly do not meet qualitative and dimensional characteristics of biocentres and biocorridors.

Highly fragmented land ownership is prevalent in the area Biotechnical and technical soil conservation measures cannot be applied without respecting property rights. Integral parts of any project of soil erosion control (its basic network) are usually line elements for soil erosion control (broad base terraces and channels etc.), which run across individual owners' fields. Therefore it is necessary to identify every owner and discuss with the project and relevant proposals. The greatest interventions with agricultural landscape are land consolidation which, apart from other less important objectives, are designed to completely eliminate or at least partly limit unfavourable effects of runoff (especially soil erosion) and thus to become one of the most important elements of territory organisation and protection. Therefore it was found suitable to design the system of the soil and water conservation in the process of land consolidation in the Czech Republic.

Optimal spatial and functional delimitation of soil erosion control practices in the landscape is one of the basic steps in the plan of comprehenshive land consolidation, in addition to the implementation of a new network of field roads and landscape features enhancing ecological stability. Soil erosion control and flood prevention practices are included in the system of public facilities within the framework of the land consolidation process (where property relations are consistently solved).

The definition of land consolidation is from Act No. 139/2002 Coll., on Reparcelling and Land Authorities and amending Act No. 229/1991 Coll., on the Arrangement of Ownership titles to Land and other Agricultural Assets, as amended. The land consolidation processes in case study area have started in 2005.

This procedure gives solutions to the whole area, both from the aspect of a new land and ownership arrangement (Figure 2) and from the aspect of soil conservation and flood prevention and improvement of environment (Figure 1).

Figure 1. Soil conservation and Flood prevention system.

Figure 2. Parcel of owners before and after land consolidation.

Recently, the process of complex land consolidation in the Czech Republic has provided a unique opportunity for improving the quality of the environment and sustainability of crop production through better soil and water conservation. The current process of the land consolidation consists of the rearrangement of plots within a given territory, aimed at establishing the integrated land-use economic units, consistent with the needs of individual land owners and land users.

Integrated territory protection can be reached by controlling runoff by means of design of terraces as a soil erosion control measures. A number of mathematical models, mostly simulation ones, to solve water-management problems have been compiled, some of which include the option of exact mathematical optimization. A certain summary of these models, including their characteristics and application possibilities, were elaborated by Kos (1992). An interesting combination of the application of a simulation and optimization model technique in the elaboration of design of a particular water-management system was described by Major, Lenton et al. (1979), a three-model approach to solve water-management systems was used by Onta, Gupta and Harboe (1991). Benedini (1988) dealt more generally with the design and possible applications of these models. Most likely, an optimization model has not been designed, which would enable to attach territory protection and the measures to eliminate the amount and accumulation of runoff in catchments areas to solving water-management problems.

The created procedure is a universal tool which can be applied for any territory. It enables to find the most suitable combination of all possible alternatives of various erosion controls and flood protection measures under given conditions of each particular site. Such sites do not always have to be ground used for farming. They may also include in forest or urban areas or site arrays in various territories.

2. METHOD

The optimization process of designing the system of *integrated territory protection* (the *IOU* system) begins with the processing of the system of organisational, agrotechnical, biotechnical and technical measures at individual sites of the case study territory. It is necessary to derive hydrograms of direct runoff from extreme rainfall events for each of these variants. Then it is necessary to elaborate the variants of terraces and other conservation measures on all sections of watercourses and variant of designs of retention protection reservoirs. Not only rivers, streams and brooks are included into the watercourse category within this procedure but also sometimes passed watercourses such as terraces, grass infiltration belts or the lines of stabilisation of concentrated runoff waterways in valley lines.

A selection of the most suitable combination of all prepared variants is listed. With respect to the fact that it is necessary to find optimal dimensions for some of the system elements, there is usually a great number, in case of a continual solving even an infinite number, of possible combinations. It is therefore necessary to use an optimalized mathematical model to find the most suitable combination. This model was created on the basis of a mixed discrete programming (Korsuň et al., 2002, Dumbrovský et al., 2006). Its basic building stones are three generally formulated partial models: *A.* partial model of protective measures at individual sites of the case study area. *B.* partial model of a watercourse. *C.* partial model of a reservoir.

It is possible to shape an *optimization model of integrated territory protection* (*OMIOU*) from these partial models for any particular territory. The partial models are repeatedly inserted into the *OMIOU* as needed so as to exactly copy the modelled system structure. It is necessary to determine in advance one criterion or more simultaneously operating optimization criteria for each optimization function. A whole range of criteria can be determined for a given purpose. These can be taken from the sphere of economy but also from those of ecology, water-management, social etc. However it is necessary to define the most suitable criteria as far as quality is concerned but also to have a chance to quantify the values of each defined criterion. On top of that, it is necessary, in case of several simultaneously operating optimization criteria, to assign each criterion its adequate weight with which it will enter the solving process and which will support its effect on the result, so called a compromise solution in competition with the other criteria.

In creating the procedure of the *IOU* system proposal optimization in connection with the process of territory organisation a requirement of a maximal protection of inhabited and other areas with the exertion of minimal means was formulated for the solving process on the level of land consolidation as one of the suitable optimization criteria. It is a criterion consisting of three simultaneously operating partial economic, but at the same time water-management and socially aimed at their impacts. Criteria include:

- minimization of the average annual damage (material damage: it is estimated that input requirements and conditions will not allow solutions which could lead to losses of human lives) originated by overland runoffs from rainfall events and then by their concentration in watercourses.
- minimization of the average annual economic losses in farming production related to the realisation of proposed protective measures on arable land.
- Minimization of the average annual expenses (the sum of expenses for running and maintenance plus the amortization of the capital goods) of the proposed conservation measures.

Seeing that in most cases they are average annual values, quantified for example in thousands of CZK per year, these criteria can be assigned the same weights 1:1:1 in reflection.

The optimization mathematical model is a system of equations, which model a given system behaviour, the variables in the equation describe a system structure and the dimensions of its individual elements. Non-equations found in each model are transformed into equations by means of additional variables in the course of the model solving process, therefore the term *equation* is used only. The above mentioned partial models were created in the modelling and calculation system *GAMS* (*General Algebraic Modelling System*) in its general form (Charamza, 1993) so it can be used to model any integrated territory protection system. The nature of the solved problems implies that the defining process of all the variables used in the model as positive variables. They can be either continuous ones which are marked x herein after or binary ones (they can take on only 0 or 1 values) marked with the symbol xB. Other symbols are used to mark variables and coefficients. Activities proceeding in time must be modelled in the whole system according uniform timekeeping.

The partial model A is aimed at terraces and other biotechnical, agrotechnical, and organisation conservation measures in the catchments area of a certain watercourse. These measures are usually designed within land consolidation to decrease overland flow of rainfall events and thus to limit the effects of soil erosion and damage in inhabited territories. The various proposals of protective measures must be elaborated in each individual case before an optimization model is designed (pre-optimization) as pragmatically created systems of various, mutually complementary interventions with the individual catchments area elements. Such a partial catchments area element could be, for example, valley and slope area above one bank of a certain watercourse section in the range from the bank line to the interstream divide line.

The part of runoff from the design rainfall events which will not be caught by the system of catchments area protective measures (*residual runoff*) will concentrate in a particular watercourse and will create a design Q runoff or flood wave. The time T of passage of the design flood wave through a watercourse will be divided into r of equally long *time intervals* (*TI*); time t of the durance of one *TI* will thus be given by the relation $t = T / r$. For the individual *TIs*, partial volumes $w\,1$ of the design flood wave are then quantified, $i = 1, 2, \ldots, r$.

In case of the application of the above mentioned optimization criterion, the following indicators must be quantified for each pre-optimization processed variant of the protective measure set on a partial catchments area element:

- its estimated effect U expressed financially as an average annual level of damage on land, growth, buildings, roads etc. which will occur after the variant has been realised (*residual damage*),
- estimated average annual economical loss E in farming production related to the realisation of the proposed measures on arable land.

- realisation costs of a particular variant and its average annual own costs *N*,
- the amount of residual runoff *Oi* into a watercourse in the individual *TI*s.

These data represent input information for the partial model *A*. In the course of the optimization process, only one – optimal – variant with the most suitable indicators will be chosen from thus prepared variants of systems of protective measures for each partial catchments area element. Residual runoffs concentrating in a watercourse runway from the watercourse adjacent partial areas protected by optimal systems of measures will cause a gradual accretion of a flood wave passing through the watercourse. The protection from damage which could be caused by this flood wave will be provided by the protective measures on the watercourse and retention protective reservoirs as mentioned later (the partial models *B* and *C*).

Binary variables can be used for modelling of individual variants of protective measure systems in each of the partial catchments area elements in a discrete way. The total number of catchments area elements will be *m*. If, for example, *n* variants of protective measure systems of a dth catchments area element are modelled by relations to binary variables $x_{B1dp} \in \{0, 1\}$, $d = 1, 2, \ldots, m$, $p = 1, 2, \ldots, n$, the effects of these measure systems for this catchments area element can be write into the model using the following equations:

the equation of protective effects (residual damage)

$$x_{Ud} = \sum_{p} U_{dp} \cdot x_{B1dp} \tag{1}$$

the equation of economic damage

$$x_{Ed} = \sum_{p} E_{dp} \cdot x_{B1dp} \tag{2}$$

the equation of own costs

$$x_{Nd} = \sum_{p} N_{dp} \cdot x_{B1dp} \tag{3}$$

the equation of residual runoff, i.e. contribution of a dth catchments area element to the flood wave volume on a particular watercourse section in i^{th} *TI*

$$x_{Oid} = \sum_{p} O_{idp} \cdot x_{B1dp} \tag{4}$$

for $i = 1, 2, \ldots, r,$

$d = 1, 2, \ldots, m,$

$p = 1, 2, \ldots, n,$

where $x\ Ud$ is the total residual damage in a dth catchments area element,

$x\ E,Ud$ is the total economic loss in a dth catchments area element,

$x\ Oid$ is the total residual runoff from a dth catchments area element in i^{th} *TI.*

Because only one of the protective measure system variants can enter the solving process, the following condition must be valid for the sum of all the binary variables of a dth catchments area element:

$$\sum_{p} x_{B1dp} = 1 \tag{5}$$

The partial model B captures the passage of the design flood wave through the watercourse sections. The sections are either left in their present state, the optimization of a river bed or a contour furrow systems design (including the building of protective dams), or the reconstruction an earlier carried out adjustment or protective dams may be required. A watercourse section can also be a water or dry protective reservoir which will be modelled in a way mentioned in the partial model C description.

Flood damage that can occur is quantified for each watercourse section during its modelling. Further, runoffs from the section are calculated in the individual *TIs* of a flood wave passage. With respect to the overland flow from the initial section profile to the last one, it is necessary to determine a time shift which will affect collisions of flood waves on the main watercourse and at the mouths of its tributaries. The mean value of the runoff volume which can be found in a section (in a river bed or also in an inundation territory) in the course of ith *TI* is at the position of the basic section variable. The values of the other variables are related to this variable: the variables of the water flowing through the section, time of concentration, the level of flood damage in the section, and the level of runoff from the section. The courses of these non-linear functions are derived from the watercourse pre-optimization variant designs. They are replaced with linear function part by part in the optimization model. The formulation of particular equations is mentioned in Chapter 5.1 of Patera, Korsuň et al. (2002).

The partial model C is outlined for a designed multipurpose water reservoir with unknown capacities of spaces protective controllable x_{OO}, protective non-controllable x_{ON} and total x_V. The necessary volumes of the spaces of dead storage $S \geq 0$ and active storage capacity $Z \geq 0$ are constant – these values result from other than protective requirements. The objective of analysis is to find its dimension which, respecting the requirement to create the spaces *S* and *Z*, with its protective spaces will ensure the reduction of culminated runoff from the reservoir to its optimal level during the passage of the design flood wave. In cases when the designed water reservoir has only a protective function, the value of the *Z* variable is zero; the values of both variables are zero $S = Z = 0$ for a dry protective reservoir.

The unknown volume of the total reservoir space is a variable, whose value which is limited from above by the maximal value V_{max} corresponding with the biggest realisable variant of the reservoir design during the pre-optimization solutions. From below it is limited by the minimal variant, still acceptable for practice, with the total volume V_{min}.

We cannot forget a situation when building a reservoir will not be acceptable due to the used optimization criteria. It is therefore necessary to introduce a binary variable $x_{B2} \in \{0, 1\}$ into the set of variable values. If this variable has a zero value, the reservoir will not enter the solving process, if $x_{B2} = 1$, the entry of the reservoir into solving is cleared. Then the volume of the total reservoir space (without evaporation and percolation) must correspond with the following conditions

$$x_V = (S + Z) \cdot x_{B2} + x_{OO} + x_{ON} \tag{6}$$

$$V_{min} \cdot x_{B2} \leq x_V \leq V_m \tag{7}$$

The equations modelling the passage of the design flood wave through a dam profile, the calculations of the volumes of individual reservoir spaces and of necessary financial means are described in Chapter 5.1 of Patera, Korsuň et al. (2002). The partial model C can be also used for already an existing reservoir with a constant volume of the total space.

The model compilation from the fore mentioned partial elements in the presented form requires the introduction of a set of concrete coefficients and variables into the model for the model equation system to copy completely a particular system of *IOU*. These coefficients and variables should be derived from the pre-optimization processed background materials. In the case of non-standard requirements of an *IOU* system structure, it is necessary to introduce other equations to the model. Such new equations would capture these requirements. The model solving process in carried out on a computer by means of some of the *GAMS* system tools.

3. MATERIAL

To verify the function and potential of the already described optimization procedure, a system of integrated territory protection was chosen that was proposed within the framework of land consolidation on the case study area between the town of Hustopeče and the village of Starovice in the Czech Republic (see Figure 1). The declining ground in this region is mostly used as arable land. Overland flow is concentrated into its main waterway, which enters the residential parts of Hustopeče. Considerable, and frequently repeating damage, is caused by soil erosion on farm crops, sediment transport from arable land and especially by flooding parts of the town.

The proposed system of integrated protection of this farming territory and town is based on a system of technical-biotechnical, organisational and agrotechnical soil erosion control measures on arable land and of two conservation measures: 1. transfer of concentrated runoff from the drainage furrow or channel *K1* in the main valley line over the terrain into the adjacent valley line and creating a channel *K2* entering watercourse, 2. building a dry protective reservoir (polder) *P1* to catch parts of runoffs from the main valley line and another polder on the channel *K2* in the adjacent valley line above the village of Starovice.

The *IOU* system design for the given territory is based on the situation which would occur during a rainstorm with hundred-year periodicity (design rainfall). The protective measures with pre-optimization were designed in ten different variants, volume and cost (own costs) functions were derived for both the polders. It is estimated that *P1* polder filling, which is a side basin for the channel *K1*, will proceed through the channel side overfall. For the individual soil erosion control measure alternatives volumes and accumulation of overland runoffs, derived from the design rainfall, in the form of runoff hydrograms from two catchments areas: from the polder *P1* catchments and from the polder *P2* catchments. The passage of runoff waves through dam profiles of both polders takes from 510 minutes in the alternatives 1 and 2 to 195 minutes in the alternatives 9 and 10. It requires limiting the culmination water passages in the river beds below the two polders: below the *P1* this passage (runoff from the *P1*), which will enter the city sewerage system in Hustopeče, should not exceed 0.125 $m^3.s^{-1}$, below the *P2* the passage limit should be, with regard to the protection of Starovice, chosen at 1.5 $m^3.s^{-1}$ at most and in variants of 1.0 and 0.5 $m^3.s^{-1}$ to determine the effect of this passage size on the *IOU* system optimal solution.

The optimization model consists of 3,506 equations with the total of 1,673 structural variables, 539 of which are binary variables. The model objective function (optimization criterion) minimises the sum of average annual values of flood damages, economic losses and biotechnical measures and polders own expenses in the proportion of 1:1:1. It is ensured that only one protection system alternative can enter the *OMIOU* optimal solution in both the catchments areas, but it can be different for each of the catchments. These alternatives are marked as *A1* and a particular alternative number for the *P1* polder catchments, the *P2* polder was allocated symbol *2* in a similar way. The polders can enter the solution but they also do not have to. The runoff wave from the *P1* polder catchments may be partly or completely transferred into the *P2* polder. Permissible maximum of water depth in the *P1* polder is 5.0 m, it is 4.34 m in the *P2* polder. The *P1* polder low outlet dimensions (the inside diameter of outlet pipeline of a round shape) d = 200 mm, there is a possibility of choice from d = 200, 300, 400, 500 or 600 mm for the *P2* polder.

4. RESULTS AND DISCUSSION

The model function and behaviour were examined first in relation to the project research objectives. Then the possibilities of experimentation on the model of the designed system were tested (Korsuň et al., 2002). The optimization process was carried out with the three above listed values of admissible maximal runoff from the *P2* polder and then in an experimental way with various runoffs from both catchments areas: with real runoffs derived from hundred-year rain storms for the individual variants of conservation measures in both the catchments areas, and with fictive multiples of these runoffs.. Variants with other changes in input conditions (e.g. without the polders entering the solving process) were calculated for the same reason. The results of these solutions are not listed here. Optimal solutions of variants No. 1, 3 and 5 correspond with the real state of the input conditions. These solution results were derived from overland flows from the grounds and from three real values of admissible maximal runoff from the *P2* polder above Starovice. The results of following experiments on the optimization model have led to a number of interesting findings. However, the most important finding is the fact that the experimental locality can be protected as required without any interference into plant production conditions, i.e. without any (on site) economic loss on the produce only by conservation measures themselves: by draining overland and hypodermic runoffs through contour furrows and channels in the *P2* polder. This protective system design is valid only provided the applied optimization criterion is kept. The resulting design can be different in the case of any change to the criterion (e.g. the changes in the weights of the three used partial criteria) or in case of the application of a different criterion.

5. CONCLUSION

The results of the practical application of the optimization procedure in designing terraces and retention reservoirs within integrated territory protection verify its functionality and applicability. In cases when it is not clear in advance which of the potential torrential rainfall could be the most dangerous, the model will provide solutions with all chosen rainfalls types for the result to comply with the territory protection requirements.

The created model can be used to find either one optimal solution or, in case it is necessary to verify the position of the optimal solutions with the changes of some input conditions and requirements, more times in more versions with variables and coefficients modified by these changes. The possibility of multiple application of this model and to obtain a whole set of optimal solutions visualises much better the character and behaviour of the designed system in reactions to modifications of the input conditions and requirements and thus enables to improve significantly the process of making decisions about the design final shape.

A great advantage of the model lies in the general formulation of its components – partial models of conservation measures at individual sites of the experimental locality, watercourse and reservoir. This should enable its problem-free application for optimization design of integrated territory protection under any conditions and at any site

REFERENCES

1. M. Benedini, 1988 Developments and possibilities of optimisation models Agric. Water Manag. 13 329 358
2. M. Dumbrovský, et al. 2003 Optimisation of the system of soil and water conservation for runoff minimizing in certain watershed in the process of land consolidation (in Czech) Final research report, NAZV-QC1292 VÚMOP, Praha
3. P. Charamza, et al. 1993 Modelling system GAMS (in Czech). MFF UK, Praha.
4. S. Korsuň, et al. 2002 Creating and verification of the model for optimisation of soil and water conservation (in Czech) Final research report, A01-NAZV-QC1292 FAST VUT, Brno
5. Z. Kos, 1992 Water management systems and their mathematical models during climate changes (in Czech) Vod. Hosp. 7 211 216
6. D. C. Major, R. L. Lenton, et al. 1979 Applied water resource systems planning. Prentice- Hall Inter., Inc., London.
7. P. R. Onta, A. D. Gupta, R. Harboe, 1991 Multistep planning model for conjunctive use of surface and groundwater resources Jour. Water Res. Plan. Manag. 6 662 678
8. A. Patera, J. Váška, J. Zezulák, V. Eliáš, S. Korsuň, et al. 2002 Floods: prognosis, water streams and landscape (in Czech) ČVUT / ČVVS Praha

CHAPTER 13

Conservation and Management of Forest Resources in India: Ancient and Current Perspectives

M. S. Umesh Babu, Sunil Nautiyal

Centre for Ecological Economics and Natural Resources, Institute for Social and Economic Change, Bangalore, India

ABSTRACT

This paper brings out the issues related to natural resource exists, management and erosion right from the ancient to modern era. The natural resource such as forests and forest products were found as a basic source of survival for living organisms during the initial period. Later, the natural resources started degrading slowly and steadily due to increased human activities. In the beginning, natural resources were the primary assets for the growth of agriculture, industry, urbanization etc. Subsequently, increased anthropogenic activities started degrading the forest cover on a large scale through extending the agricultural land scape and other developmental activities in the country. The importance of traditional knowledge regarding resource conservation, protection of sacred grooves, endangered species etc. started declining in modern India. Innovativeness and policy framework for resource conservation and utilization became unfriendly in addition to gradual decline of community management. The key findings of the study point to major gaps existing such as between line departments and institutions in the promotion and management of forests, human welfare, agriculture, animal husbandry etc. Moreover, people tend to be more inclined towards the utilization of natural resources rather than conservation. From an observational and practical point of view, the natural resource has been declining rapidly in India. Therefore, this study strongly suggests that a due attention be directed towards developing people-friendly and environment friendly conservation strategies.

Keywords: Theory, Practice, Natural Resources, Approaches, Management

1. INTRODUCTION

The management of Natural Resources is a scientific approach for utilization and conservation of land, water bodies, soil systems, plants and animals to enable the people to fulfill the requirement so that along with the present generation the future generations would also be benefited. Therefore, the management of natural resources is combination of scientific and technical understanding of the same. This study mainly focuses on conservation and management of forest resources in India.

Considering that the indigenous communities happen to be the holders and carriers of traditional ecological knowledge and wisdom regarding the use and management of natural resources, their active participation in the various biodiversity programmes is absolutely essential to ensuring their sustainability and cost-effectiveness. In other words, it is important that their prior consultation, participation, and consent procedures (communication with each other) are ensured for implementing such programmes mainly due to the fact that the approach of indigenous communities towards assessing the value of natural resources (both in terms of conservation and utilization) is quite unique as compared to the present day conservation and economists [1] .

2. A BRIEF HISTORY

Conservation Principles as Explained in Atharva Veda: Atharva Veda (12.1.11) hymns believed to have been composed at around 800 BC, somewhere amidst deep forests, read: "O Earth! Pleasant be thy hills, snow-clad mountains and forests; O numerous coloured, firm and protected Earth! On this earth I stand, undefeated, unslain, unhurt."

Utilization and Regeneration Principles: Another hymn from Atharva Veda (12.1.35) reads: “Whatever I dig out from you, O Earth! May that has quick regeneration again; may we not damage the vital habitat and heart.”

About 10,000 years ago, people started adopting an agrarian way of life and over time this new type of livelihood system spread everywhere with the population beginning to expand faster than ever before. In the course of time, a sedentary agricultural life led to the evolution of villages, cities and eventually states all of which were highly dependent on fresh water resources[1].

As per the Vedic tradition village all the villages came under three main categories, namely, Mahavan, Tapovan and Shrivan[2]. From third century AD onwards, town-building began in the northern regions of India with ecology and the nature of human and animal habitat changing over period; for instance, the dry lands, where people lived by rearing animal herding, became lands of great scarcity and danger whenever rains failed, with wolves and thieves attacking people. Long before the times referred to by the Sanskrit texts, wild animals were a major source of meat across various places of the Harappan civilization. However, in certain parts of western India, seeds were found mostly in ancient houses. While people hunted a wide variety of wild animals, they reared only a few varieties. It is significant to note that the use of iron tools and fire often referred to in the ancient Sanskrit texts is considered as being mainly responsible for replacing forests with farmlands, and nature with culture[2]. Thus, approaches towards NRM can be explained, in terms of historic, functional and futuristic perspectives (Table 1).

Natural resources such as land, water bodies, biodiversity and genetic resources, biomass resources, forests, livestock and fisheries?are the basic sources of human survival and prosperity. However, it is observed that they have been steadily degrading since 5000 B.C and the unprecedented pace of their erosion is the root cause for the present environmental uncertainties. The over-exploitation of natural resources in the context of an ever growing population has resulted in various problems such as land degradation, vegetation loss, pollution and so on. Such problems need to be addressed seriously using traditional knowledge, a proven treasure for natural resource management, as the main basis. One of the major concerns in this endeavor is to rehabilitate the degraded and vulnerable lands. In this respect, an account of the existing historical and traditional aspects of natural resource management, particularly forest resources is presented here in the Indian context.

Table 1. NRM from different perspectives?historical, functional and futuristic?and their relation to traditional conservation methods.

Attributes of Natural Resource Management (NRM)	Historical Perspective	Functional Perspective	Futuristic Perspective
Traditional Ecological Knowledge (TEK) regarding NRM	Was given a high importance.	Apart from TEK, certain new interventions have been made for conservation of natural resources. However, TEK has started losing its prime role in the management of natural resources across various regions.	TEK is in jeopardy, thus needs to be given a high priority for proper documentation.
Conservation and Management of Natural Resources	The resources were used for fulfilling the basic needs on a sustainable basis.	Resource exploitation is high and as a result, resources are depleting very fast.	Need to rethink from the perspective of growing population on the one hand, and the declining resources on the other.
Dependence on Natural Resources	Sustainable utilization for only meeting the basic requirements.	Increased for various needs without thinking needs of the future generation.	Depends on the availability of various ecological resources
Supply and Demand for Natural Resources	It was on a sustainable basis.	Not in an equilibrium.	Lesser chances of a sustainable ecological resource flow, if the current trend were to continue.
Innovativeness in Resource conservation	Through the process of trial and error with TEK as the main basis.	Various innovations have been made, but a majority of them are not being practiced.	There is a need for a proper evaluation of all the innovations made so far with appropriate strategies have to be devised for resource utilization and conservation for future generations.
Role of institutions in resource management	Community participation	Institutions have been developed without a proper thought given to most of the cases.	Has to play a major role within an integrated framework.
Policy inventions in resource utilization	Various traditional ecological methods introduced for resource utilization.	The policy has become the main tool for resource utilization.	Existing policies need to be revisited for proper implementation.

3. ANCIENT VIEWPOINT

According to study [2] , Indo-Aryans (circa 3000 B.C) were occupying the country as an Indian frontier during the pre-Indian age. The Aryans were a nomadic agricultural tribe quite different from their Iranian brethren peculiar to the Indo-Aryans of the Rig-Veda. The Indian civilization evolved around the five river basins namely the Indus Valley, Sind and Western Rajputana, Gangetic provinces & delta, Narmada and Peninsula. The Aryans were located around the Hindu Kush and Sutlej basins/regions. Although rich in soils, the flourishing communities were still primitive with respect to their agricultural practices their period approximately lasted from 2000-1500 B.C.

Certain scientific inquiries required for a proper understating of the Vedas and the Vedic rituals [2] , can be located in the elaboration of Vedangas, namely 1) Kalpa which includes geometry so far as it is applied to the construction of sacrificial altars; 2) Siksha or phonetics; 3) Chandas or meter; 4) Vyakarana or grammar; 5) Nirukta or study of words; 6) Jyotisha or astrology. These Vedangas find a brief treatment in the Brahmanas or Upanishads acquiring a necessary scientific cast relevant even to the present age.

Moreover, Pali literature is undoubtedly the most important source with regard to the topography ancient Indian topography. Similarly the great literature of the early Buddhist period is the main source of the historical and geographical information about ancient India, followed by Jaina and Brahmanical sources to a certain extent. For later periods, we have abundant geographical and archaeological sources and literary sources, as well as the accounts of the classical geographers and the travelogues of the Chinese pilgrims, while the geographic information contained in Pali and Sanskrit, Buddhist literature is considerably important [3] .

Grammarian Panini refers to this period while bringing the history of South India into touch with that of the North. If the great epic Ramayana, could be trusted, the great forest extended up to the Pampa Saras on the north bank of the Tungabhadra near modern Hampi, though the Saras (or Tank) must have been forgotten under the name as the author of the Tamil Ramayana refers to it as Pampanadi. The advent of Agastia introduced the reclamation of the jungle into arable land, and was reputed author of the first Tamil grammar [2] .

According to study [3] India was bounded on the north by the stupendous Himalayan mountain ranges and on the other three sides by the mighty beas and the ocean, constituting a distinct geographical unit. The sheer vastness of the country with its infinite variety of fauna and flora, races and languages, religions and cultures justly lent itself to be called a great subcontinent. The word "India" is derived from the river Sindhu. The Chinese also knew the ancient name of India as Shin-tnh or Sindbu. It should, however, be noted that the Vedic "Sapta Sindhavat" and the Persian "Hi(n)du" corresponded only to a particular geographical area lying to the north-west of India.

3.1. Forest Resources as Explained in Epic Ramayana (300 BC-200 AD)

In Valmiki Ramayana, two Sanskrit words viz., "Vana" and "Aranya" are used while referring to forests. The term "Vana" is used to denote sub-forestation, a kind of cultivated forest with a clustering of desirable plants, conserved for a purpose. As the legend goes, Rama, Sita and Lakshmana had settled down in Chitrakoota forest because of its rich biodiversity. Strangely, Lanka supported two kinds of forests—the natural woodland of its own and the cultivated forest of the Royal Majesty. The Indian forests are enigmatically known and described as the land of "glowing" also rich in medicinal plants.

The narrative style of all these forest types is supposed to evoke four predominant sentiments or "rasas" (i.e. moods) such as "Santa" (Tranquility), "Madhura" (Sweetness), "Roudra" (Fury) and "Bibhatsa" (repulsion)[3]. The example of importance of tree is mentioned in Box 1.

3.2. Forest Destruction in Ancient India

According to study [4] , the destruction of forests for valuable wood on the plains causes many of the hills lose their forest cover resulting in the rainfall rushing quickly down their bare slopes which in turn, prevents the growth of new trees. Further the cultivation of potato crop, introduced from England, leads then to a further destruction of trees. Those living on the hill clear their potato fields now and then by burning a ring round the sterns of the great trees, while converting the mountain sides into terraces. As a result within a few years, the barks and leaves would drop off the branches, with the forest standing bleached and ruined. Some of the trees would rot on the ground, like giants fallen in a confused fight, while others still remaining upright with white trunks and skeleton-like arms. In the ranks, green potato crop mark the spot where forest has been slain and buried. Several of the ruder hill tribes follow an even more wasteful mode of tillage. Bereft of either ploughs or cattle, they burn down the jungle, exhausting the soil by a quick succession of crops, raised by a within a year or two, the whole settlement moves on to a fresh patch of jungle, while repeating the same process.

There is a little-known legend related to Vijayadashami festival associated with Mahabharata. The Pandavas underwent a period of exile, i.e., 12 years of dwelling in the forest followed by a year of incognito exile. It is said that the particular Shami tree chosen by the Pandavas stood inside a cremation ground. The Pandavas wrapped their weapons in a white cloth and concealed the same on the same tree, making the weapons look like a dead body. Virata Parva, Chapter 5 of the Mahabharata epic, mentions that on the southern bank of River Yamuna in Viratanagara, Arjuna hide his bow "Gandiva" on a Shami tree. Meanwhile, the Kauravas had invaded that area, suspecting that the Pandavas resided there. Upon completing their exile period, the Pandavas headed straight to the battle field, and won the contest comprehensively. Even to this day, across various parts of Indian, people exchange Shami leaves on Vijayadashami, while wishing each other success in their own ventures and efforts.

There is a little-known legend related to Vijayadashami festival associated with Mahabharata. The Pandavas underwent a period of exile, *i.e.*, 12 years of dwelling in the forest followed by a year of incognito exile. It is said that the particular Shami tree chosen by the Pandavas stood inside a cremation ground. The Pandavas wrapped their weapons in a white cloth and concealed the same on the same tree, making the weapons look like a dead body. Virata Parva, Chapter 5 of the Mahabharata epic, mentions that on the southern bank of River Yamuna in Viratanagara, Arjuna hide his bow "Gandiva" on a Shami tree. Meanwhile, the Kauravas had invaded that area, suspecting that the Pandavas resided there. Upon completing their exile period, the Pandavas headed straight to the battle field, and won the contest comprehensively. Even to this day, across various parts of Indian, people exchange Shami leaves on Vijayadashami, while wishing each other success in their own ventures and efforts.

Box 1. The legend of Shami tree (Prosopis Cineraria)[4].

Habitat destruction is identified as the main threat to biodiversity conservation. Under diverse natural conditions, over a billion people in rural and urban areas live in harmony under a democratic system in India. However, their pressing needs of food, fibre, shelter, fuel and fodder combined with a compelling need for economic development exert an enormous pressure on the natural resources and the resultant loss and fragmentation of natural habitats affect all animal and plant species. Therefore, it is imperative to prevent not only any further habitat loss in the immediate future, but also to restore a substantial fraction of the wilderness that has been lost in the past. Various plant and animal species are on the decline largely due to habitat fragmentation and overexploitation, e.g. Habitats of Great Indian Bustard in Madhya Pradesh, Gujarat and Rajasthan, and the Lion-tailed Macaque in the Western Ghats [5] .

The predominant causes behind the dwindling of forest wealth are over-exploitation, overgrazing, illegal encroachments, unsustainable practices, forest fires, and environment-unfriendly development projects in the forest areas. The collection of forest products including fuel wood, timber etc, are observed to be much beyond the regenerative capacity of our forests [7] .

The two main direct threats to biodiversity conservation are habitat degradation and loss of biodiversity for example refer Box 2. Pollution and invasive species are also significant threats whose adverse effects are clearly discernible in relation to freshwater ecosystems. The biodiversity loss can be attributed to economic growth and increasing consumption, poverty, capacity constraints, lack of environmental safeguards, lack of comprehensive land-use policies and plans, undervaluation, lack of grassroots level support for conservation, and global climate change.

3.3. Degradation of Forest Resources

An ever expanding population pressure, poverty and a weak institutional framework have often been viewed as the predominant causes underlying deforestation and degradation in the developing and developed countries Ta- ble 2 and Table 3. An ever growing human and livestock population and the dependence on the forest products essential for sustenance and development invariably exert enormous pressures on the forest resources like fuel wood, fodder, timber, paper, etc., which, in turn, trigger an irreversible deforestation process. The over exploitation of forest resources as compared to its incremental and regenerative capacities tends to escalate the forest

(Source: [4]).

The ancient Sanskrit poets speak of the southern tableland as buried under forest with *sdl*, ebony, *sissu*, teak, and other great tree species that are still found in abundance. The Ghats, in particular, are endowed with a magnificent vegetation cover. However, the increased cultivation of crops has driven the jungle to the fields of wheat, and many other kinds of smaller grain or millets, tobacco, cotton, sugar-cane, and pulses, which are spread over the open country. The black soil of Southern India is proverbial for its fertility as also the lowlands between the Ghats and the sea rival including even Lower Bengal with their fruit-bearing palms, rice harvests, and rich succession of crops. The tableland is, however, very susceptible to droughts; and the people have devised a varied system of irrigation, (depending on the region) such as wells, tanks, artificial lakes (formed by damming up the mouths of river valleys), etc. They thus store rain water during a few months of the northern and southern monsoons.

Box 2. A scenario of forest resources in India.

Table 2. Forest cover across the World from 1990-2010.

Region	Total forest cover (in MHa)			Change (in MHa)
	1990	2000	2010	
Africa	749	709	674	−75
Asia	576	570	593	17
Europe	989	998	1005	16
North and Central America	708	705	705	−3
Oceania	199	198	191	−8
South America	946	904	864	−82
World	4168	4085	4033	−135

Source: http://www.earth-policy.org/indicators/C56/forests_2012 (Accessed 19.03.2015).

Table 3. Indicative estimation of species loss due to deforestation across the world.

Ecological zone	Forest cover area (in MHa)		Deforestation rate (in MHa)	Relative species richness (Species number)	Indicative species loss (% loss)
	1980	1990	1981-90	1990	1981-90
			Africa		
Tropical rain forest	91	87	4.7	30700	2.0
Moist deciduous forest	274	251	22.5	13000	2.5
Dry and very dry zone	162	151	10.8	1900	1.0
Upland formations	38	35	2.9	12300	2.5
			Asia and Pacific		
Tropical rain forest	199	177	21.6	40400	4.3
Moist deciduous forest	48	42	6.6	7600	4.3
Dry and very dry zone	46	41	4.7	1600	1.6
Upland formations	53	47	5.8	13500	3.8
			Latin America and Caribbean		
Tropical rain forest	474	454	19.4	57900	1.6
Moist deciduous forest	326	294	31.8	13700	3.0
Dry and very dry zone	52	46	6.2	1600	1.9
Upland formations	138	122	16.2	18400	4.0

Source: http://www.ciesin.columbia.edu/docs/002-471/002-471.html (Accessed 19.03.2015).

depletion and degradation process Table 3. At present the human population exceeded 7 billion which is growing every second. The sheer force of our population where the human species is the dominant factor affecting land, air, water, soil, and species. Figure 1 depicts the relationship between human population and extinction of species between 1800 and 2010 [6] . In fact, an excessive deforestation has impacted not only the local environment, but also the global environment as also the socio-economic and development processes in the developing countries as forests generate a lot of employment opportunities in the primary, secondary, and tertiary sectors besides being a source of subsistence for the poorest of poor in the agricultural economies [7] .

Population-driven pressures on croplands, pasturelands and forest lands coupled with rural poverty have triggered a massive migration to urban and industrial centres for wage employment. Further, the pace of urbanization has always been faster than the pace of population growth in India. More importantly, the process of politicization has always been more pronounced than urbanization, since the 1980s [7] become negative growth towards forest resource conservation.

3.4. Knowledge Dissemination

Local Ecological Knowledge based on prolonged observations and experiences has evolved over time through oral transfer down the generations. As per the understanding from FGD, farmers need to apply scientific knowledge of resource conservation. Based on the discussion, the traditional knowledge has been declining not only from generation to generation but also is not being transferred effectively. The scientist has to focus on capturing the declining knowledge to link with scientific knowledge. Such initiatives can certainly benefit the societies in the coming days. The changes in the developmental policies and acts also do have negative implications for the traditional system besides being unfavorable to medicinal plants. For example, tribal express that the regeneration of medicinal plants is not taking place in their native places and that their diversity is also declining. This shows the lack of coordination between the local communities and the government while framing policies or rules for conservation. It would be better if researchers could involve tribal people as partners in natural resource conservation rather than just as informers in their studies.

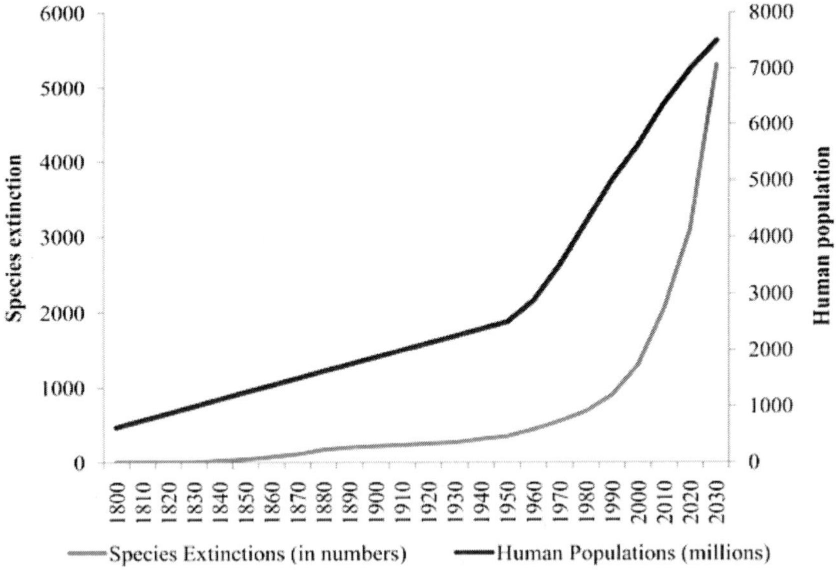

Source: Redrawn from Glikson, A. (2012) [6] . Is another mass extinction event on the way? The Conservation.

Figure 1. Rise in human population and extinction of species between 1800 and 2030.

3.5. Sacred Groves in India

The sacred groves dedicated to local deities or ancestral spirits are protected by local communities through social norms and taboos that incorporate spiritual and ecological values. Preserved over many generations, sacred groves represent native vegetation in a natural or near-natural state rich in biodiversity and harbour many rare species of plants and animals. However, the forces of modernization are depleting these sacred groves, weakening the traditional institutions that protect them. Fortunately, thousands of sacred groves still remain intact mainly because many villages continue to observe traditional practices. Most of the sacred groves in India are associated with almost 40,000 endogamous groups within the Hindu caste system and other major religions such as Buddhism and Islam and traditional tribal groups. However, to date, there has been no comprehensive survey of sacred groves conducted in India, and as such, their exact number and the area covered under them are unknown. Nevertheless studies have found 13,720 sacred groves reported across various regions of the country. The studies inform us that the number of sacred groves in India is likely to be between 100,000 and 150,000[5].

3.6. Sacred Trees as Part of Resource Conservation

According to Hindu mythology, the 14 jewels (fourteen precious objects) which emerged from the depths of the sea (they vary according to the texts) included: horse Uchchaihshravas (symbol of sovereignty), white elephant Airavata (symbol of royalty), throne (symbol of kingship and power), fire (symbol of life), Tulsi plant (repre- senting physician Dhanvantari), chariot, goddess Lakshmi (bestower of abundance) Parijata tree (divine tree), Surabhi (cow), Soma (intoxicating juice derived from sacred plants), Varuna (the sea god), precious stones, tree of life, divine damsels and A lakshmi, goddess of misfortune etc.

Trees being nature's major processors of solar energy besides yielding flowers, fruits, wood and medicine, essential for our existence, are worshipped by Hindus with a deep sense of gratitude and reverence. Many believe that they are conscious like humans and feel pleasure and pain. The Indian sages and seers have eulogized the significance of asvattha or peepal (Ficus religiosa), gular (Ficus glomerata), neem (Azadirachta indica), bel (Aegle marmelos), barged or banyan (Ficus bengalensis), Asoka (Sereca indica), Emile (Phyllanthus emblica), Arjuna (Terminalia Arjuna) and many other tree species which had acquired social and religious sanctity in the ancient times. Due to their ecological value and efficacious properties, trees continue to be used in the religious and social ceremonies by Hindus (for example, see Box 3 and Table 4).

Folio from *Baramasa* (12 months of the year) set. Bundi School, 1675-1700 A.D.

Fairs and festivals are important parts of Indian life. There are many *vratas* (*observations*) and festivals related to natural forces. Teej is one of them. Teej is celebrated at the onset of monsoon in the month of July-August. On this occasion, women, dressed colorfully, carry the image of Goddess Parvati in a procession to the outskirts of a village or town. They spend the whole day in the company of nature playing, singing, dancing and swinging. At some places, *neem*, a (useful medicinal tree) and Banyan are also worshipped.

Box 3. Month of sravana (rainy season-July-August)[6].

Table 4. List of trees conserved and worshipped through festivals and vratas in India[7].

Name of the Tree	Related Festival or Vrata	Time of Celebration and Rituals
Amala	Amala Ekadasi	11th day of Phalguna Sukla; people bathe with Amala twigs soaked in water and also consume it; worship it along with Radha-Krishna.
Amra or Mango	Amra-puspa Bhaksana Vrata	1st day of Chaitra Sukla; people consume mango blossoms and worship Kamadeva.
Asoka	Asoka Pratipada	1st day of Chaitra Shukla; only women worship the Tree; they also observe fast seeking longevity.
Bakula	Bakula Amavasya	Bakula flowers are offered to the manes, seeking their blessings.
Vata or Bargad	Vata Savitri Vrata	Jyestha Purnima or Amavasya day; having fasted for the three previous days, married women worship the bagged tree by circumambulating, while tying the sacred protective thread around the tree (raksa Sutra), and listen to the sacred Savitri-Satyavan story; some women stay awake during the night before completing the vow by feeding brahmins; in the western parts of India, devout women observe this vow for five consecutive years after marriage.
Bilva or Bel	Bilva Tri-ratri Vrata	On Tuesday of Jyestha Purnima when the constellation is Jyestha, people worship the bel tree for three consecutive nights as per Hemadri's injunctious in the Skanda Purana; the vow comprises bath with water mixed with mustard seeds, partaking of sacred sattvic food (havisyanna), adorning the tree with two pieces of red cloth and placing the image of Uma-Mahesvara beneath it; Homa is performed with 1008 Bolivia leaves offered; Brahmins are fed.
Bilva or Bel	Sravana Krisna Ekadasi	Involves the ceremonial offering of water to the bel tree.
Bilva or Bel Bhadra	Shukla Chaturthi	Offering of trifoliate leaves of bel to Lord Ganesa.
Bilva or Bel	Bilva Nimantrana	Asvina shukla sasthi; invoking the tree-goddess and worshipping the Devi.
Bilva or Bel	Bilva Saptami	Asvina shukla saptami; a twig of bel with two fruits is offered to Devi.
Bilva or Bel	Bilva Navami	Asvina shukla navami; bel leaves are offered to Lord Siva.
Karavira or Kaner or Oleander (Neriumindicum)	Karavira Vrata	Jyestha shukla prathama tithi; kaner roots and branches are soaked in water and adorned with a red cloth; offerings of seven cereals (sapta dhanya) and fruits are made followed by fasting; Savitri, Satyabhama, and others performed this vrata when they were in trouble.
Kadali or Kela	Kadali Vrata	Vaisakha, Magha or Kartika shukla chaturdasi; a banana tree is planted and nurtured till it bears fruits; for the welfare of one's family, a person should worship the tree with flowers, fruits, etc and circumambulate it.
Kadali or Kela	Yaksa-samantaka	Kadali Vrata: A golden banana tree is worshipped and offered to a brahmin on any auspicious day.
Kevada or Screw Pine (Panadanusodoratissimus)	Kevada Teej	Bhadra shukla tritiya; As a way of soliciting unbroken married life, women offer Kevada leaves to Lord Siva.
Neem	Sitala Puja	Chaitra navaratras; goddess Sitala who is said to reside in the neem tree is propitiated ritually; Pat Gosain festival in Bengal means neem tree worship; neem leaves are consumed on Vaisakha shukla saptami.

[6]http://bmtrainingprog.files.wordpress.com/2010/08/14-vandana-prapanna.pdf, accessed: 17.12.2011.

[7]http://www.salagram.net/Sacred-trees.html, accessed: 17.12.2011.

4. AN EMPIRICAL ANALYSIS UNDERTAKEN ACROSS THE WESTERN GHATS OF INDIA

The Western Ghats (WG), region a valuable natural resource pool of the country, plays a crucial role in shaping the topography, climate pattern and the environmental system of the nation besides serving as a living laboratory of biodiversity. Several ecosystems within the Western Ghats landscape play a major role in connecting the ridge between the Western Ghats and Eastern Ghats, the two important flora and fauna regions of the country. Thus, the Western Ghats landscape is important not only for the people and biodiversity of the region, but also the social and ecological systems of the low lying areas as ecosystem services that flow from WG provide sustainability to the low landscapes of the entire peninsular India. Human beings derive many useful benefits from the existing biodiversity and it is a well known fact that a large number of people, directly or indirectly continuous to depend on forest resources for their livelihoods. Keeping in view the ever increasing population and an actual decrease in the forest areas, the Indian Government allocates funds under the Special Central Assistance (SCA) and Hilly Areas Development Programme for conserving and improving the present Western Ghats in Karnataka and other states. This programme, initially started during the Fifth Five Year Plan (1974-79), is still continuing. Across various plan periods, the focus of the program has varied. During the sixth plan, the watershed development programme was taken up, keeping in view the significance of ecological restoration and conservation of biodiversity of the region. Later, during the eighth five year plan, an integrated approach towards watershed development was initiated, while during the Ninth Five Year Plan, a greater emphasis was given to awareness creation, maintenance, preservation and restoration of biological diversity. Further, during the Tenth Five Year Plan, the emphasis was on the preservation of biodiversity and rejuvenation of the hill ecology with an equal emphasis given to economic activities and livelihood security of the inhabitants under the environmental conservation programme. The major conservation activities are depicted in Table 5.

The Western Ghats Development Programme is currently being implemented across 171 taluks covered by viz. Maharashtra (63 taluks), Karnataka (40 taluks), Kerala (32 taluks), Tamil Nadu (33 taluks) and Goa (3 taluks) with some of the original talukas sub-divided. In Karnataka WGDP has been implemented across all the 40 taluks coming under 11 districts of the state.

To conserve the natural resources, the Government has formed farmers organization called Executive Committee (E.C.) involving the local elected representatives, self help groups and user groups of the proposed watershed development area. The GP President happens to be President of E.C. with one watershed department official as Secretary of the E.C for facilitating the smooth functioning of the proposed watershed development activities with a view to for enhancing ecosystem services to the local people. This committee holds the sole responsibility of distributing WGDP benefits to the selected beneficiaries such as landless, labors, women groups, SC, ST communities, marginal farmers etc. (Figure 2).

4.1. Green Cover Management

The implementation of forest schemes such as agroforestry, social forestry and other forest activities is the second largest and important scheme under WGDP in Karnataka. The department of watershed in each district concerned executes monitor forestry works. Assistant Conservator of Forests (ACF) is the implementing officer in the Watershed Department and works under the guidance of District Watershed Development Officer (DWDO). Under this programme, the department raises plantations in the designed watershed areas and distributes seedlings to farmers to be planted in their farmyards and homestead lands. The department raises saplings which are suitable for the local environment and plant the same accordingly for the prescribed year. Under this scheme, the department considers both public and private lands and supplies seedlings/plants to beneficiaries free of cost. The benefits accruing from the plantations during the harvest period are shared by farmers equitably. Views of the villagers/beneficiaries are taken into consideration by the department officials prior to planting in the watershed areas and public lands and are mainly done on bunds and banks of streams and rivers for arresting rain water and also for avoiding runoffs. More importantly, it is the forestry related activities that mainly characterize the ecological restoration besides facilitating development under WGDP in Karnataka. The government allots 20 per cent of the WGDP budget for promoting forest development related activities in the Western Ghats. However, it needs to undertake more activities to fulfill the actual needs.

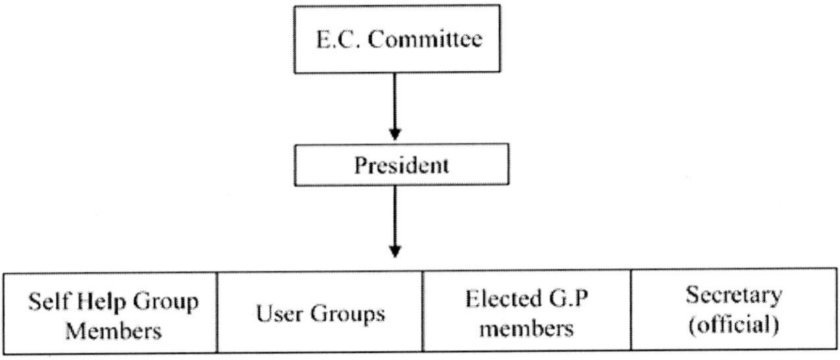

Figure 2. Ground level organizational chart for sustainable development of natural resources.

Table 5. Conservation approaches for Western Ghats development.

	Watershed works	
1	Agriculture & Soil Conservation works	The main objective of this scheme is to undertake soil conservation measures in watershed areas, and the activities covered under this scheme, include land development activities like levelling, bench terracing, land reshaping, amalgamation of paddy fields, land reclamation and contour bunding for increasing fertility and productive potential of agricultural fields; Water harvesting and erosion prevention structures like check dams, vented dams, nala bunding, boulder bunds and gully checks, diversion channels, filter strips for regulating the flow of water and diverting excess water.
2	Forestry	This scheme focuses on conservation, preservation and development of biodiversity in the watershed area; activities under this scheme include the development of forest nurseries, conversion of grazing lands into forested areas; Planting of tree species on either side of roads, Mangroves cultivation, Planting of tree species like bamboo, azardecta indica etc on farm land bunds; Protection of forest areas and planting of tree species for green manure and afforestation.
3	Agro forestry/Social Forestry	This scheme aims at raising plantations in the beneficiaries land holdings coming under the watershed programme with a view to meeting the growing demand for timber, fuel, wood fodder etc., thereby reducing the pressure on the forest area. Planting of nurseries along roadsides, river and canal banks and on the village common lands comes under this scheme.
4	Spice Board	This scheme includes the distribution of I.P sets, sprinkler sets, construction of farm ponds and gravity irrigation in the beneficiaries lands coming under the watershed area.
5	Horticulture	The scheme serves the purpose of converting waste land into horticultural plots and adoption of rain fed crops on dry lands; taking up inter and mixed cropping activities.
6	Foot bridges/Hanging Bridges/vented dam-cum-foot bridges/PRED	The objective of this scheme is to provide rural connectivity to small habitations of Western Ghats. Under this scheme roads and foot bridges are provided for the people residing in the steep hilly terrains, (the Western Ghats region).
	D) Livelihood activities for asset-less persons	
7	Animal husbandry, Supply of Giriraja, birds/calfs/Development of small animals, Supply of Sewing machines/Masonry, Fodder, Carpentry kits/Kitchen garden Animal health camp/ Fodder development	This scheme includes development programmes for landless farmers coming under the watershed programme like artificial insemination for upgrading cattle, fodder cultivation on plots or farm lands, supply of carpentry kits and many other household kits for landless farmers; conducting of animal health camps and supplying of birds for the beneficiaries.
8	Livestock management including Goatry, Piggery.	Income generating activities for landless farmers—development of piggery and Goatry for meat and fibre.
	E) Production system & micro enterprises	
9	Crop diversification, crop demonstration, Nutrition management/Organic fertiliser etc.	Demonstration regarding crop diversification is given to the beneficiaries in their own lands. Demonstration classes are conducted for generating awareness regarding benefits associated with and using organic fertilizers rather than synthetic fertilizers.

4.2. Critical Empirical Field Observations

The analysis highlights the schemes and programmes that are performing well and also the ones which are not. There are some schemes which need very innovative methods of implementation so as to be more effective, while some others need to be re-worked in terms of their focus and resource allocation. There is also an urgent need for a change in the approach and attitude on the part of the forest department towards the local communities as well as the programmes and schemes keeping in view the significance of community participation. Although there have been attempts made from time to time to involve the local communities in many of the programmes like Special Development Plan, Greening of Urban Areas, Eco tourism, Raising of seedlings for public distribution, Development of degraded forest etc, the extent of their active participation leaves much to be desired. It is important to note that Participatory approaches for resource governance have proved to be effective in terms of yielding favorable results such as halting the degradation of environment in different parts of the world. An important issue which needs to be addressed is the burgeoning of the staffs at higher levels and the reverence of it at lower levels.

4.3. Man, Animal and National Parks

Bandipur National Park is rich in biodiversity and wild animal Species. There are two major issues related to the wildlife division; 1) Man-animal conflict; 2) Habitat improvement of these two issues, man-animal conflict is the major one because, a majority of the people (75%) of the region have a negative attitude towards the park.

However the forest department has taken up some major steps towards reducing man-animal conflicts by putting up a solar fencing around the forest boundary and also by digging up deep trenches to stop the movement of animals into farm fields. Japan International Cooperation Agency is providing 30 per cent of the cost incurred for the solar fencing (Rs. 2 lakh per kilometer), across the 135 Sq. km boundary, but the Government is not providing the required annual maintenance cost of about Rs. 30,000/km.

The compensation amount given to farmers constitutes 1/10th of their loss caused by wild animals, even as the forest officers feel that the compensation amount should be increased to 1/3rd of the total cost incurred and that the timely release of such funds is necessary.

Habitat improvement is another essential aspect concerning this protected area. The luxuriant growth of Lantana species has occupied the maximum area (more than 60%) of BNP. It is one of the major weeds that has invaded this forest ecosystem, replacing the habitat of the native vegetation cover. In view of this development, the Government has initiated a scheme called "Eradication of Lantana", but it is not being properly carried out because of an insufficient allocation of funds and focusing on smaller areas.

According to Divisional Conservator of Forest (DCF), approximately Rs. 25,000 - 30,000/Ha is required for Lantana removal in the national park, but the current allocation is very inadequate. The tremendous growth of Lantana is closing in on many corridors which the wild animals use for moving around. However, the challenge is that, within the framework of habitat improvement, the forest department has to provide food, water and shelter to wildlife. Lantana growth is a threat to the native vegetation cover leading to a reduction in grass availability for the wild herbivores. Therefore, the involvement of local people is the necessary in conservation and management of biodiversity of the region. However, the forest department has reported a relatively few number of cases of poaching, smuggling and hunting of wild animals due to the presence of strict rules and regulations.

Further, there should be a separate contingency grant/fund in place in case of any human or wildlife injury/ loss.

4.4. Issues That Need to Focus

There seems to be no proper cooperation and co-ordination between the forest department and other line departments with regard to forest management. The research wing of the forest department, is weak and hence, it should associate with other research organizations for better outputs. The policy implementation should be region- specific with a better recruitment policy at all levels. As there are more personnel at the top and few at the ground level there is no balance observed in the execution of work. The critical suggestion based on observations includes the following and Figure 3 represents the better management process:

1) Quality Planting Material (QPM) should be planted.

2) Accountability and Transparency should be ensured at various levels.

3) Proper maintenance is required.

4) Computerization of records is essential.

5) Global Position System (GPS) should take a keen interest in monitoring the works handed over by the forest department.

6) There should be no irregular release of funds that affects the plantation activities right from germination to seedling transplantation stages.

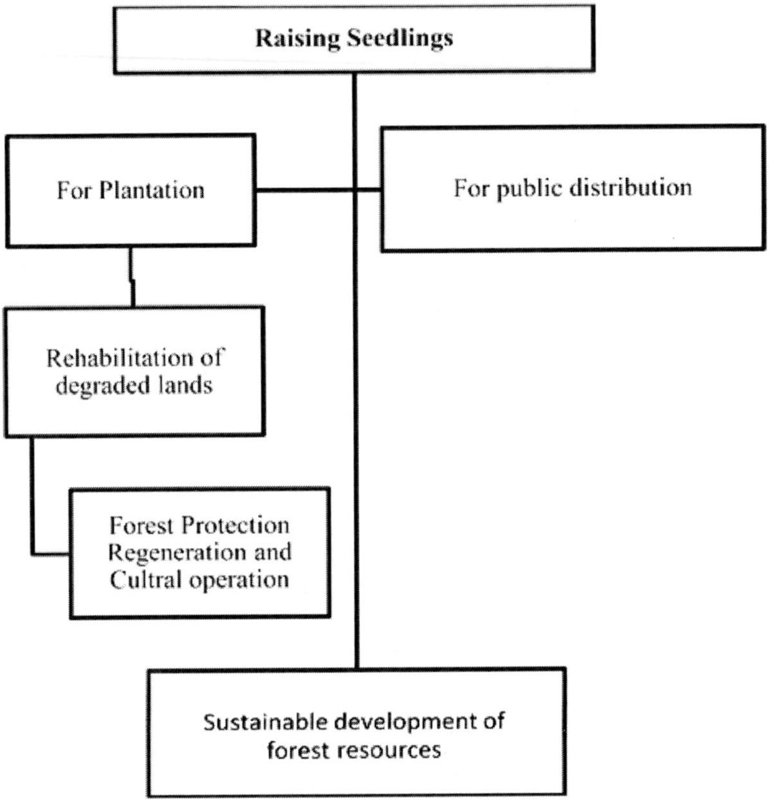

Figure 3. Sustainable approaches for resource conservation.

7) For pre-monsoon period, activities like nursery development might be a problem due to irregular release of funds which can be avoided.

8) There is no timely tender process in place.

9) Inadequate staff at the field level.

10) The department should encourage villagers to form VFCs and the watchers should work under the guidance of VFCs.

11) Basic information should be given for forming VFCs, like how to conduct meetings, who should be involved in meetings and training programmes and benefits of forming VFCs.

12) Trainings should be given to villagers and VFCs regarding conservation and management of forest or plantations.

13) Protection should be provided by a committee consisting of members from the forest department and villages.

5. NEED TO RETHINK POLICIES WITH REGARD TO DEVELOPMENT OF PROTECTED AREAS

Various studies support conservation oriented policies in respect of the protected areas as they constitute cornerstones in conserving biodiversity [8] - [11] . However, some significant research studies point out that a further debate is required concerning the process and relevance of establishing protected areas which tend to demean human existence [12] - [15] . A variety of external driving factors including climate change, have brought about a shift in terms of social and ecological processes across the landscapes protected for biodiversity conservation [11] [16] . Further, in view of the fact that these changing socio-economic and ecological processes are more visible in the protected areas Figure 4, it is necessary that agencies, managers, policy makers, scientists and the public do a rethink in terms of devising innovative policy responses for biodiversity conservation and sustainable socio-economic development. This is especially important in respect of the protected areas where people have been living in harmony with nature for centuries.

Thus, for effective implementation of the conservation policies, scientific and technological research should support the policy making processes. When science and policy differ, the outcome, in the form of communication, is often problematic [16] . The requirement is?that the scientific information is as per the requirements of policy demand and that it is easily accessible to policy makers and decision takers. On the other, policy makers and decision takers should formulate appropriate policy responses in such a way that it is easily understandable for researchers so as to be able to provide available scientific information based on their deliberations [17] . This approach could help evolve effective science-policy responses for protected area management (i.e. in terms of impact, causes, effects, adaptation, conservation and management of biodiversity) at local, regional, national and international levels.

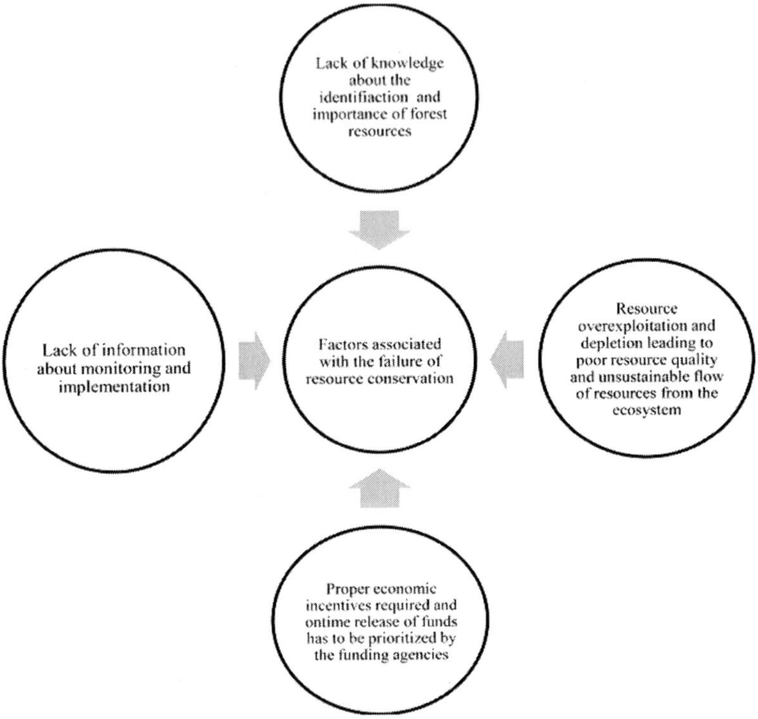

Figure 4. Factors responsible for resource conservation failure.

6. CONTEMPORARY APPROACHES FOR RESOURCE MANAGEMENT

The growing global concern for the conservation of natural resources has resulted in the formulation of long- term perspective plans. Forest resources facilitate the maintenance of ecological balance, biodiversity, enhancement of the quality of environment by checking soil erosion, water retention and conservation, regulation of water cycle, acting as carbon sinks which balance carbon dioxide and oxygen in the atmosphere and the greenhouse gas effects, etc.

Biodiversity and ecosystems are a natural basis ensuring sustainable resource use, including forestry, farms, renewable energy, urban land use, fisheries and other coastal & marine resources [17] . There has been a growing awareness of the business impacts of biodiversity loss and degradation of ecosystems. Conservation measures that combine land development, land conservation, and revenue generation provide a functional protection for conserving resources. Further, conservation projects are created through a process of ecologically based planning and design, whereby planners assess a site's natural resources and environmental context and use this knowledge for conserving portions of the site with a high resource value while designing a development process that minimizes the environmental negative externalities [18] .

Although conservation oriented projects appear to have a fair potential for protecting biodiversity and ecosystem services, there have been few attempts made to evaluate their actual conservation benefits. In practice, the success of such conservation projects is typically measured in terms of the percentage of the total site area set aside as a protected land, but this indicator does not explain whether a given site's conservation values are being protected functionally. Therefore, it is argued that *traditional knowledge is very crucial to contemporary natural resource management and conservation.*

6.1. Socioeconomic Implications

The development of the local economy, while managing Common Pool Resources (CPRs), has become an integral part of the sustainable development policy in many of the developing countries over the past few years [19] . This initiative has emerged largely due to a strong realization of disillusionment with the performance of the centralized management policy in terms of providing sufficient incentives to the resource users for managing the local resources on a sustainable basis. Participatory resource management is often seen as an appropriate solution to reducing rural poverty and resource degradation, while granting community property rights over the local level natural resources would ensure an equitable and sustainable use of environmental resources [19] . More precisely, when the responsibility of allocating natural resources for conservation is delegated to local organizations, communities tend to use these resources for the collective welfare of the community.

Poverty, property rights and the distributive consequences of community-based resource management are becoming increasingly a major subject of discourse with respect to the local level collective action. A long held perception is that common property resource appropriators can create and sustain the local management institutions for ensuring an equitable access to, and income from resource management. Further, CPR related literature claims that since poor people are heavily dependent on natural resources, they derive higher incomes from CPRs [19] . More importantly, there is a less attention given to the fact that a mere institutional change of resource management might not help certain sections of the community, especially the poor and marginalized people, as they are often the most vulnerable sections of the community.

6.2. Innovativeness towards Resource Conservation

An innovative community is not one founded to produce and distribute products invented by a single individual. An innovative community has the means to stimulate, nurture and develop innovation among its people. The lessons learned from open-source development can be transferred to other existing innovative user communities of physical products as well. Farmers in the developing countries possess quite a sophisticated knowledge of agriculture and natural resource management which is recognized as being more eco-friendly and sustainable (Singh, online). Indigenous Knowledge (IK) is dynamic, in terms of changing mechanisms of creativity and innovativeness and contact with other local through international knowledge systems. Indigenous Knowledge Systems (IKS) may appear simple to an outsider, but they represent mechanisms that ensure the minimal livelihoods for the local people mainly because the structure and content of traditional knowledge are intimately linked to the local bio-resources and ecosystems [20] .

Most natural-resource planners recognize genetic diversity and its underlying processes as essential components of an ecosystem and species stability, adaptability and conservation, but there is rarely an explicit provision for the conservation of genetic diversity in management planning and decision-making [21] . The conservation of genetic diversity is essential for many reasons. Given the current worldwide pervasiveness of habitat fragmentation, integrating the conservation of essential ecosystem processes with human population needs is urgent [21] . There is a need for promoting social and economic infrastructure development to realize the potential benefits of the market economy, under natural resource policies that incorporate the environment as well as local people's rights and economies. There is a need for reforming and strengthening of forest governance and the forest-sector by way of ensuring a sustainable forest resource use and forest conservation through promoting modern scientific approaches Figure 5. Because, forest development with a mere industrial emphasis and a narrow-minded resource use, can further weaken the connection between local people and forests.

While every site is unique, challenges involved in managing the same tend to differ from country to country and even region to region or site to site [22] . Social and ecological dynamics and the human dependence on the capacity of ecosystems in terms of generating essential services and the vast importance of ecological feedbacks for societal development, suggest an interconnection between social and ecological systems. A social system includes economy, actors and institutions in mutual interactions [22] . More importantly, power, politics, knowledge, social differentiation, economic development, livelihood quality (often understood as both social and environmental), and ecological resilience have emerged of late particularly salient aspects of human-environmental issues.

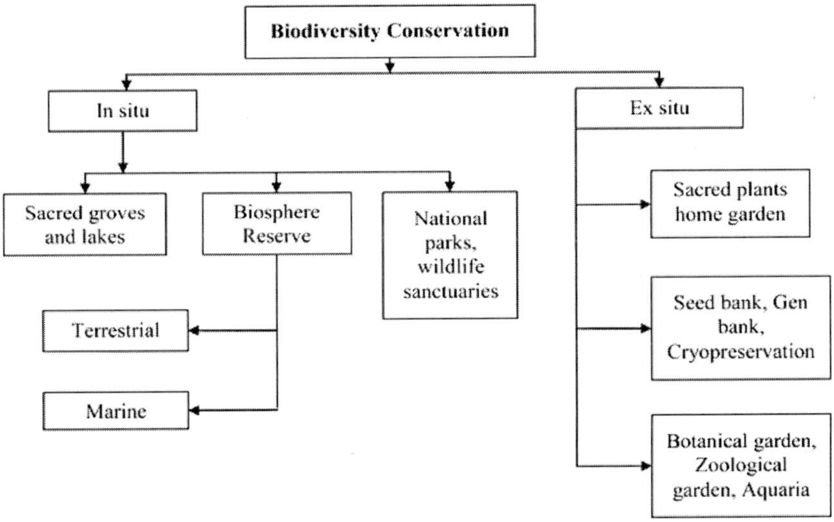

Source: http://www.tutorvista.com/content/biology/biology-iv/biodiversity/india-biodiversity conservation.php (Accessed 19.03.2015).

Figure 5. Approaches for biodiversity conservation (in situ and ex situ).

6.3. The Need: Integrated Natural Resource Management Policy

The need of the hour and a possible solution, particularly in view of a possible threat looming large over the future generation, is to think of natural resources as one integrated piece of asset on this planet. These resources should be governed by uniform use policy, be it land, water or air or even forest or minerals. In fact, natural resource management with conservation and resource enhancement and preservation and pollution abatement as its primary goals, should be the focus of a new set of policies by the government. An integrated natural resource management policy can be a directive principle underlying the policy for states to administer with various departments Figure 6 and the Centre to monitor and evaluate the programmes. An umbrella legislation like the well-drafted EPA (Environmental Protection Act) 1986, can go a long way in making a thematic conservation strategy for resource utilization and regeneration possible. People need to be the focus of any such policy [23] .

Fundamental changes, necessary for preventing a large scale destruction of the nations (and the world's) natural resources, can occur only if people could enjoy greater incentives so as to be able to rethink and reform their behavior towards the environment [23] .

The failure to view natural resources and the environment as one whole entity can have far reaching adverse consequences. Ecology entails an interrelated existence of living beings and natural resources with environmental justice as the touching stone of resource conservation. Considering that resource conservation is a necessary condition for ensuring environmental justice, reorienting the legal regime towards this goal becomes significant in terms of policy and effects. The interlinkage between water (jal), soil (jameen) and forest (jungle), calls for a long-term plan for developing forests, stopping the expansion of deserts, conserving soil fertility and nurturing groundwater through rain water harvesting, especially for meeting the challenges of droughts. The lack of integrated approaches and definitive standards in this sphere allows different statutory bodies to adopt different criteria and policies while exercising overlapping controls with communication gaps so as to escape public accountability through mutual accusations [23] .

Unless conservation and preservation become the main aim of policy-making, environment in general and resources in face a possible threat of extinction. Nature as a harmoniously unified entity, cannot tolerate a proverbial situation of "too many cooks spoiling the broth" [24] .

7. CONCLUSIONS

This paper highlights the significance of natural resources and their utilization right from ancient to modern times in addition to an inalienable relationship between human beings and natural biodiversity. NRM refers to the management of natural resources such as land, water, soil, plants and animals, with a specific focus on how management affects the overall quality of life. Indigenous people happen to be the carriers of ancestral knowledge and wisdom regarding biodiversity. Their active participation in biodiversity conservation programs is essential for resource management to be more effective and cost-efficient. A comprehensive NRM approach is very critical to human survival and prosperity. Forests worldwide cover around 3.9 billion hectares, while the existing forest resources are divided into six categories globally, namely, Moist Tropical, Montane Sub Tropical, Dry Tropical, Montane Temperate Forests, Sub Alpine and Alpine.

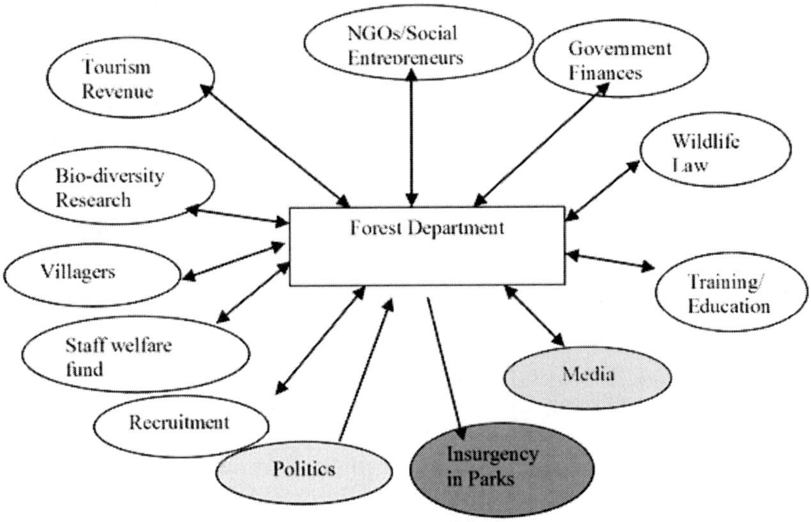

Source: www.iimahd.ernet.in/egov/ifip/susan-sharma.htm (Accessed 20.06.2014).

Figure 6. Approaches for biodiversity conservation (in situ and ex situ).

Under diverse natural conditions, over a billion people in rural and urban areas live in harmony under a democratic system in India. However, the loss and fragmentation of natural habitats invariably affect all animal and plant species. The predominant causes underlying the declining forest cover are over-exploitation, overgrazing, encroachments, unsustainable practices, forest fires, and non eco-friendly development projects in the forested areas. Forest resources help protect ecological systems besides enhancing the quality of environment by way of controlling soil erosion and water retention. But, increasing population and commercial demand for wood have dented the natural forest cover nurtured by our ancestors.

Considering that local knowledge is no longer receiving its due attention in modern India, the loss of natural biodiversity has doubled with conservation becoming much more challenging. Along with science, local technologies and people's knowledge regarding ethno forestry have an important role to play in biodiversity conservation and sustainability. A balance between local and formal institutions can result in empowerment, security and opportunities for local people. Also equity of knowledge provides an opportunity for local people to participate in the management of local resources with global implications.

Conservation management is created through a process of ecologically based planning and design, whereby planners assess the state of natural resources in the environmental context and use their knowledge for conservation to minimize the environmental impacts.

Participatory resource management is often seen as an appropriate approach to reducing rural poverty and resource degradation. A long held perception is that common property resource appropriators can create local management institutions that can ensure an equitable access to the locally based natural resources. India was one of the first nations in the world to have established a professional forest service system under the provisions of the Forest Act 1865. The first forest policy of 1894 had upheld the right of the state to an exclusive control over the forest resources.

There is a need for promoting social and economic infrastructure development for realizing the potential benefits of the market economy through natural resource policies that incorporate the environment as well as local people's rights and local economies. Farmers in the developing countries possess a fairly sophisticated knowledge of agriculture and natural resource management which is recognized as being more eco-friendly and sustainable. Reforming and strengthening of forest governance and the forest industry sector is essential for ensuring a sustainable forest resource use and forest conservation through promoting the modernization of forest conservation. Because, development with an industrial emphasis coupled with a reckless resource use can further weaken the interrelationship between the local people and forest resources.

M. S. Umesh Babu, Sunil Nautiyal Since ecology is all about an interrelated existence of living beings and natural resources, an integrated natural resource management policy can be an ideal directive principle of the State policy for states to administer and the Centre to monitor and evaluate.

REFERENCES

1. Sobrevila, C. (2008) The Role of Indigenous Peoples in Biodiversity Conservation: The Natural but Often Forgotten Partners. The World Bank, Washington DC.
2. Aiyangar, K.S. (1941) Ancient India and South Indian History & Culture: Papers on Indian History and Culture; India to A.D. 1300. Oriental Book Agency.
3. Bimala, C.L. (1888) Historical Geography of Ancient India. Societe Asiatique De Paris France.
4. Hunter, W.W. (1883) A Brief History of the Indian People. 3rd Edition, Trubner and Company, London.
5. Karthikeya, S. (2007) What Are the Threats to Biodiversity. Centre for Education and

Environment. http://www.vigyanprasar.gov.in/Radioserials/Threat_to_biodiversity_-_draft_paper.pdf
6. Glikson, A. (2012) Is Another Mass Extinction Event on the Way? The Conservation. http://theconversation.com/is-another-mass-extinction-event-on-the-way-5397
7. Gulati, S.C. and Suresh, S. (online) Population Pressure and Deforestation in Institute of Economic Growth, New Delhi. http://www.corecentre.co.in /database /docs/docfiles/population_pressure.pdf
8. Baron, J.S. (2004) Research in National Parks. Ecological Applications, 14, 3-4.
9. Zeng, H., Daniel, Z.S. and Wu, X.B. (2005) Human Disturbance on Landscapes in Protected Areas: A Case Study of Wolong Nature Reserve. Ecological Research, 20, 487-496.
10. Ninan, K.N., Jyothis, S., Babu, P. and Ramakrishnappa, V. (2007) The Economics of Biodiversity Conservation-Valuation in Tropical Forest Ecosystems. Earthscan Publications, London.
11. Baron, J.S., Gunderson, L., Allen, C.D., Fleishman, E., McKenzie, D., Meyerson, L.A., Oropeza, J. and Stephenson, N. (2009) Option for National Parks and Reserves for Adapting to Climate Change. Environmental Management, 44, 1033-1042.
12. Kothari, A., Suri, S. and Singh, N. (1995) People and Protected Areas. Rethinking Conservation in India. The Ecologist, 25, 88-194.
13. Kothari, A., Pathak, N., Anuradha, R.V. and Taneja, B. (Eds.) (1998) Communities and Conservation: Natural resource management in South and Central Asia. Sage Publications, New Delhi.
14. Bruner, A.G., Gullison, R.E., Rice, R.E. and da Fonesca, G.A.B. (2001) Effectiveness of Parks in Protecting Tropical Biodiversity. Science, 291, 125-128.
15. Lasgorceix, A. and Kothari, A. (2009) Displacement and Relocation of Protected Areas: A Synthesis and Analysis of Case Studies. Economic & Political Weekly, 49, 37-47.
16. Klabbers, J.H.G., Swart, R.J., Janssen, R., Vellinga, P. and Van Ulden, A.P. (1996) Climate Science and Climate Policy: Improving the Science/Policy Interface. Mitigation and Adaptation Strategies for Global Change, 1, 73-93.
17. Van den Hove, S. (2007) A Rationale for Science-Policy Interfaces. Future, 39, 807-826.
18. Jeffrey, C.M., Lassoie, P.J. and Bedford, L.B. (2007) Conserving Biodiversity and Ecosystem Function through Limited Development: An Empirical Evaluation. Conservation Biology, 22, 70-79.
19. Adhikari, B. (2002) Property Rights and Natural Resource: Socio Economics Heterogeneity and Distributional Implications of Common Property Resource and Management. EEE Working Papers Series-No. 3, University of York, UK.
20. David R. Downes, D.R. and Laird, S.A. (1999) Community Registries of Biodiversity Related Knowledge: The Role of Intellectual Property in Managing Access and Benefit. http://www.ciel.org/Publications/CommunityRegistries.pdf
21. Boshier, D. and Amaral, W. (2004) Threats to Forest Ecosystems and Challenges for the Conservations and Sustainable Use of Forest Genetic Resources. In: Vinceti, B., Amaral, W. and Meilleur, B., Eds., Challenges in Managing Forest Genetic Resource for Livelihoods: Examples from Argentina and Brazil, International Plant Genetic Resources Institute, Rome.
22. Shioya, M. (online) Socio-Ecological Resilience and Adaptive Governance towards Sustainable Resource Management (Case Studies from Malaysia and Japan). http://umconference.um.edu.my/upload/163-1/Paper%20122.pdf
23. Bhat, S. (2010) Natural Resources Conservation Law. Sage Publications Pvt. Ltd, New Delhi, 13.
24. Ball, S. and Bell, S. (1994) Environmental Law. 2nd Edition, Universal Book Traders, New Delhi, 254.

NOTES

[1]http://www.iwawaterwiki.org/xwiki/bin/view/Articles/BRIEFHISTORYOFWATERANDHEALTHFROMANCIENTCIVILIZATIONSTOMODERNTIMES. Accessed: 19.11.2011.

[2]http://www.indiaenvironmentportal.org.in/files/Forestry%20in%20ancient%20India.pdf. Accessed: 19.11.2011.

[3]http://gattflysnest.blogspot.in/2010/05/biodiversity-in-epic-ramayana_19.html. Accessed: 10.11.2011.

[4]http://en.wikipedia.org/wiki/Flora_of_the_Indian_epic_period. Accessed: 15.12.2011.

[5]http://www.sacredland.org/sacred-groves-of-india/. Accessed: 05.01.2012.

[6]http://bmtrainingprog.files.wordpress.com/2010/08/14-vandana-prapanna.pdf. accessed: 17.12.2011.

[7]http://www.salagram.net/Sacred-trees.html. accessed: 17.12.2011.

CHAPTER 14

Forest Management Challenges for Sustaining Water Resources in the Anthropocene

Ge Sun [1,*] and James M. Vose [2]

[1] *U.S. Department of Agriculture, Forest Service, Southern Research Station, Eastern Forest Environmental Threat Assessment Center, Raleigh, NC 27606, USA*

[2] *U.S. Department of Agriculture, Forest Service, Southern Research Station, Center for Integrated Forest Science, Department of Forestry and Environmental Resources, Raleigh, NC 27695, USA*

ABSTRACT

The Earth has entered the Anthropocene epoch that is dominated by humans who demand unprecedented quantities of goods and services from forests. The science of forest hydrology and watershed management generated during the past century provides a basic understanding of relationships among forests and water and offers management principles that maximize the benefits of forests for people while sustaining watershed ecosystems. However, the rapid pace of changes in climate, disturbance regimes, invasive species, human population growth, and land use expected in the 21st century is likely to create substantial challenges for watershed management that may require new approaches, models, and best management practices. These challenges are likely to be complex and large scale, involving a combination of direct and indirect biophysical watershed responses, as well as socioeconomic impacts and feedbacks. We discuss the complex relationships between forests and water in a rapidly changing environment, examine the trade-offs and conflicts between water and other resources, and propose new management approaches for sustaining water resources in the Anthropocene.

Keywords: climate change; disturbance; drought; forest hydrology; modeling; urbanization; watershed management

1. INTRODUCTION

According to the International Union of Geological Sciences (IUGS), we are officially in the Holocene ("entirely recent") epoch, which began 11,700 years ago after the last major ice age. However, since the 1990s, some scientists have argued that the Earth entered an "Anthropocene" epoch beginning from the industrialization in the 1800s [1,2]. Anthropocene represents anthropo, for "man", and cene, for "new". The Anthropocene epoch is dominated by humans and is characterized by mass extinctions of plant and animal species, water and soil pollution, and an altered atmosphere. We are living in an environment which is significantly different from the Holocene [1,2]. For example, the concentration of atmospheric CO_2 during the preindustrial period was 270–275 ppm, but it has exceeded 400 ppm today [3]. The world population has reached to 7.3 billion in contrast to merely 1.0 billion in 1800. The world's urban population (3.9 Billion) has grown more than four times during the past 60 years. Global cropland area expanded from about 4 million km^2 in the 1800s to 15 million km^2 in the 1990s at the expense of forest, shrub, and grasslands.

Today about 31% of the land surface or 4 billion ha is covered by forests [4]. About one half of the primary forests on Earth have disappeared from land conversion, and 16 million hectares of the remaining forests are lost each year. At the same time, forests have been increasingly recognized for their important services, such as water supply and provision of food, medicinal, and forest products, as well as other recreational, cultural, aesthetic, and spiritual benefits [5]. For example, on a global scale, forests contribute ~50% of terrestrial net primary production and store ~45% of terrestrial carbon [5]. Over half of the water supply in the U.S. flows from forestlands [6,7]. It is estimated that tropical and temperate forests worldwide provide ecosystem goods and services of $23.32 trillion per year [8].

Great progress has been made in understanding the complex interactions among forests, water, climate change and humans during the past century (Figure 1). Forest conservation and sustainable forest management practices around the world have slowed forest losses [9]; however, serious future challenges are emerging in the Anthropocene. For example, forest managers face global environmental threats from a warming climate [10,11], and rapid urbanization and demographic changes are increasing the demands of forest ecosystem services such as timber supply, clean water [12], and recreation opportunities. To meet these unprecedented challenges, we propose that land managers will require new thinking and innovative approaches for sustainable forest management in the 21st century [13]. Based on many past successes, there is an expectation by land managers and the public that we have sufficient knowledge and tools to keep watersheds functioning and capable of providing and sustaining ecosystem services into the future [10]. Recent research suggests rapid and substantial progress in our knowledge of watershed sensitivity to rapidly changing conditions [13,14]; however, critical knowledge gaps exist in applying forest watershed sciences to sustain ecosystem services in a new environment [10]. In particular, we lack a mechanistic understanding of hydrological responses to the combined effects of climate change (especially climate extremes) and human disturbances such as urbanization and land use change. Without a mechanistic understanding, our modeling tools and management approaches developed in the past may not fit the future environment.

In this communication, we examine how new emerging global environmental threats interact with forest water resources and ecosystem functions in the Anthropocene. We discuss how watershed ecohydrological science [13], the study of interactions between hydrological processes (*i.e.*, water quantity and quantity) and ecological processes (*i.e.*, vegetation dynamics) under a changing environment, can help forest managers achieve forest sustainability for the benefits of current and future generations.

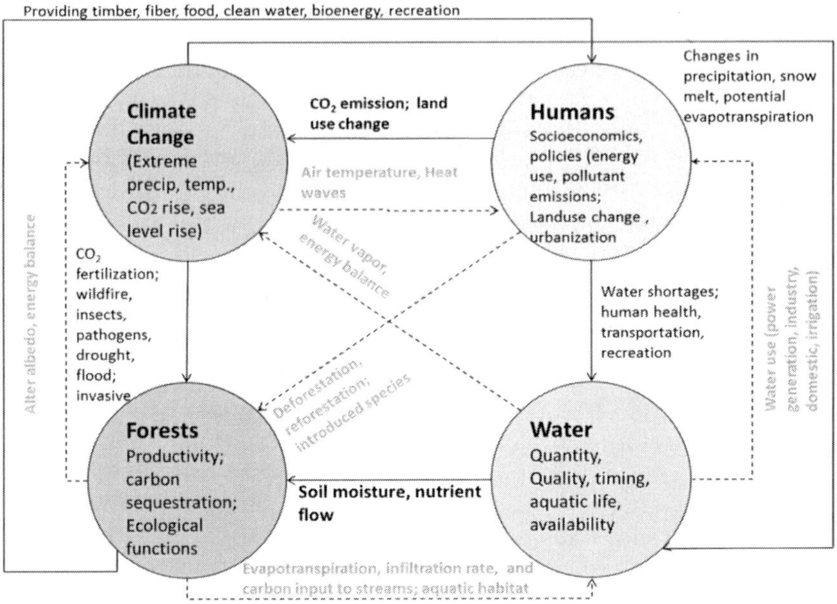

Figure 1. Complex interactions among forests, water resources, climate change, and humans in the Anthropocene. The solid lines represent impacts of stressors while the dotted lines represent feedbacks.

2. EMERGING GLOBAL ENVIRONMENTAL THREATS TO FOREST WATER RESOURCES

2.1. Climate Change: Warming Temperature, Increasing Storms, and Sea Level Rise

Anthropogenic climate change refers to the changes of meteorological variables such as air temperature and precipitation over an extended period in terms of their average and/or variability. Elevated concentrations of atmospheric carbon dioxide (CO_2) concentrations and other greenhouse gases are the causes of climate change [3,15]. Since 1750, atmospheric concentrations of CO_2 have increased by about 40%, nitrous oxide by 20%, and methane by about 150%. The global average temperatures increased by 0.85 °C (about 1.6 °F) between 1880 and 2012 [15]. Annual precipitation has increased over the mid-latitude terrestrial areas of the Northern Hemisphere at a rate from 1.44 to 3.82 mm per decade. While annual precipitation trends over other areas have been less significant, the temporal variability of precipitation has increased. Increased ocean temperatures along with the melting of glaciers and ice caps have contributed to an observed rise in global sea level of approximately 0.2 m between 1901 and 2010 [15].

Climate change is hydrological change, thus it has direct and indirect impacts on forest ecosystems [16] through altering the amount and timing of water and energy movement and availability [13,17,18]. One of the most observable changes is hydrologic intensification: the increased frequency of hydrologic extremes such as low and high flows. For example, studies have detected both decreasing and increasing flows in the southern U.S. and the changes were attributed, at least in part, to greater precipitation variability [19]. While changes in annual mean (or totals) values in metrics such as streamflow and groundwater recharge are important, a greater challenge is posed by changes in hydrologic extremes. Climate change implies that the stationarity of ecosystem functions is a false assumption. The lack of stationarity amplifies the challenges because our reference points for developing and applying management responses may no longer be appropriate. The past no longer serves as an appropriate analog for the future and this non-stationarity is likely to amplify in the future. Many of the tools (e.g., models), guidelines, and best management practices have been developed based on historical (and soon to be obsolete) hydrologic conditions and disturbance regimes. A key question is whether existing approaches and tools for protecting and enhancing water resources will be sufficient to mitigate or adapt to future conditions.

2.2. Population Growth, Urbanization, Land Use Change, and Demographic Change

Population growth is a strong driver of urbanization, land use change, and water supply stress. By 2050, the world population is projected to be 9.6 billion [20] and majority of the total population is expected to live in urban areas. For example, 80% of the population lives in urban areas in the United States and urban population has exceeded 50% in China. In 1950 there were fewer than a dozen Mega (population >10 million) cities worldwide. Today, there were almost 40 Mega Cities (population >10 million) [21].

Population expansion over the next century is expected to occur primarily in less-developed regions placing more pressure on forest ecosystems to provide essential ecosystem services. By 2025 it is projected that there will be 50 Mega Cities with the fastest growth occurring in Africa and Asia. Urban expansion is usually characterized as increasing impervious surface areas and losing agricultural and forest lands bring many well-recognized environmental consequences such as water shortages [21,22], water and air pollution [23], and urban heat island [24]. In particular, urbanization affects watershed microclimate, surface water dynamics, groundwater recharge, stream geomorphology, biogeochemistry, and stream ecology [23,25]. We lack knowledge of the impacts of urbanization on ecosystem structure and function, society, and culture under future climate change [22] and how forest management can play a role in an urbanizing world to reduce the negative aspects of urbanization [26].

Securing adequate and reliable water resources for large cities has become one of top priorities for policy makers, city planners, and land managers worldwide [27]. Forested watersheds are often the most important sources of clean water for city inhabitants. As noted previously, there is an expectation that we have sufficient knowledge and tools to keep watersheds functioning and capable of providing and sustaining ecosystem services into the future [21]. While substantial progress has been made in research approaches and our understanding of watershed sensitivity to rapidly changing environmental conditions [28], it is largely unknown whether this new knowledge will be sufficient or effective in changing management practices because linking this new knowledge to existing modeling tools, Best Management Practices (BMPs), and guidelines is not straightforward and often requires challenging existing dogma and revising long-standing approaches [29].

3. CHALLENGES TO FOREST WATER MANAGEMENT

3.1. Rapid and Complex Environmental Changes Are Difficult to Understand

The Anthropocene represents a relatively recent period in the earth's history where human-caused changes are dramatically altering the structure and functions of natural and social systems [1]. Over the past few decades, these changes have accelerated and are expected to accelerate even more rapidly in the future [1]. Changes in the earth's climate have significant impacts on forest water resources by altering the ecohydrological processes [30] such as plant growth rate and water use efficiency and consequently water balances [28]. Direct effects include the influences of altered precipitation amount, timing, and variation, and changes in temperature and elevated atmospheric CO_2 concentration [31,32]. Indirect effects include vegetation responses to these direct changes and other disturbances and stressors, such as fire, insect outbreaks, tree mortality [33,34] and sea level rise that are indirectly altered by direct effects. These changes are complex and often occur in combination. An even greater challenge will be new combinations and interactions that we have not observed. Changing conditions may favor (or tolerate) new invasive species that may increase wildfire risk and/or permanently alter hydrologic processes. For example, wetter conditions, fire suppression, and the maturation of much of the forest following widespread harvests during the 20th century in the southern U.S. have resulting in forest "mesophication", a process of shifting species dominance to more xeric conditions. Mesophication caused an increase in evapotranspiration and an decrease in water yield [19].

It is difficult to predict how forest ecosystems will respond to traditional forest management practices in a novel environment. For example, fertilization under drought may do harm to plantation forests and increase vulnerability to drought [35,36]. The traditional practices of ditching to grow trees in coastal lowlands may need to be revisited under sea level rise to maximize economic and ecological benefits of intensively managed plantation forests [37,38]. At the large watershed scale, climate change impacts may be masked by management effects. For example, deforestation (deforestation) generally increases (decreases) streamflow, but the influences of this management practice could be offset by increasing or decreasing precipitation and greater evapotranspiration due to climate warming [39,40].

3.2. Extreme Events Challenge Existing Modeling Tools

Sophisticated simulation models have been widely used in forest ecohydrological research and watershed management since the 1990s when personal computers became available [41,42,43,44]. However, the ability to predict the impacts of extreme events presents considerable challenges to existing models [45]. Performance of watershed-scale models, lumped models in particular, is often evaluated on data that have been averaged in space and time [41] and this precludes evaluation of performance of extreme events [46,47,48] such as drought and flooding events. Where finer resolution evaluations have been conducted, model performance of most hydrologic models is often poor, especially for drought conditions. The impacts of high rainfall events on streamflow are easier to model because once soils are saturated, hydrologic responses are driven primarily by physical features of the watershed. If these characteristics are well defined, then flood characteristics (amount, timing, location, *etc.*) can be predicted with relative certainty. However, in mountainous terrain, large storms may increase landslide risk and understanding and modeling the biophysical controls on landslide risk in space and time are difficult [49].

Predicting the impacts of drought on streamflow is especially challenging for a variety of reasons. First, there is often a temporal disconnect between meteorological drought and hydrologic drought that varies based on physical properties (e.g., soil depth, soil water holding capacity, topography, *etc.*) [50,51]. Secondly, physical responses interact with chemical and biological responses [52]. For example, recent warming trends and more prolonged and frequent droughts have increased wildfire frequency and intensity in the western U.S. [53]. Wildfire events can have short term (e.g., peakflow, flooding, landslide) and long term (e.g., geomorphology, land cover shift) consequences to watershed hydrology [54,55]. Droughts have also accelerated the spread and intensity of insect (*i.e.*, Mountain Pine Beetles) attacks and tree die-off [56] in the western U.S. that kill canopy trees, altering stand structure, changing the energy balance of the land surface and affecting many hydrologic processes [57,58,59]. Consequently, predicting the future impacts of climate warming and hydrologic drought on watershed hydrology is difficult. While there is growing information about how drought interacts with hydraulic architecture and stomatal responses [60,61,62], we know very little (and hence, can't model) about how drought impacts root structure and function. Furthermore, in mixed species stands drought does not affect trees equally. Some tree species may be more or less affected by drought through better resistance to drought relative to other species, or through enhanced competitive ability during or after drought (*i.e.*, resilience) [63].

Hydrologic models will need to be able to account for these interactions and responses at the species-level. This expectation requires models that couple leaf-level physiology, above- and belowground whole-tree responses, root dynamics and soil water access, stand level responses, and physical hydrology [64]. Generalized empirical models will have limited utility, as conditions are likely to exceed the data used to develop empirical relationships and non-linearity should be expected [65,66]. Unfortunately, process-based models require a large amount of parameters and input variables and can be only applied at intensively studied research sites, and thus have limited use to answer regional questions [44].

3.3. Challenges to Existing Best Management Practices (BMPs) and Modeling Tools

Extreme climates, such as drought, ice storms, heat waves, are often large scale (*i.e.*, region, continental, global) environmental stressors [67], but their impacts are observable at a range of spatial scales (*i.e.*, tree, stand, basin). In contrast, management responses are typically focused on a limited area, especially in areas with mixed ownership, and usually at the stand scale. For example, climate change, movement of invasive species, water withdrawals, wildfires are global or regional in scale, and cross ecosystem and geographic and political boundaries. Dealing with these large scale issues requires management responses that are also large scale [10]. However, coordinated and large scale management activities are rare. This is especially true where private land is predominant and coordination is especially difficult. Forest BMPs must be designed site specific to suit local watershed physical conditions such as topography, geology and soils, drainage patterns, but they also should consider future climate and hydrologic conditions [29].

Many of models available are not capable of providing the information needed to assist water managers [68]. For example, water managers need information on streamflow amount and quality at a weekly resolution at a spatial scale specific to the water intake or storage reservoir for the water treatment facility. However, models are usually generic and need specific parameters for a certain watershed with unique characteristics and management conditions. Watershed managers also need to know how changing landscape conditions, forest type, and climate interact to determine risks and vulnerabilities, and evaluate management actions to offset them. Hence, models and tools need to be dynamic and account for varying land uses, species and structure [69], and disturbances at fine spatial (e.g., tree) and temporal (e.g., storm event) scales. Empirical rainfall-runoff models built from historical data may not be applicable under future climate change conditions when plant growing season length, forest structure and species composition, and plant water use efficiency [70] have changed over time. The ability of current models to provide this type of information is extremely limited and the models do not match the needs of water managers in space and time.

4. WHAT IS NEEDED TO MEET THE CHALLENGES OF INCREASING DEMAND FOR FOREST WATER RESOURCES?

4.1. Anticipate and Manage for Extreme Stressors

As discussed in the above section, anticipating and managing for extreme stressors poses a great challenge for land managers [71] due to high uncertainty and the lack of effective tools to manage forests at the appropriate scales. Recent mega-droughts, heat waves, and large wildfires portend a future with large-scale changes to ecohydrologic processes and forest functions [72]. In some cases, management activities such as thinning and tree species conversion can increase resilience to these extreme events [11], but increasing resistance on a small subset of stands may be futile unless management occurs at the scale and intensity that will be required. Recognizing that extreme events will alter forests ecosystems, land managers may need to consider management actions that can help facilitate transitions to new and perhaps novel conditions.

4.2. Develop Flexible Modeling Tools in Anticipating Novel Conditions

Modeling tools must be able to account for complexity at the correct spatial and temporal scales [73] and they must include variables required to evaluate management options such as fertilization [74], prescribed burning, and thinning [75]. We should also expect that watershed ecosystems will continue to respond to global change in complex ways characterized as nonlinear and threshold response, some of which may be novel and unprecedented. For example, climate warming may increase evaporation potential but the rise of atmospheric CO_2 concentration may increase water use efficiency thus reduce water use by trees at the leaf level for some species [76]. The end results on hydrology and ecosystem dynamics at the scale of a watershed with mixed land use and variable precipitation patterns can be unpredictable [31]. As these changes occur, it is uncertain whether existing models will be capable of predicting hydrologic responses at the appropriate spatial and temporal resolution. This is especially true for empirical models, as many of the changes will likely be outside the range of data used to generate relationships. Hydrologic "process-based models" usually also include a high level of empirical simplifications and calibrations that will also challenge the performance of existing models.

Forest ecosystem functions and forest uses by people are changing in Anthropocene. Land managers require new tools to detect change using updated knowledge and adjust management approaches accordingly. The rapid pace of change will amplify the need to detect ecosystem responses over wide spatial scales in both natural and managed forests. Significant advances in remote sensing-based change recognition [77] and tracking system holds promise for monitoring some forest conditions in near real time (*i.e.*, ForWarn) [78]; however, it is unclear if this technology will be appropriate for quantifying and analyzing the effectiveness of management actions. Hence, a combination of remote sensing and "on-the-ground" management platforms such as FIA [79], NEON, and other large-scale networks will be critical for change detection. In addition, new knowledge needs to be rapidly incorporated into management actions and user friendly predictive models, such as the Water Supply Stress Index (WaSSI) model (http://www.wassiweb.sgcp.ncsu.edu/), a water and carbon accounting model that has been used to project climate change impacts on water and carbon resources across the United States [7,12,80]. Climate change mitigation and adaptation management guides such as the Template for Assessing Climate Change Impacts and Management Options (TACCIMO) (www.forestthreats.org/taccimo) provide rapid updates to the "state-of-the-science" from the published literature thus connects forest planning to peer-reviewed climate change science; Such a system delivers information from peer-reviewed publication findings describing effects and management options and interactive maps of climate projections and models that provide insight into climate influences on natural resources. Additional information could be provided from land managers as they observe changes in the forests that they manage. New approaches to collecting and disseminating information such as citizen science networks and "crowd sourcing" approaches could accelerate the pace of experiential knowledge, observations, and data collection to supplement information from the published literature. Finally, improvements in data collection, storage, and access systems over the past few decades has created massive amounts of readily available data that has facilitated large scale "big data" analyses of hydrologic trends [81]. These new approaches are likely to provide significant insight into the interactions between large scale drivers such as climate change, and smaller scale controls such as land use and management intensity. We also recommend that models be available as open source that can be modified as new data and understanding become available.

4.3. Be Realistic about Forest Management Options

Active management including implementing Best Management Practices (BMPs) is critical to offset increasing environmental threats to forest ecosystem services [82]. In the past, improvement in forest condition with forest management in many parts of the world has been tremendously successful [9]. Among the best examples, is the recovery of heavily cutover lands in the eastern U.S. Reforestation and the implementation of best management practices has restored these forests and associated ecosystem services [82,83]. Forest coverage in China has been increasing thanks to large scale reforestation campaigns that aims at reducing soil erosion and protecting remaining forests in the last three decades [84]. The socioeconomic (*i.e.*, poverty reduction through food subsidies through the Grain for Green Project) and ecological benefits (*i.e.*, soil erosion control, carbon sequestration) of these policy-driven ecological restoration efforts are well documented [85]. However, tradeoffs of ecosystem services of reforestation and unintended environmental consequences such as decline in deep soil moisture (*i.e.*, soil desiccation) and water yield reduction in some arid regions [86,87] are emerging.

Can we manage forests in the face of future threats to continue to provide the level of existing ecosystem services? Or, more importantly, can we provide and sustain ecosystem services required in the future? We contend that large scale and extreme changes in social-economic and biophysical conditions will preclude the ability to sustain many ecosystem services. We contend that it will be difficult to "manage" our way out of future threats due to the large uncertainty of environmental conditions. Traditional approaches to forest conservation and management that assume a constant climate; stable forest dynamics, and socioeconomic, and demographic conditions; and rely heavily on historical reference conditions will be inadequate [88] to meet future demand on forest ecosystem services. Instead, new approaches that focus on anticipating and guiding ecological responses to change, are urgently needed to ensure the full value of forest ecosystem services for future generations. For example, warmer and drier conditions are increasing the frequency and size of wildfires throughout many areas of the world [89]. How to manage the threats from large area of wildfires is debatable although comprehensive strategies have been proposed [90,91,92]. There are proposed management options that may be implemented to minimize the impacts of drought on water quantity and quality [10]. For example, reducing leaf area by thinning and regenerating cut or planting native tree species that use less water than exotic species may help reduce water stress and increase water availability to tree growth [93], aquatic systems in forest streams, and downstream water supply for people. However, as with other natural disturbances, droughts are difficult to prepare for because they are unpredictable. Management actions such as thinning [75] and prescribed burning [94] are typically not implemented at a scale or intensity to offset climatic driving forces. Similarly, the growing conflict between managing for carbon *vs.* managing for water [95] will only increase as efforts to mitigate CO_2 emissions using bioenergy promote management of fast growing species [96,97]. In short, managers should prepare for growing conflicts among management priorities and the need to articulate the limitations of forest management for providing ecosystem services in the future. The disparity between winners and losers will widen, and trade-offs will need to be carefully evaluated.

5. SUMMARY

In the remainder of the twenty-first century, humans will likely be impacted by the degradation of ecosystem services, and the potential loss of the planet's ability to recover [98]. The forestry community is facing large global environmental and socioeconomic challenges, such as climate change and urbanization, to meet the ever increasing demand for ecosystem services from forest ecosystems. Traditional watershed management is facing new challenges as rapid and compounded environmental, economic, and social change contribute to an increasingly uncertain future [19]. Our knowledge of ecohydrological response to extreme stressors (e.g., drought) is lacking, so current modeling tools may be insufficient to project the impacts of climate change on ecosystem functions at the scales needed in forest management. Future forest water resource management must consider the trade-offs of forest ecosystem services and coupled nature–human systems. Existing forest Best Management Practices (BMPs) [82] should be revisited to mitigate and adapt to the negative impacts of natural and anthropogenic disturbances that are expected to increase in the Anthropocene.

CONFLICTS OF INTEREST

The authors declare no conflict of interest.

REFERENCES

1. Steffen, W.; Grinevald, J.; Crutzen, P.; McNeill, J. The Anthropocene: Conceptual and historical perspectives. Philos. Trans. R. Soc. A 2011, 369, 842–867.
2. Steffen, W.; Crutzen, P.J.; McNeill, J.R. The Anthropocene: Are humans now overwhelming the great forces of nature. Ambio 2007, 36, 614–621.
3. IPCC. Summary for policymakers. In Climate Change 2014: Impacts, Adaptation, and Vulnerability. Part A: Global and Sectoral Aspects; Contribution of Working Group ii to the Fifth Assessment Report of the Intergovernmental Panel on Climate Change; Field, C.B., Barros, V.R., Dokken, D.J., Mach, K.J., Mastrandrea, M.D., Bilir, T.E., Chatterjee, M., Ebi, K.L., Estrada, Y.O., Genova, R.C., et al, Eds.; Cambridge University Press: Cambridge, UK; New York, NY, USA, 2014; p. 32.
4. FAO. Global Forest Resources Assessment 2015: How Have the World's Forests Changed? FAO: Rome, Italy, 2015.
5. Bonan, G.B. Forests and climate change: Forcings, feedbacks, and the climate benefits of forests. Science 2008, 320, 1444–1449.
6. Brown, T.C.; Hobbins, M.T.; Ramirez, J.A. Spatial distribution of water supply in the coterminous United States. J. Am. Water Resour. Assoc. 2008, 44, 1474–1487.
7. Sun, G.; Caldwell, P.; Noormets, A.; McNulty, S.G.; Cohen, E.; Myers, J.M.; Domec, J.C.; Treasure, E.; Mu, Q.Z.; Xiao, J.F.; et al. Upscaling key ecosystem functions across the conterminous United States by a water-centric ecosystem model. J. Geophys. Res. Biogeosci. 2011, 116.
8. De Groot, R.; Brander, L.; van der Ploeg, S.; Costanza, R.; Bernard, F.; Braat, L.; Christie, M.; Crossman, N.; Ghermandi, A.; Hein, L.; et al. Global estimates of the value of ecosystems and their services in monetary units. Ecosyst. Serv. 2012, 1, 50–61.
9. MacDicken, K.G. Global forest resources assessment 2015: What, why and how? For. Ecol. Manag. 2015, 352, 3–8.
10. Vose, J.M.; Klepzig, K.D. Climate Change Adaptation and Mitigation Management Options: A Guide for Natural Resource Managers in Southern Forest Ecosystems; CRC Press: Boca Raton, FL, USA, 2014; p. 476.
11. Ford, C.R.; Laseter, S.H.; Swank, W.T.; Vose, J.M. Can forest management be used to sustain water-based ecosystem services in the face of climate change? Ecol. Appl. 2011, 21, 2049–2067.
12. Sun, G.; McNulty, S.G.; Myers, J.A.M.; Cohen, E.C. Impacts of multiple stresses on water demand and supply across the southeastern United States. J. Am. Water Resour. Assoc. 2008, 44, 1441–1457.
13. Vose, J.M.; Sun, G.; Ford, C.R.; Bredemeier, M.; Otsuki, K.; Wei, X.H.; Zhang, Z.Q.; Zhang, L. Forest ecohydrological research in the 21st century: What are the critical needs? Ecohydrology 2011, 4, 146–158.
14. Vogel, R.M.; Lall, U.; Cai, X.M.; Rajagopalan, B.; Weiskel, P.K.; Hooper, R.P.; Matalas, N.C. Hydrology: The interdisciplinary science of water. Water. Resour. Res. 2015, 51, 4409–4430.
15. Stocker, B.D.; Roth, R.; Joos, F.; Spahni, R.; Steinacher, M.; Zaehle, S.; Bouwman, L.; Xu-Ri; Prentice, I.C. Multiple greenhouse-gas feedbacks from the land biosphere under future climate change scenarios. Nat. Clim. Chang. 2013, 3, 666–672.
16. Melillo, J.M.; Richmond, T.; Yohe, G.W. Climate Change Impacts in the United States: The Third National Climate Assessment; Global Change Research Program: Washington, DC, USA, 2014.
17. Felzer, B.; Sahagian, D. Climate impacts on regional ecosystem services in the United States from cmip3-based multimodel comparisons. Clim. Res. 2014, 61, 133–155.
18. Sun, S.L.; Sun, G.; Caldwell, P.; McNulty, S.; Cohen, E.; Xiao, J.F.; Zhang, Y. Drought impacts on ecosystem functions of the us national forests and grasslands: Part ii assessment results and management implications. For. Ecol. Manag.2015, 353, 269–279.

19. Vose, J.M.; Martin, K.L.; Barten, P.K. Applications of forest hydrologic science to watershed management in the 21st century. In Forest Hydrology; Amatya, T., Williams, L., de Jong, C., Eds.; CABI: Cambridge, UK, in press.
20. UN. World Urbanization Prospects; the 2014 Revision; UN: New York, NY, USA, 2014.
21. Li, E.J.; Endter-Wada, J.; Li, S.J. Characterizing and contextualizing the water challenges of megacities. J. Am. Water. Resour. Assoc. 2015, 51, 589–613.
22. McDonald, R.I.; Green, P.; Balk, D.; Fekete, B.M.; Revenga, C.; Todd, M.; Montgomery, M. Urban growth, climate change, and freshwater availability. Proc. Natl. Acad. Sci. USA 2011, 108, 6312–6317.
23. O'Driscoll, M.; Clinton, S.; Jefferson, A.; Manda, A.; McMillan, S. Urbanization effects on watershed hydrology and in-stream processes in the southern United States. Water-Sui 2010, 2, 605–648.
24. Zhou, D.C.; Zhao, S.Q.; Zhang, L.X.; Sun, G.; Liu, Y.Q. The footprint of urban heat island effect in China. Sci. Rep.2015, 5.
25. Paul, M.J.; Meyer, J.L. Streams in the urban landscape. Annu. Rev. Ecol. Syst. 2001, 32, 333–365.
26. Sun, G.; Lockaby, B.G. Water quantity and quality at the urban-rural interface. Urban Rural Interfaces Link. People Nat.2012, 29–48.
27. Sun, G.; Michelsen, A.M.; Sheng, Z.P.; Fang, A.F.; Shang, Y.Z.; Zhang, H.L. Featured collection introduction: Water for megacities challenges and solutions. J. Am. Water Resour. Assoc. 2015, 51, 585–588.
28. Vose, J.M.; Miniat, C.F.; Luce, C.H.; Asbjornsen, H.; Caldwell, P.V.; Campbell, J.L.; Grant, G.E.; Isaak, D.J.; Loheide, S.P.L., II; Sun, G. Ecohydrological implications of drought for forests in the United States. For. Ecol. Manag. 2016. in press.
29. Marion, D.A.; Sun, G.; Caldwell, P.V.; Miniat, C.F.; Ouyang, Y.; Amatya, D.M.; Clinton, B.D.; Conrads, P.A.; Gull Laird, S.; Dai, Z.; et al. Managing forest water quantity and quality under climate change. In Climate Change Adaption and Mitigation Management Optionsa Guide for Natural Resource Managers in Southern Forest Ecosystems; Vose, J.M., Klepzig, K.D., Eds.; CRC Press/Taylor and Francis: Boca Raton, FL, USA, 2014; pp. 249–306, p. 58.
30. Parolari, A.J.; Katul, G.G.; Porporato, A. An ecohydrological perspective on drought-induced forest mortality. J. Geophys. Res. Biogeosci. 2014, 119, 965–981.
31. Tor-Ngern, P.; Oren, R.; Ward, E.J.; Palmroth, S.; McCarthy, H.R.; Domec, J.C. Increases in atmosphericco(2) have little influence on transpiration of a temperate forest canopy. New Phytol. 2015, 205, 518–525.
32. Hanson, P.J.; Wullschleger, S.D.; Norby, R.J.; Tschaplinski, T.J.; Gunderson, C.A. Importance of changing CO2, temperature, precipitation, and ozone on carbon and water cycles of an upland-oak forest: Incorporating experimental results into model simulations. Glob. Chang. Biol. 2005, 11, 1402–1423.
33. Allen, C.D.; Breshears, D.D.; McDowell, N.G. On underestimation of global vulnerability to tree mortality and forest die-off from hotter drought in the anthropocene. Ecosphere 2015, 6, 1–55.
34. Allen, C.D.; Macalady, A.K.; Chenchouni, H.; Bachelet, D.; McDowell, N.; Vennetier, M.; Kitzberger, T.; Rigling, A.; Breshears, D.D.; Hogg, E.H.; et al. A global overview of drought and heat-induced tree mortality reveals emerging climate change risks for forests. For. Ecol. Manag. 2010, 259, 660–684.
35. Ward, E.J.; Domec, J.C.; Laviner, M.A.; Fox, T.R.; Sun, G.; McNulty, S.; King, J.; Noormets, A. Fertilization intensifies drought stress: Water use and stomatal conductance of pinus taeda in a midrotation fertilization and throughfall reduction experiment. For. Ecol. Manag. 2015, 355, 72–82.
36. Bartkowiak, S.M.; Samuelson, L.J.; McGuire, M.A.; Teskey, R.O. Fertilization increases sensitivity of canopy stomatal conductance and transpiration to throughfall reduction in an 8-year-old loblolly pine plantation. For. Ecol. Manag.2015, 354, 87–96.

37. Lohmus, A.; Remm, L.; Rannap, R. Just a ditch in forest? Reconsidering draining in the context of sustainable forest management. Bioscience 2015, 65, 1066–1076.
38. Amatya, D.M.; Gregory, J.D.; Skaggs, R.W. Effects of controlled drainage on storm event hydrology in a loblolly pine plantation. J. Am. Water Resour. Assoc. 2000, 36, 175–190.
39. Wei, X.H.; Liu, W.F.; Zhou, P.C. Quantifying the relative contributions of forest change and climatic variability to hydrology in large watersheds: A critical review of research methods. Water-Sui 2013, 5, 728–746.
40. Liu, W.F.; Wei, X.H.; Liu, S.R.; Liu, Y.Q.; Fan, H.B.; Zhang, M.F.; Yin, J.M.; Zhan, M.J. How do climate and forest changes affect long-term streamflow dynamics? A case study in the upper reach of poyang river basin. Ecohydrology2015, 8, 46–57.
41. Sun, S.L.; Sun, G.; Caldwell, P.; McNulty, S.G.; Cohen, E.; Xiao, J.F.; Zhang, Y. Drought impacts on ecosystem functions of the us national forests and grasslands: Part i evaluation of a water and carbon balance model. For. Ecol. Manag. 2015, 353, 260–268.
42. Tian, S.Y.; Youssef, M.A.; Sun, G.; Chescheir, G.M.; Noormets, A.; Amatya, D.M.; Skaggs, R.W.; King, J.S.; McNulty, S.; Gavazzi, M.; et al. Testing drainmod-forest for predicting evapotranspiration in a mid-rotation pine plantation. For. Ecol. Manag. 2015, 355, 37–47.
43. Sun, G.; Riekerk, H.; Comerford, N.B. Modeling the hydrologic impacts of forest harvesting on florida flatwoods. J. Am. Water Resour. Assoc. 1998, 34, 843–854.
44. Ollinger, S.V.; Aber, J.D.; Federer, C.A. Estimating regional forest productivity and water yield using an ecosystem model linked to a gis. Landsc. Ecol. 1998, 13, 323–334.
45. Amatya, D.M.; Sun, G.; Rossi, C.G.; Ssegane, H.S.; Nettles, J.E.; Panda, S. Forests, land use change, and water. In Impact of Climate Change on Water Resources in Agriculture; Rodrigues, R., Ed.; CRC Press/Taylor & Francis Group: Boca Raton, FL, USA, 2015.
46. Sun, S.L.; Chen, H.S.; Ju, W.M.; Yu, M.; Hua, W.J.; Yin, Y. On the attribution of the changing hydrological cycle in poyang lake basin, China. J. Hydrol. 2014, 514, 214–225.
47. Hao, L.; Zhang, X.Y.; Gao, J.M. Simulating human-induced changes of water resources in the upper xiliaohe river basin, China. Environ. Eng. Manag. J. 2011, 10, 787–792.
48. Hao, L.; Sun, G.; Liu, Y.Q.; Qian, H. Integrated modeling of water supply and demand under management options and climate change scenarios in Chifeng city, China. J. Am. Water Resour. Assoc. 2015, 51, 655–671.
49. Band, L.E.; Hwang, T.; Hales, T.C.; Vose, J.; Ford, C. Ecosystem processes at the watershed scale: Mapping and modeling ecohydrological controls of landslides. Geomorphology 2012, 137, 159–167.
50. Geris, J.; Tetzlaff, D.; McDonnell, J.; Soulsby, C. The relative role of soil type and tree cover on water storage and transmission in northern headwater catchments. Hydrol. Process. 2015, 29, 1844–1860.
51. Adams, H.R.; Barnard, H.R.; Loomis, A.K. Topography alters tree growth-climate relationships in a semi-arid forested catchment. Ecosphere 2014, 5, 1–16.
52. Hanson, P.J.; Amthor, J.S.; Wullschleger, S.D.; Wilson, K.B.; Grant, R.F.; Hartley, A.; Hui, D.; Hunt, E.R.; Johnson, D.W.; Kimball, J.S.; et al. Oak forest carbon and water simulations: Model intercomparisons and evaluations against independent data. Ecol. Monogr. 2004, 74, 443–489.
53. Flannigan, M.D.; Krawchuk, M.A.; de Groot, W.J.; Wotton, B.M.; Gowman, L.M. Implications of changing climate for global wildland fire. Int. J. Wildland Fire 2009, 18, 483–507.
54. Bladon, K.D.; Emelko, M.B.; Silins, U.; Stone, M. Wildfire and the future of water supply. Environ. Sci. Technol. 2014, 48, 8936–8943.
55. Ice, G.G.; Neary, D.G.; Adams, P.W. Effects of wildfire on soils and watershed processes. J. For. 2004, 102, 16–20.

56. Adams, H.D.; Luce, C.H.; Breshears, D.D.; Allen, C.D.; Weiler, M.; Hale, V.C.; Smith, A.M.S.; Huxman, T.E. Ecohydrological consequences of drought- and infestation-triggered tree die-off: Insights and hypotheses. Ecohydrology 2012, 5, 145–159.
57. Bearup, L.A.; Maxwell, R.M.; Clow, D.; McCray, J.E. Hydrological effects of forest transpiration loss in bark beetle-impacted watersheds. Nat. Clim. Chang. 2014, 4, 481–486.
58. Langhammer, J.; Su, Y.; Bernsteinova, J. Runoff response to climate warming and forest disturbance in a mid-mountain basin. Water-Sui 2015, 7, 3320–3342.
59. Pugh, E.; Gordon, E. A conceptual model of water yield effects from beetle-induced tree death in snow-dominated lodgepole pine forests. Hydrol. Process. 2013, 27, 2048–2060.
60. Domec, J.C.; Noormets, A.; King, J.S.; Sun, G.; McNulty, S.G.; Gavazzi, M.J.; Boggs, J.L.; Treasure, E.A. Decoupling the influence of leaf and root hydraulic conductances on stomatal conductance and its sensitivity to vapour pressure deficit as soil dries in a drained loblolly pine plantation. Plant Cell Environ. 2009, 32, 980–991.
61. Noormets, A.; McNulty, S.G.; DeForest, J.L.; Sun, G.; Li, Q.; Chen, J. Drought during canopy development has lasting effect on annual carbon balance in a deciduous temperate forest. New Phytol. 2008, 179, 818–828.
62. Domec, J.C.; Palmroth, S.; Ward, E.; Maier, C.A.; Therezien, M.; Oren, R. Acclimation of leaf hydraulic conductance and stomatal conductance of pinus taeda (loblolly pine) to long-term growth in elevated CO2 (free-air CO2enrichment) and n-fertilization. Plant Cell Environ. 2009, 32, 1500–1512.
63. Ford, C.R.; Hubbard, R.M.; Vose, J.M. Quantifying structural and physiological controls on variation in canopy transpiration among planted pine and hardwood species in the southern appalachians. Ecohydrology 2011, 4, 183–195.
64. Tague, C.L.; McDowell, N.G.; Allen, C.D. An integrated model of environmental effects on growth, carbohydrate balance, and mortality of pinus ponderosa forests in the southern rocky mountains. PLoS ONE 2013, 8, e80286.
65. Zhou, G.; Wei, X.; Chen, X.; Zhou, P.; Liu, X.; Xiao, Y.; Sun, G.; Scott, D.F.; Zhou, S.; Han, L.; et al. Global pattern for the effect of climate and land cover on water yield. Nat. Commun. 2015, 6.
66. Sun, G.; Alstad, K.; Chen, J.Q.; Chen, S.P.; Ford, C.R.; Lin, G.H.; Liu, C.F.; Lu, N.; McNulty, S.G.; Miao, H.X.; et al. A general predictive model for estimating monthly ecosystem evapotranspiration. Ecohydrology 2011, 4, 245–255.
67. Sun, G.; Caldwell, P.V.; McNulty, S.G.; Georgakakos, A.P.; Arumugam, S.; Cruise, J.; McNider, R.T.; Terando, A.; Conrads, P.A.; Feldt, J.; et al. Impacts of climate change and variability on water resources in the southeast USA. In Climate of the Southeast United States: Variability, Change, Impacts, and Vulnerability, NCA Southeast Technical Report; Ingram, K.T., Dow, K., Carter, L., Anderson, J., Eds.; Island Press: Washington, DC, USA, 2013; p. 31.
68. Amatya, D.M.; Douglas-Mankin, K.R.; Williams, T.M.; Skaggs, R.W.; Nettles, J.E. Advances in forest hydrology: Challenges and opportunities. Trans. ASABE 2011, 54, 2049–2056.
69. Brantley, S.T.; Miniat, C.F.; Elliott, K.J.; Laseter, S.H.; Vose, J.M. Changes to southern appalachian water yield and stormflow after loss of a foundation species. Ecohydrology 2015, 8, 518–528.
70. Liu, Y.B.; Xiao, J.F.; Ju, W.M.; Zhou, Y.L.; Wang, S.Q.; Wu, X.C. Water use efficiency of China's terrestrial ecosystems and responses to drought. Sci. Rep. 2015, 5.
71. Patel-Weynand, T.; Peterson, D.L.; Vose, J.M. Effects of Climatic Variability and Change on Forest Ecosystems: A Comprehensive Science Synthesis for the U.S.; USDA Forest Service Pacific Northwest Research Station: Portland, OR, USA, 2012; p. 265.
72. Matyas, C.; Sun, G. Forests in a water limited world under climate change. Environ. Res. Lett. 2014, 9.

73. Porporato, A.; Feng, X.; Manzoni, S.; Mau, Y.; Parolari, A.J.; Vico, G. Ecohydrological modeling in agroecosystems: Examples and challenges. Water. Resour. Res. 2015, 51, 5081–5099.
74. Amatya, D.M.; Rossi, C.G.; Saleh, A.; Dai, Z.; Youssef, M.A.; Williams, R.G.; Bosch, D.D.; Chescheir, G.M.; Sun, G.; Skaggs, R.W.; et al. Review of nitrogen fate models applicable to forest landscapes in the southern US. Trans. ASABE2013, 56, 1731–1757.
75. Sun, G.; Caldwell, P.V.; McNulty, S.G. Modelling the potential role of forest thinning in maintaining water supplies under a changing climate across the conterminous United States. Hydrol. Process. 2015.
76. Cheng, L.; Zhang, L.; Wang, Y.P.; Yu, Q.; Eamus, D.; O'Grady, A. Impacts of elevated CO2, climate change and their interactions on water budgets in four different catchments in Australia. J. Hydrol. 2014, 519, 1350–1361.
77. Hansen, M.C.; Potapov, P.V.; Moore, R.; Hancher, M.; Turubanova, S.A.; Tyukavina, A.; Thau, D.; Stehman, S.V.; Goetz, S.J.; Loveland, T.R.; et al. High-resolution global maps of 21st-century forest cover change. Science 2013, 342, 850–853.
78. Norman, S.P. Highlights of Satellite-Based Forest Change Recognition and Tracking Using the Forwarn System; U.S. Department of Agriculture, Forest Service, Research & Development, Southern Research Station: Asheville, NC, USA, 2013; p. 30.
79. Coulston, J.W.; Wear, D.N.; Vose, J.M. Complex forest dynamics indicate potential for slowing carbon accumulation in the southeastern United States. Sci. Rep. 2015, 5.
80. Caldwell, P.V.; Sun, G.; McNulty, S.G.; Cohen, E.C.; Myers, J.A.M. Impacts of impervious cover, water withdrawals, and climate change on river flows in the conterminous US. Hydrol. Earth Syst. Sci. 2012, 16, 2839–2857.
81. Rice, J.S.; Emanuel, R.E.; Vose, J.M.; Nelson, S.A.C. Continental U.S. Streamflow trends from 1940 to 2009 and their relationships with watershed spatial characteristics. Water Resour. Res. 2015, 51, 6262–6275.
82. Cristan, R.; Aust, W.M.; Bolding, M.C.; Barrett, S.M.; Munsell, J.F.; Schilling, E. Effectiveness of forestry best management practices in the United States: Literature review. For. Ecol. Manag. 2016, 360, 133–151.
83. Boggs, J.; Sun, G.; McNUlty, S.G. Effects of timber harvest on water quantity and quality in small watersheds in the piedmont of North Carolina. J. For. 2015, 114, 27–40.
84. Sun, G.; Zhou, G.Y.; Zhang, Z.Q.; Wei, X.H.; McNulty, S.G.; Vose, J.M. Potential water yield reduction due to forestation across China. J. Hydrol. 2006, 328, 548–558.
85. Lu, Y.H.; Fu, B.J.; Feng, X.M.; Zeng, Y.; Liu, Y.; Chang, R.Y.; Sun, G.; Wu, B.F. A policy-driven large scale ecological restoration: Quantifying ecosystem services changes in the loess plateau of China. PLoS ONE 2012, 7, e31782.
86. Cao, S.X.; Sun, G.; Zhang, Z.Q.; Chen, L.D.; Feng, Q.; Fu, B.J.; McNulty, S.; Shankman, D.; Tang, J.W.; Wang, Y.H.; et al. Greening China naturally. Ambio 2011, 40, 828–831.
87. Liang, W.; Bai, D.; Wang, F.Y.; Fu, B.J.; Yan, J.P.; Wang, S.; Yang, Y.T.; Long, D.; Feng, M.Q. Quantifying the impacts of climate change and ecological restoration on streamflow changes based on a budyko hydrological model in China's Loess Plateau. Water Resour. Res. 2015, 51, 6500–6519.
88. Golladay, S.W.; Martin, K.L.; Vose, J.M.; Wear, D.N.; Covich, A.P.; Hobbs, R.J.; Klepzig, K.D.; Likens, G.E.; Naiman, R.J.; Shearer, A.W. Achievable future conditions as a framework for guiding forest conservation and management. For. Ecol. Manag. 2016, 360, 80–96.
89. Liu, Y.Q.; Stanturf, J.; Goodrick, S. Trends in global wildfire potential in a changing climate. For. Ecol. Manag. 2010, 259, 685–697.
90. Moritz, M.A.; Batllori, E.; Bradstock, R.A.; Gill, A.M.; Handmer, J.; Hessburg, P.F.; Leonard, J.; McCaffrey, S.; Odion, D.C.; Schoennagel, T.; et al. Learning to coexist with wildfire. Nature 2014, 515, 58–66.
91. Stephens, S.L.; Agee, J.K.; Fule, P.Z.; North, M.P.; Romme, W.H.; Swetnam, T.W.; Turner, M.G. Managing forests and fire in changing climates. Science 2013, 342, 41–42.

92. North, M.P.; Stephens, S.L.; Collins, B.M.; Agee, J.K.; Aplet, G.; Franklin, J.F.; Fule, P.Z. Reform forest fire management. Science 2015, 349, 1280–1281.
93. Grant, G.E.; Tague, C.L.; Allen, C.D. Watering the forest for the trees: An emerging priority for managing water in forest landscapes. Front. Ecol. Environ. 2013, 11, 314–321.
94. Vose, J.M.; Elliott, K.J. Oak, fire, and global change: What might the future hold? Fire Ecol. 2016. in prerss.
95. Jackson, R.B.; Jobbagy, E.G.; Avissar, R.; Roy, S.B.; Barrett, D.J.; Cook, C.W.; Farley, K.A.; le Maitre, D.C.; McCarl, B.A.; Murray, B.C. Trading water for carbon with biological sequestration. Science 2005, 310, 1944–1947.
96. King, J.S.; Ceulemans, R.; Albaugh, J.M.; Dillen, S.Y.; Domec, J.C.; Fichot, R.; Fischer, M.; Leggett, Z.; Sucre, E.; Trnka, M.; et al. The challenge of lignocellulosic bioenergy in a water-limited world. Bioscience 2013, 63, 102–117.
97. Vose, J.M.; Miniat, C.F.; Sun, G.; Caldwell, P.V. Potential implications for expansion of freeze-tolerant eucalyptus plantations on water resources in the southern United States. For. Sci. 2015, 61, 509–521.
98. Steffen, W.; Persson, A.; Deutsch, L.; Zalasiewicz, J.; Williams, M.; Richardson, K.; Crumley, C.; Crutzen, P.; Folke, C.; Gordon, L.; et al. The anthropocene: From global change to planetary stewardship. Ambio 2011, 40, 739–761.

CHAPTER 15

Resolving Conflicts between Agriculture and the Natural Environment

Andrew J. Tanentzap[1] *, Anthony Lamb[2] , Susan Walker[3] , Andrew Farmer[4]

[1] *Ecosystems and Global Change Group, Department of Plant Sciences, University of Cambridge, Cambridge, United Kingdom,*

[2] *Conservation Science Group, Department of Zoology, University of Cambridge, Cambridge, United Kingdom,*

[3] *Landcare Research, Dunedin, New Zealand,*

[4] *Institute for European Environmental Policy, London, United Kingdom*

ABSTRACT

Agriculture dominates the planet. Yet it has many environmental costs that are unsustainable, especially as global food demand rises. Here, we evaluate ways in which different parts of the world are succeeding in their attempts to resolve conflict between agriculture and wild nature. We envision that coordinated global action in conserving land most sensitive to agricultural activities and policies that internalise the environmental costs of agriculture are needed to deliver a more sustainable future.

INTRODUCTION

Agriculture is the most dominant land use on Earth and will remain so as the world's population and global food demand rise [1] (Figure 1). This poses a fundamental challenge to the natural environment, which requires access to many of the same resources as agricultural producers and is harmed by agriculture's external impacts [2,3]. Using land resources for agriculture prioritises crop and livestock production over the many other benefits that natural ecosystems provide and that agriculture displaces, including biodiversity, carbon storage, and water purification [4]. Governments also heavily support agricultural production in ways that are often bad for the environment, such as rewarding output or lowering input prices [5,6]. Very little is spent on actively mitigating the environmental impacts (Figure 1).

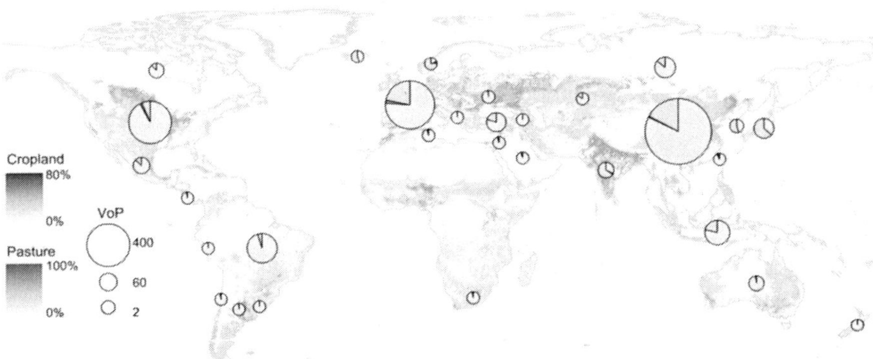

Figure 1. Financial support to farmers from taxpayers and consumers associated with agricultural policies as a proportion of the total value of agricultural production (VoP) at the farm gate.

We distinguished between policies aiming to achieve specific environmental objectives beyond those required by regulation (purple segments) and all other types of support irrespective of their influence on farm production or income (orange segments), e.g., market price support, monetary transfers based on output, input subsidies. Sizes of symbols were scaled to VoP in hundreds of millions of United States dollars. Data are from the most recent year available (2008–2013) and described in S1 Text. Green and brown shading in the background is the percent of cropland and pasture at 5' resolution in the year 2000 [7]. Financial support data were preferentially sourced from the Organisation for Economic Cooperation and Development (OECD) (n = 21 countries), supplemented from the World

Trade Organization (WTO) where possible for additional countries (n = 12). Data used to make this figure are provided in S1 Data.

Here, we summarise how agriculture impacts the natural environment and then scrutinise the success of government agricultural policies that we believe offer the most promise for safeguarding nature while feeding a growing human population. We use economic and political theory to help explain why some measures are more prone to succeed than others, recognising that scientific understanding of a problem does not mean it will be solved [8]. Throughout, we touch upon examples from centres of agricultural production.

A CULTIVATED PLANET

Agriculture's demand for land drives conversion of natural habitats, and this is arguably its largest environmental cost. Approximately 40% of ice-free land (4,300 million ha) is already covered by crops or used to raise livestock [7], and an additional 2.7–4.9 million ha of cropland annually, the size of a small European nation, may be required to feed the world's population by 2030 [1,9]. Conversion of land for agriculture is estimated to account for 80% of global deforestation [10], and ca. 53% of terrestrial species assessed as threatened by the International Union for Conservation of Nature (IUCN) [11] are negatively impacted by agriculture (Figure 2). Land conversion also reduces the size of the terrestrial carbon sink. Global simulation models predict that 24% and 10% less carbon is stored in vegetation and soil, respectively, than if present-day landscapes retained their natural vegetation [12].

We counted the total number of species classified as either critically endangered, endangered, or vulnerable by the IUCN Red List in different taxonomic units and the number that were assessed to be threatened by at least one agricultural activity [11]. Agricultural activities were annual and perennial nontimber crops, wood and pulp plantations, livestock farming and ranching, logging and wood harvesting, abstracting of surface water (agricultural use), abstraction of ground water (agricultural use), and agricultural and forestry effluents. Many other species may become threatened with agricultural expansion (S2 Fig).

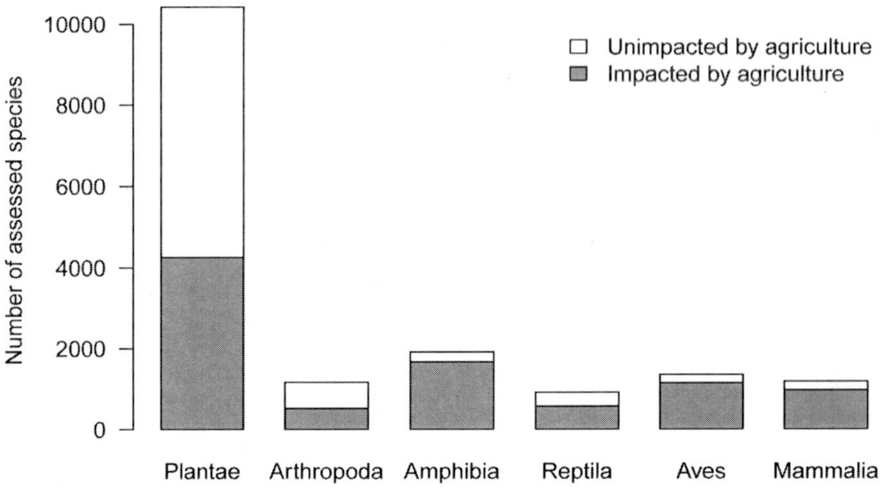

Figure 2. Estimated numbers of threatened species negatively impacted by agriculture.

Agriculture also contributes more to other forms of environmental degradation than any other economic sector. Between 30%–35% of global greenhouse gas (GHG) emissions come from agriculture [3], and crop irrigation accounts for 70% of the world's freshwater withdrawals [13]. The use of synthetic nitrogen fertiliser has increased nearly 21-fold since 1950, and more nitrogen is now added to agricultural soils than from natural processes [14]. Virtually all human-derived nitrogen is lost to the atmosphere or receiving waters [14,15]. This eutrophication suffocates aquatic ecosystems [16]. Phosphorus from fertiliser and livestock manure [14,17], pesticides [18–20], and nanoparticles [21,22] are also exacerbating environmental pollution. Additionally, poor agricultural management causes soil degradation, reducing agricultural productivity and creating further demand for nutrients, water, and land conversion [23,24].

Solutions for limiting agriculture's environmental costs are increasingly well understood. These include reducing land clearance by improving yields of existing crops, minimizing excess nutrient and water use, and shifting production from livestock towards crops [3]. Implementing these practices for only a handful of commodities in a subset of the world can have a large environmental benefit [25]. The success of these practices, however, depends on how well policy puts them into practice.

Challenges to Limiting Agriculture's Environmental Costs

Designing and implementing policies to reduce agriculture's environmental costs is difficult. Like other industries, farmers and the businesses reliant upon them, such as input suppliers, processors, and retailers, have a rational self-interest in maximising economic returns and oppose policies that threaten their viability [26,27]. In developed economies, this has historically allowed them to organise into focused lobby groups with considerable advantage over diffuse public interests seeking to safeguard the environment [28,29]. Governments rarely benefit from challenging agri-food lobbies over environmental issues that lack immediate returns for the public good. These lobbies reward government for less demanding environmental policies [30,31], and tailoring policies appropriate for the range of physical and social conditions that farms occupy is costly [26,27]. Concerns about foodborne illness (including genetically modified organisms), environmental impacts (e.g., pesticide use), and animal welfare have ceded some power in the policy arena of developed countries from producers to consumers [32], and increasingly to large retailers, but existing governance structures remain highly susceptible to farmer interests despite spatiotemporal variation in power relations [29]. For example, corporate lobbying efforts in the agricultural sector were deemed effective by 52% of respondents in a 2013 survey of nearly 600 European Union (EU) decision makers, and >80% effectiveness was reported in Denmark, Latvia, and the United Kingdom [33]. Policymakers therefore face major challenges in aligning the private interests of the agri-food sector with public interest in environmental goods across the diversity and extent of farm entities.

In addition to political challenges, measures that do succeed in reducing the environmental harms of agriculture in one place can perversely increase them elsewhere. This is because, in a global market, local policies or technologies that cause agricultural production to carry a greater opportunity cost than using land for environmental goods will displace production to regions where these costs are lower [34]. For example, regulations limiting fertilisers can reduce local water pollution and GHG emissions. But because such measures may reduce yields (i.e., tonnage of production per unit area), food demand will be met elsewhere, leading to intensification and land clearance that potentially creates a net environmental loss [35]. Incentivising low-intensity systems, seen as beneficial to biodiversity conservation in regions such as Europe, may similarly limit production and displace environmental impacts while also lacking the full benefits of natural ecosystems [36]. Efforts to limit the environmental costs of agriculture must therefore consider the environmental impacts beyond farm boundaries as well as the consequences for food production.

Potential Resolutions to the Agriculture–Nature Conflict

Agricultural policy with explicit environmental objectives is generally organized around three approaches (Table 1 and expanded upon in S1 Table; see Box 1 for an exception):

1. regulations, such as limits on pesticide use or abstracting water, which can be enforced through penalties and conditions placed on financial support to farmers;

2. community-based approaches, which support farmers and local stakeholders to work collectively in addressing environmental impacts; and

3. economic instruments, which pay farmers directly or create markets for adopting practices that minimise environmental impacts and provide noncommodity outputs beyond those required by existing regulation; tariffs can also be used to internalise environmental costs.

Table 1. Successes associated with different policy approaches and the local[1]obstacles to their implementation.

Approach	Success story	Obstacles
Regulations	Critical Counties programme launched in 2008 by Brazilian government suspended access to credit for farmers in counties with the highest deforestation rates. This stimulated collective actions to conserve native vegetation, resulting in deforestation rates dropping to <20% of previous 10-yr average [40].	Complexity, enforceability, cost of implementation. Where compliance is linked to payments, as in EU's Common Agricultural Policy, clear environmental standards and operational guidance required at farm level [41]. Payments must also incorporate regional and/or sectorial variation in the cost of compliance.
Community-based	Landcare Australia is among the best known partnerships between communities, government, and organisations. Established by a $360 million programme in 1989 and engaging about 30% of farming community at its height [42]. Notable successes improving water quality, reducing soil degradation, restoring habitat [43].	Delivering social and/or short-term economic benefits to individual farmers; lack of these benefits has limited uptake of Landcare practices in Australia [44,45].
Economic instruments	Conservation Reserve Program (CRP) established in 1985 by US government provides farmers with annual rental payments for removing land most sensitive to agricultural impacts from production for ≥10 years. Currently, >10 million ha enrolled at annual cost of ca. US$2 billion [46]. CRP has eliminated millions of tonnes of GHG emissions and fertiliser additions and established new habitat supporting tens of millions of birds and other wildlife [47]. Gains arisen despite conversion of non-cropland into production predicted over area equal to 20% of CRP land [48].	Can perform poorly if farmers paid for actions rather than actual environmental outcomes associated with their land [49]. For example, temporary retirement schemes emerged initially in EU to reduce overproduction. As no environmental outcome was explicitly targeted, farmers tended to retire least profitable land that was often the poorest quality, limiting environmental benefits [50,51].

[1]Globally, the major obstacle for all approaches is to avoid displacing production and causing environmental degradation elsewhere.

Box 1. Abandoning the Agricultural Sector to Market Forces

New Zealand radically altered its agricultural policy in 1984 to embrace minimal government intervention. It now relies exclusively on regulation to minimise the environmental impacts of agriculture while providing among the least support to domestic producers out of all developed countries (ca. 0.5% of the value of production; S1 Text). Consequently, the natural environment has been left vulnerable to the pressures of market forces, which are rapidly accelerating production and land conversion with increasing demand for dairy products in particular. The rate of conversion of indigenous grassland to exotic pasture in the South Island has increased by 67% from the period between 1990–2001 to 2001–2008 [52], and New Zealand leads the world in livestock emissions per capita, exceeding the global total across developed countries by more than 13-fold [53].

The nonprescriptive Resource Management Act (RMA) is the primary national legislation intended to oversee natural and physical resources, with land regulation devolved to elected district councils. Although the RMA tasks councils with "maintaining" biodiversity in balance with resource development, they generally expedite the latter. Fewer than 5% of agricultural sector consents requiring biodiversity or ecosystem services maintenance were recently found to comply with conditions [54]. Limited resources are often blamed for poor enforcement, but political interference also occurs [55]. Other regulations, such as the Wildlife Act and Native Plant Protection Act, are rarely invoked, and the latter does not apply to landowners [55].

New Zealand also has a dedicated judicial system to contest decisions made under the RMA. While this system has delivered several notable "wins" for the environment (e.g., [56]), many local actions escape scrutiny. In practice, biodiversity protection by rural landowners occurs mainly on small, voluntary, set-asides of residual unproductive land (known as covenants) and through patchy predator-trapping and stock fencing activities, supported by limited local and national funds. It appears that the laissez-faire approach to the agricultural sector does little to safeguard the environment.

Farmers will be motivated to accept one or more of these approaches where they incur small socioeconomic costs relative to the benefits that they receive from achieving environmental objectives [37,38]. Eliminating financial support for farmers to maximise agricultural production is important in helping to reduce the socioeconomic costs of adopting environmental practices. Reforms in developed economies since the 1980s are now supporting producer incomes more through direct payments than by subsidising production or protecting domestic prices [2,39] (Box 2).

Even with farmer acceptance, success of any policy approach will ultimately depend on whether farmers maintain or boost yields on existing agricultural lands rather than convert new natural lands to production. Halting local land conversion can be achieved with strict enforced regulations (Table 1), but in a globalized market, changes in the local supply of a given cash crop will lead to changes in demand and supply in other distant locations [59]. Trends in cereal production in 19 countries between 1990 and 2005 clearly show this link between local actions and global consequences. These countries managed to boost cereal yields on existing lands while removing cropland from production, but without fully meeting local demand. Per capita cereal imports consequently increased by 12-fold in these countries compared with those lacking land retirement policies [60], clearly suggesting that production was displaced elsewhere. Where subsistence agriculture is important, limiting its local production may similarly boost reliance on cash crops produced abroad [61]. National governments implementing any of the three main policy approaches need to begin working in a coordinated way to combat land use displacement.

Box 2. Environmental Harms of Agricultural Price Support

Price support policies, such as import tariffs or minimum commodity prices, encourage overproduction in the absence of supply management. This strains natural resources and makes marginal land artificially profitable [57]. In the US, the federal government historically managed the dairy industry by guaranteeing to purchase milk if its market price fell below a set value. Modelling revealed that this would drive abandonment and reversion to natural forest across thousands of hectares of land in the state of Wisconsin if milk prices were decreased by only 5%–15%, with millions saved in water treatment and government expenditure [58]. The US has now adopted a strategy of paying producers should their margins fall beneath national averages, likely with similar environmental consequences.

Finding the Right Mixture of Solutions

Maintaining the status quo with agricultural policy is unsustainable for the environment [3,35,62]. On their own, regulations and community approaches will offer little help if damaging the environment remains financially lucrative because the associated penalties are trivial, governments continue to subsidise production (Figure 1, S1 Fig), and farmers are not accountable for the environmental costs of their actions [63–65] (Box 3). Economic instruments, such as payments for public goods, are also contentious, especially for biodiversity conservation [66–68]. Positive outcomes often arise only from certain actions, such as enhancing specific food resources or nesting habitat [69], and catch-all interventions may avoid targeting rare species [70,71]. Removing production subsidies and making farmers financially responsible for their environmental impacts, such as with markets for public goods, can help local outcomes but must avoid curbing intensification because this will encourage land conversion elsewhere to meet demand for food [35].

Box 3. Balancing Large-Scale Incentives for the Rural Poor with Strict Penalties for Agri-business

China—the world's largest agricultural producer—offers a stark contrast in government policy to that implemented by New Zealand described in Box 2. The country's economy is heavily controlled by centralised government and relatively free from the political influence wielded by farmers lobbies.

Government can therefore act largely in its own self-interest, which may align more often with the public good. This has allowed the Chinese government to implement environmental policies on scales that would be virtually impossible elsewhere in the world. The Grain to Green Project is one example, having converted nearly 9.1 million hectares of cropland into forest whilst incentivising farmers to participate with grain and cash subsidies [72]. This permanent land retirement scheme has reduced soil erosion and increased carbon sequestration at landscape scales [73]. China is also adopting a stricter approach to large industrial agriculture. In 2014, the country's top legislature revised the national environmental protection law for the first time in 25 years. The revisions included harsher penalties to deter companies from accepting fines rather than investing in pollution control [74]. Whether enforcement follow suits remains to be seen.

Ruhl [26] argued that successful agricultural policies will target large industrial operations that can be readily regulated while incentivising changes in production practices by small farmers without compromising their livelihoods. Recent developments in China are therefore promising, but the country's agricultural sector still causes major environmental harm, including poor water quality in almost all rivers and lakes [75], and nearly 40% more GHG emissions per tonne of nitrogen fertiliser than in Europe [76]. Subsidies on inputs such as fertilisers and pesticides are largely to blame as they promote overuse. Fertilisers are estimated to be overused by 30%–60% [77], yet nearly CMY¥108,000 million (US$17,000 million) was spent in 2012 on just one scheme to help farmers purchase more of these inputs [78]. Removing harmful financial subsidies would help greatly in aligning the country's agricultural and environmental aims.

We assert that conflict between agriculture and the environment will be best resolved by policies dedicating high-quality habitat towards nature conservation, while encouraging intensive production on existing farmland with stringent limits on environmental impacts [35,36,79–81]. Measures that make farmland itself more benign—so called "land-sharing" approaches—also deliver local environmental and social benefits [82,83], but can reduce agricultural yields and cause land conversion elsewhere, effectively displacing environmental burdens [80]. Intensifying production on existing farmland may thus be a better option because it avoids displacing impacts and can allow other land to be freed for nature conservation and restoration, with no net loss in agricultural production.

Our consideration of policy instruments approaches and the obstacles to their success (Table 1) suggests that our proposed resolution would practically entail a combination of punitive measures, such as taxes that reduce existing overuse of farm inputs [84], with positive incentives linked to measurable outcomes [49], such as for permanently removing land from production. Taxes on farm inputs can improve upon existing payments to limit input use because they do not explicitly cap nutrient, pesticide, and water use. Without input limits, some farmers may find it profitable to intensify existing production, despite the added taxes, while others might profit more from reducing resource consumption. Across the landscape, yields will be maintained, and potentially increased without additional land conversion. For example, water surcharges in Mexico are used to pay landowners to retain forested watersheds. Displaced land conversion from exploitation of communal properties and changing market pressures reduced the environmental benefits of this programme by only ca. 4% [85]. By enabling farms to respond differently to the same policy, input taxes carry an additional advantage over traditional regulations that hold all farmers to the same standard and ignore sectorial diversity [86]. An aligned but more audacious approach would be to replace taxes on income with taxes on environmental consumption. The idea is to shift from taxing the production of private wealth toward taxing the consumption of public environmental goods, thus promoting economic growth from sustainable use rather than depletion of natural capital. Brown et al. [55] have proposed that such a system would most heavily tax intensive land uses that generate the most negative environmental impacts, while areas of intact ecological function would entitle the owners to a rebate. This approach might encourage farmers to work intensively with the most productive land but set aside and restore less productive areas with net gains for the environment.

Evidence at national levels suggests that investments in land retirement and input-limiting schemes, consistent with our suggested policy approach, do deliver local environmental benefits. To illustrate this, we examined trends in two environmental responses, farmland bird populations and GHG emissions from synthetic fertilisers, across countries with similar agricultural patterns, i.e., dominated by livestock and arable crops (see S1 Text for methods). We calculated the annual, per hectare monetary investment in a specific environmental outcome beyond that required by regulation from a detailed inventory of government policies compiled by the OECD (annotated in S1 Data). As expected, because habitat loss is a major threat to farmland and grassland biota, we found that farmland birds benefited as more money was invested in retiring agricultural land (Figure 2). We also found that total emissions of N_2O declined with increasing investment in reducing fertiliser use (Figure 3). While cursory in nature, these results suggest that targeted policies have the potential to deliver positive environmental outcomes.

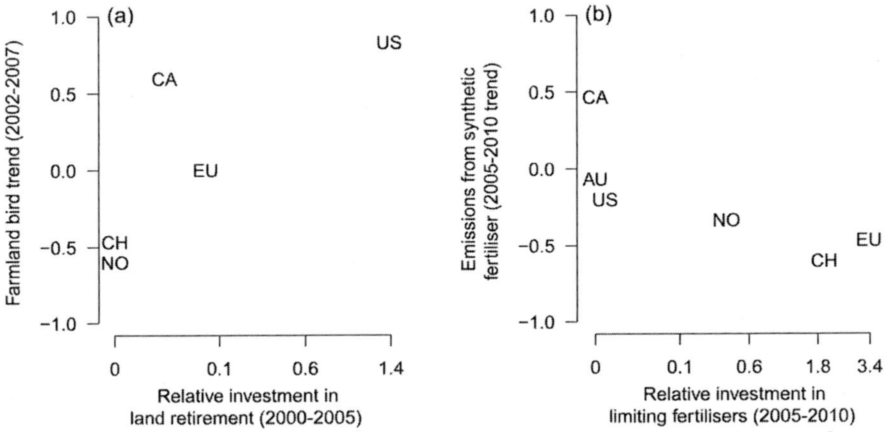

Figure 3. Improvements in environmental performance of agriculture correlate with investments in agri-environment schemes (AES).

Temporal trend in (**A**) farmland bird populations and (**B**) GHG emissions from synthetic fertiliser were calculated over time for each country using Kendall's τ (see S1 Text for full methods). Investments in AES were scaled by the ratio of investments in AES relative to financial transfers to agricultural producers for all other purposes so as to account for the fact that environmental improvements can be offset by policies that maximise production. Values were then summed across 5-y periods with a pre-emptive 2-y lag for land retirement. Habitat quality will take at least one growing season to improve following land retirement and any changes in bird populations from improved breeding success will thus largely be measurable the following year, i.e., two years later [87,88]. Spearman's rank correlation: $\rho = 0.95, p = 0.005$ and $\rho = -0.93, p = 0.008$ for (**A**) and (**B**), respectively. Points are responses in individual countries: AU = Australia [data available for (**B**) only], CA = Canada, EU (aggregated together), NO = Norway, CH = Switzerland, and US. Data used to make this figure are provided in S1 Data.

Measures delivering environmental benefits must also be coupled with mechanisms that promote intensification to avoid displacing production elsewhere, and they must do so with fewer external costs. Research suggests increased yields and better environmental practices are possible by improving nutrient and water management and breeding more efficient crops [3]. Reductions in food waste and human diets with less livestock and dairy consumption can also help meet rising food demands without increasing resource use [89,90]. Dietary change is, however, contentious. Governments may need to consider policies that tax livestock products and/or subsidize the development of livestock-free protein sources, e.g., [91], in order to dislodge entrenched norms.

Here, we have laid out a vision for limiting the environmental impacts of agriculture, yet at least two major challenges remain. First, winning political support for policy change remains a fundamental requirement to any of our proposed reforms. Ultimately, democratically elected governments, while not immune to the lobbying power of agribusiness, will generally respond to electoral pressures [31]. Electorates may express little desire for policy change if they are poorly informed, unmotivated, or misled by industry about limiting the environmental costs of agriculture. Second, implementation of our policy mixture will vary spatially. For example, successful land retirement depends on features of the surrounding landscape, such as habitat connectivity [92], and can reduce yields in regions where the most productive land also has the most valuable environmental assets. Nonetheless, improvements in the management of land resources in developing countries [93], new technologies that optimally allocate farm inputs to reduce overuse [94], advances in crop breeding [95], and tools for better conservation planning [96] give us hope that innovations exist for meeting the need for more sustainable and intensive agriculture.

ACKNOWLEDGMENTS

We thank Andrew Balmford, David Coomes, and staff at the OECD for discussion.

REFERENCES

1. Tilman D, Balzer C, Hill J, Befort BL (2011) Global food demand and the sustainable intensification of agriculture. Proc Natl Acad Sci USA 108: 20260–20264. pmid:22106295
2. Warren J, Lawson C, Belcher K (2007) The agri-environment. Cambridge: Cambridge University Press. 232 p.
3. Foley JA, Ramankutty N, Brauman KA, Cassidy ES, Gerber JS, et al. (2011) Solutions for a cultivated planet. Nature 478: 337–342. pmid:21993620
4. Millennium Ecosystem Assessment. Ecosystems and human well-being: synthesis. Washington, DC: World Resources Institute; 2005. 155 p.
5. James AN, Gaston KJ, Balmford A (1999) Balancing the Earth's accounts. Nature 401: 323–324. pmid:16862091
6. Myers N, Kent J (2001) Perverse subsidies: how tax dollars can undercut the environment and the economy. Washington, DC: Island Press. 277 p.
7. Ramankutty N, Evan AT, Monfreda C, Foley JA. (2008) Farming the planet: 1. Geographic distribution of global agricultural lands in the year 2000. Global Biogeochem Cycles 22: GB1003.

8. Lawton JH (2007) Ecology, politics and policy. J Appl Ecol 44: 465–474.
9. Lambin EF, Meyfroidt P (2011) Global land use change, economic globalization, and the looming land scarcity. Proc Natl Acad Sci USA 108: 3465–3472. pmid:21321211
10. Kissinger G, Herold M, de Sy V (2012) Drivers of deforestation and forest degradation: a synthesis report for REDD+ policymakers. Lexeme Consulting, Vancouver. 46 p.
11. International Union for Conservation of Nature [IUCN] (2014) The IUCN Red List of Threatened Species. Version 2014.3. http://www.iucnredlist.org via the Internet. Accessed 2015 March 14.
12. Bondeau A, Smith PC, Zaehle S, Schaphoff S, Lucht W, et al. (2007) Modelling the role of agriculture for the 20th century global terrestrial carbon balance. Glob Chang Biol 13: 679–706.
13. Food and Agriculture Organization of the United Nations [FAO]. (2015) AQUASTAT database. Available: http://www.fao.org/nr/water/aquastat/data/ via the Internet. Accessed 14 March 2015.
14. Bouwman L, Goldewijk KK, Hoek KWVD, Beusen AHW, Van Vuuren DP, et al. (2013) Exploring global changes in nitrogen and phosphorus cycles in agriculture induced by livestock production over the 1900–2050 period. Proc Natl Acad Sci USA 110: 20882–20887. pmid:21576477
15. Galloway JN, Aber JD, Erisman JW, Seitzinger SP, Howarth RW, et al. (2003) The nitrogen cascade. BioScience 53: 341–356.
16. Diaz RJ, Rosenberg R (2008) Spreading dead zones and consequences for marine ecosystems. Science 321: 926–929. pmid:18703733
17. MacDonald GK, Bennett EM, Potter PA, Ramankutty N (2011) Agronomic phosphorus imbalances across the world's croplands. Proc Natl Acad Sci USA 108: 3086–3091. pmid:21282605
18. Rohr JR, Schotthoefer AM, Raffel TR, Carrick HJ, Halstead N, et al. (2008) Agrochemicals increase trematode infections in a declining amphibian species. Nature 455: 1235–1239 pmid:18972018
19. Henry M, Béguin M, Requier F, Rollin O, Odoux J-F, et al. (2012) A common pesticide decreases foraging success and survival in honey bees. Science 336: 348–350. pmid:22461498
20. Hallmann CA, Foppen RPB, van Turnhout CAM, de Kroon H, Jongejans E (2014) Declines in insectivorous birds are associated with high neonicotinoid concentrations. Nature 511: 341–343. pmid:25030173
21. Priester JH, Ge Y, Mielke RE, Horst AM, Moritz SC, et al. (2012) Soybean susceptibility to manufactured nanomaterials with evidence for food quality and soil fertility interruption. Proc Natl Acad Sci USA 109: E2451–E2456. pmid:22908279
22. Santiago-Martín A, Constantin B, Guesdon G, Kagambega N, Raymond S, et al. (2015) Bioavailability of engineered nanoparticles in soil systems. J. Hazard. Toxic Radioact. Waste in press.
23. Pimentel D, Harvey C, Resosudarmo P, Sinclair K, Kurz D, et al (1995) Environmental and economic costs of soil erosion and conservation benefits. Science 267: 1117–1123. pmid:17789193
24. Lal R (1997) Degradation and resilience of soils. Phil Trans R Soc Lond B 352: 997–1010.
25. West PC, Gerber JS, Engstrom PM, Mueller ND, Brauman KA, Carlson KM, et al. (2014) Leverage points for improving global food security and the environment. Science 345: 325–328. pmid:25035492
26. Ruhl JB (2000) Farms, their environmental harms, and environmental law. Ecol Law Q 27: 263–349.
27. Walker S, Brower AL, Stephens RTT, Lee WG (2009) Why bartering biodiversity fails. Conserv Lett 2: 149–157.
28. Olson M (1965) The logic of collective action. Cambridge: Harvard University Press. 208 p.

29. Daugbjerg C, Swinbank A (2012) An introduction to the "new" politics of agriculture and food. Policy Soc 31: 259–270.
30. Eskridge WN Jr (1988) Politics without romance: implications of public choice theory for statutory interpretation. Va Law Rev 74: 275–338.
31. Bellemare MF, Carnes N (2015) Why do members of congress support agricultural protection? Food Policy 50: 20–34.
32. Grant WP (2012) Can political science contribute to agricultural policy? Policy Soc 31: 271–279.
33. .Burson-Marsteller (2013) A guide to effective lobbying in Europe: the view of policy-makers. Available: http://burson-marsteller.nl/wp-content/uploads/2013/06/Lobbying-survey-Report.pdf. Accessed 1 June 2015.
34. Meyfroidt P, Rudel TK, Lambin EF (2010) Forest transitions, trade, and the global displacement of land use. Proc Natl Acad Sci USA 107: 20917–20922. pmid:21078977
35. Balmford A, Green R, Phalan B (2012) What conservationists need to know about farming. Proc Roy Soc B 279: 2714–2724.
36. Merckx T, Pereira HM (2015) Reshaping agri-environmental subsidies: From marginal farming to large-scale rewilding. Basic and Applied Ecology 16: 95–103.
37. Lastra-Bravo XB, Hubbard C, Garrod G, Tolón-Becerra A (2015) What drives farmers' participation in EU agri-environmental schemes? Results from a qualitative meta-analysis. Environmental Science & Policy 54: 1–9.
38. Wilson GA, Hart K (2000) Financial imperative or conservation concern? EU farmers' motivations for participation in voluntary agri-environmental schemes. Environment and Planning A 32: 2161–2185.
39. Organisation for Economic Co-operation and Development [OECD] (2004) Agriculture and the environment: lessons learned from a decade of OECD work. Paris: OECD. 36 p.
40. Nepstad D, McGrath D, Stickler C, Alencar A, Azevedo A, et al. (2014) Slowing Amazon deforestation through public policy and interventions in beef and soy supply chains. Science 344: 1118–1123. pmid:24904156
41. European Court of Auditors (2008) Is cross compliance an effective policy? Special report no 8/2008. Luxembourg: Office for Official Publications of the European Communities. 57 p.
42. Curtis A, Lockwood M (2000) Landcare and catchment management in Australia: lessons for state-sponsored community participation. Soc Nat Resour 13: 61–73.
43. OECD (2013) Providing agri-environmental public goods through collective action. Paris: OECD. 306 p.
44. Curtis A, Lacy TD (1998) Landcare, stewardship and sustainable agriculture in Australia. Environ Values 7: 59–78.
45. Pannell DJ, Marshall GR, Barr N, Curtis A, Vanclay F, et al. (2006) Understanding and promoting adoption of conservation practices by rural landholders. Aust J Exp Agric 46: 1407–1424.
46. Stubbs M (2014) Conservation Reserve Program (CRP): status and issues. Washington, DC: Congressional Research Service. 24 p.
47. United States Department of Agriculture (2011) CRP benefits. http://www.fsa.usda.gov/Internet/FSA_File/united_states.pdf via the Internet. Accessed 14 March 2015.
48. Wu J (2000) Slippage effects of the conservation reserve program. Am J Agric Econ 82: 979–992.
49. Franks JR, McGloin A (2007) Joint submissions, output related payments and environmental co-operatives: Can the Dutch experience innovate UK agri-environment policy? Journal of Environmental Planning and Management 50: 233–256.
50. Sotherton NW (1998) Land use changes and the decline of farmland wildlife: An appraisal of the set-aside approach. Biol Conserv 83: 259–268.
51. Robinson GM, Lind M (1999) Set-Aside and environment: a case study in southern England. Tijdschr Econ Soc Geogr 90: 296–311.

52. Weeks ES, Walker SF, Dymond JR, Shepherd JD, Clarkson BD (2012) Patterns of past and recent conversion of indigenous grasslands in the South Island, New Zealand. N Z J Ecol 37: 127–138.
53. Caro D, Davis SJ, Bastianoni S, Caldeira K (2014) Global and regional trends in greenhouse gas emissions from livestock. Climatic Change 126: 203–216.
54. Brown MA, Clarkson BD, Stephens RT, Barton BJ (2014) Compensating for ecological harm–the state of play in New Zealand. N Z J Ecol 38: 139–146.
55. Brown MA, Stephens RTT, Peart R, Fedder B (2015) Vanishing nature: facing New Zealand's biodiversity crisis. Auckland: Environmental Defence Society. 208 p.
56. Forest & Bird (2013) 2013 Annual Report. http://www.forestandbird.org.nz/files/file/FB%20Annual%20Report%202013%20V4%20low%20res.pdf via the Internet. Accessed 14 March 2015.
57. Abler D (2004) Multifunctionality, agricultural policy, and environmental policy. Agricultural and Resource Economics Review 33: 8–17.
58. Plantinga AJ (1996) the effect of agricultural policies on land use and environmental quality. Am J Agr Econ 78: 1082–1091.
59. MacDonald GK, Brauman KA, Sun S, Carlson KM, Cassidy ES, Gerber JS, et al. (2015) Rethinking agricultural trade relationships in an era of globalization. BioScience 65: 275–289.
60. Rudel TK, Schneider L, Uriarte M, Turner BL, DeFries R, Lawrence D, et al. (2009) Agricultural intensification and changes in cultivated areas, 1970–2005. Proc Natl Acad Sci USA 106: 20675–20680 pmid:19955435
61. Murray BC (2008) Leakage from an avoided deforestation compensation policy: concepts, empirical evidence, and corrective policy options. Raleigh: Nicholas Institution for Environmental Policy Solutions. 32 p.
62. Pe'er G, Dicks LV, Visconti P, Arlettaz R, Báldi A, et al. (2014) EU agricultural reform fails on biodiversity. Science 344:1090–1092. pmid:24904142
63. Pretty JN, Brett C, Gee D, Hine RE, Mason CF, et al. (2000) An assessment of the total external costs of UK agriculture. Agric Syst 65: 113–136.
64. Pretty J, Brett C, Gee D, Hine RE, Mason CF, et al. (2001) Policy challenges and priorities for internalizing the externalities of modern agriculture. Journal of Environmental Planning and Management 44: 263–283.
65. Balmford A, Bruner A, Cooper P, Costanza R, Farber S, et al. (2002) Economic reasons for conserving wild nature. Science 297: 950–953. pmid:12169718
66. Kleijn D, Sutherland WJ (2003) How effective are European agri-environment schemes in conserving and promoting biodiversity? J Appl Ecol 40: 947–969.
67. Kleijn D, Baquero RA, Clough Y, Díaz M, De Esteban J, et al. (2006) Mixed biodiversity benefits of agri-environment schemes in five European countries. Ecol Lett 9: 243–254. pmid:16958888
68. Whittingham MJ (2007) Will agri-environment schemes deliver substantial biodiversity gain, and if not why not? J Appl Ecol 44: 1–5.
69. Baker DJ, Freeman SN, Grice PV, Siriwardena GM (2012) Landscape-scale responses of birds to agri-environment management: a test of the English Environmental Stewardship scheme. J Appl Ecol 49: 871–882.
70. Kleijn D, Berendse F, Smit R, Gilissen N (2001) Agri-environment schemes do not effectively protect biodiversity in Dutch agricultural landscapes. Nature 413: 723–725. pmid:11607029
71. 71.71.Vickery JA, Bradbury RB, Henderson IG, Eaton MA, Grice PV (2004) The role of agri-environment schemes and farm management practices in reversing the decline of farmland birds in England. Biol Conserv 119: 19–39.
72. State Forestry Administration of China (2014) Forestry in China. http://english.forestry.gov.cn/uploads/Information_Services/Latest_Publication/Forestry_in_China.pdf via the Internet. Accessed 14 March 2015.
73. Liu J, Li S, Ouyang Z, Tam C, Chen X (2008) Ecological and socioeconomic effects of China's policies for ecosystem services. Proc Natl Acad Sci USA 105: 9477–9482 pmid:18621700

74. National People's Congress of the People's Republic of China. (2014) China's legislature adopts revised Environmental Protection Law. http://www.npc.gov.cn/englishnpc/news/Legislation/2014-04/25/content_1861275.htm via the Internet. Accessed 14 March 2015.
75. Sun B, Zhang L, Yang L, Zhang F, Norse D, et al. (2012) Agricultural non-point source pollution in china: causes and mitigation measures. Ambio 41: 370–379. pmid:22311715
76. Zhang W, Dou Z, He P, Ju X-T, Powlson D, et al. (2013) New technologies reduce greenhouse gas emissions from nitrogenous fertilizer in China. Proc Natl Acad Sci USA 110: 8375–8380. pmid:23671096
77. Ju X-T, Xing G-X, Chen X-P, Zhang S-L, Zhang L-J et al. (2009) Reducing environmental risk by improving N management in intensive Chinese agricultural systems. Proc Natl Acad Sci USA 106: 3041–3046. pmid:19223587
78. OECD (2015) Producer and consumer support estimates database. Available http://www.oecd.org via the Internet. Accessed 14 March 2015.
79. Hodgson JA, Kunin WE, Thomas CD, Benton TG, Gabriel D (2010) Comparing organic farming and land sparing: optimizing yield and butterfly populations at a landscape scale. Ecol Lett 13:1358–1367. pmid:20825453
80. Phalan B, Onial M, Balmford A, Green RE (2011) Reconciling food production and biodiversity conservation: land sharing and land sparing compared. Science 333: 1289–1291. pmid:21885781
81. Godfray HCJ, Garnett T (2014) Food security and sustainable intensification. Phil Trans R Soc Lond 369: 20120273.
82. 82.82.Fischer J, Abson DJ, Butsic V, Chappell MJ, Ekroos J, Hanspach J, et al. (2014) Land sparing versus land sharing: moving forward. Conserv Lett 7: 149–157.
83. Frishkoff LO, Karp DS, M'Gonigle LK, Mendenhall CD, Zook J, Kremen C, et al. (2014) Loss of avian phylogenetic diversity in neotropical agricultural systems. Science 345: 1343–1346. pmid:25214627
84. Mueller ND, Gerber JS, Johnston M, Ray DK, Ramankutty N, et al. (2012) Closing yield gaps through nutrient and water management. Nature 490: 254–257. pmid:22932270
85. Alix-Garcia JM, Shapiro EN, Sims KRE (2012) Forest conservation and slippage: evidence from Mexico's national payments for ecosystem services program. Land Econ 88: 613–638.
86. Stavins RN, Whitehead BW (1992) Pollution charges for environmental protection: a policy link between energy and environment. Annual Review of Energy and the Environment 17: 187–210.
87. Newson SE, Johnston A, Renwick AR, Baillie SR, Fuller RJ (2012) Modelling large-scale relationships between changes in woodland deer and bird populations. J Appl Ecol 49: 278–286.
88. Chamberlain D E, Fuller RJ, Bunce RGH, Duckworth JC, Shrubb M (2000) Changes in the abundance of farmland birds in relation to the timing of agricultural intensification in England and Wales. J Appl Ecol 37: 771–788.
89. Bajželj B, Richards KS, Allwood JM, Smith P, Dennis JS, et al. (2014) Importance of food-demand management for climate mitigation. Nature Clim Change 4: 924–929.
90. Tilman D, Clark M (2014) Global diets link environmental sustainability and human health. Nature 515: 518–522. pmid:25383533
91. Pandya R (2014) Milk without the moo. New Sci 222: 28–29.
92. Gilroy JJ, Edwards FA, Uribe CAM, Haugaasen T, Edwards DP (2014) Surrounding habitats mediate the trade-off between land-sharing and land-sparing agriculture in the tropics. J Appl Ecol 51: 1337–1346.
93. Hobbs PR, Sayre K, Gupta R (2008) The role of conservation agriculture in sustainable agriculture. Phil Trans R Soc Lond B 363: 543–555.
94. Chen X-P, Cui Z-L, Vitousek PM, Cassman KG, Matson PA, et al. (2011) Integrated soil–crop system management for food security. Proc Natl Acad Sci USA 108: 6399–6404. pmid:21444818

95. Tester M, Langridge P (2010) Breeding technologies to increase crop production in a changing world. Science 327: 818–822. pmid:20150489
96. Bateman IJ, Harwood AR, Mace GM, Watson RT, Abson DJ, et al. (2013) Bringing ecosystem services into economic decision-making: land use in the United Kingdom. Science 341: 45–50. pmid:23828934
97. Anderson K, Nelgen S (2013) Updated national and global estimates of distortions to agricultural incentives, 1955 to 2011. Available: http://www.worldbank.org/agdistortions via the Internet. Accessed 2015 June 1.
98. Anderson K, Kurzweil M, Martin W, Sandri D, Valenzuela E (2008) Measuring distortions to agricultural incentives, revisited. World Trade Rev 7: 1–30.

CHAPTER 16

Modelling of Climate Conditions in Forest Vegetation Zones as a Support Tool for Forest Management Strategy in European Beech Dominated Forests

Ivo Machar [1,*], Veronika Vlckova [2], Antonin Bucek [3], Vit Vozenilek [4], Lubomir Salek [5] and Lucie Jerabkova [6]

[1]*Department of Development Studies, Faculty of Science, Palacky University, 17 listopadu 12, 771 46 Olomouc, Czech Republic*

[2]*Department of Transport Telematics, Faculty of Transportation Sciences, Czech Technical University, Konviktska 20, 110 00 Prague, Czech Republic*

[3]*Department of Forest Botany, Faculty of Forestry and Wood Technology, Mendel University, Geobiocoenology and Dendrobiology, Zemedelska 3, 613 00 Brno, Czech Republic; bucek @mendelu.cz*

[4]*Department of Geoinformatics, Faculty of Science, Palacky University, 17 listopadu 12, 771 46 Olomouc, Czech Republic*

[5]*Department of Forest Management, Faculty of Forestry and Wood Sciences, Czech University of Life Sciences, Kamycka 129, 165 00 Prague, Czech Republic*

[6]*Department of Geography, King's College London, King's Building, Strand London, London WC2R 2LS, UK*

ABSTRACT

The regional effects of climate change on forest ecosystems in the temperate climate zone of Europe can be modelled as shifts of forest vegetation zones in the landscape, northward and to higher elevations. This study applies a biogeographical model of climate conditions in the forest vegetation zones of the Central European landscape, in order to predict the impact of future climate change on the most widespread tree species in European deciduous forests—the European beech (*Fagus sylvatica* L.). The biogeographical model is supported by a suite of software applications in the GIS environment. The model outputs are defined as a set of conditions - climate scenario A1B by the Special Report on Emission Scenarios (SRES) for a forecast period, for a specified geographical area and with ecological conditions appropriate for the European beech, which provide regional scenarios for predicted future climatic conditions in the context of the European beech's environmental requirements. These predicted changes can be graphically visualized. The results of the model scenarios for regional climate change show that in the Czech Republic from 2070 onwards, optimal growing conditions for the European beech will only exist in some parts of those areas where it currently occurs naturally. Based on these results, it is highly recommended that the national strategy for sustainable forest management in the Czech Republic be partly re-evaluated. Thus, the presented biogeographical model of climate conditions in forest vegetation zones can be applied, not only to generate regional scenarios of climate change in the landscape, but also as a support tool for the development of a sustainable forest management strategy.

Keywords: biogeographical model; climate change scenarios; *Fagus sylvatica*; sustainable forest management

1. INTRODUCTION

Increasingly, severe and more frequent drought events are expected to become a major risk for forest ecosystems under predicted climate change [1]. The changes in climate conditions are expected to result in changes in the forest structure [2]. We know that the current form of forest management in Central Europe, with its preference for conifer monocultures, will not be sustainable in the future [3,4]. A substantial increase in air temperature, along with a simultaneous reduction in precipitation during the growing season in Central and Southern Europe, will increase the likelihood of long and intensive summer droughts, which may have severe effects on vegetation in natural and managed ecosystems [5]. Timber production from these ecosystems also needs to be maintained because timber is a renewable material and, as such, it represents an environmentally viable material option. In the face of change and uncertainty [6], a forester needs to know the types of forests that will be most suitable under future climatic conditions. However, forests develop slowly, and during periods of rapid climate change they are likely to be in disequilibrium with the environment, leading to changed competition and possibly mortality [7]. Conversions to more drought-adapted forest types have been recommended in order to prevent climate change-induced forest dieback, and this should be the aim of any adaptive management, especially in areas where monocultures of drought-sensitive Norway spruce (*Picea abies* L. Karst) were promoted in the past. Although a complete forest conversion takes up to 120 years [8], a gradual process of introducing drought-tolerant species into drought-sensitive conifer monocultures provides immediate benefits, with the short-term enhancement of biodiversity, and the maintenance of timber production. This process includes planning for future long-term conditions.

The European beech (*Fagus sylvatica* L.) is the dominant tree species of the potential natural vegetation, from planar to montane vegetation zones in temperate Central Europe [9], and it has a key role in the European forest transition and adaptation strategies [10]. In the future, the European beech may occupy a larger area of the lower uplands and mountains, while its area in the lowlands and lower uplands may be restricted due to higher competitive pressure from the oak [11]. European beech distribution has been at equilibrium with the climate on its north-eastern leading edge, but not on its north-western leading edge [12]. The annual gross primary production of European beech is primarily affected by spring temperatures and, more irregularly, by summer water stress [13]. Based on climate change scenarios, by 2100, we can expect an expansion of climatically suitable habitats for the European beech towards northeastern Europe and higher elevations, especially in the Alps and the Pyrenees. At the same time, a sharp decline in suitable areas for the European beech in Western Europe can be expected [14]. The changing climate may also affect the European beech's life history. In the next few decades, the European beech's maximum lifespan could be shortened at higher elevations in Central Europe because of its faster growth, and in the Mediterranean mountains (e.g., Apennines) due to drought-induced mortality [15]. Several studies of European beech forests point to particular sensitivities to meteorological summer drought conditions [16,17,18,19,20], indicating that the growing season's precipitation totals, which are often used to characterize the local growing conditions of stands of trees, are only subordinate determinants to above-ground productivity in European beech forests. Sensitivity to drought can also be reduced with an appropriate thinning strategy [21]. However, some studies show that European beech dominated forests are not severely prone to negative functional climate change effects and, therefore, their key functions, such as welfare and recreation, can be maintained [22].

In lowland areas, the European beech is often associated with the oak (*Quercus* sp.). Each of these species have different ecological requirements and the dominance of the European beech has begun to decline in locations where the mean annual temperature exceeds 11.2 °C and the annual precipitation drops below 510 mm [23]. It is predicted that in Serbia (the European beech's most southeastern occurrence in Europe), by the end of the 21st century, approximately 90% of the current European beech forests will be outside their 20th century bioclimatic niche, and approximately 50% of European beech forests will be in a bioclimatic niche where mass European beech mortality has been recorded [24].

Although the European beech shows sensitivity to drought in forests where it is mixed with the Norway spruce, it can mitigate this limitation, as became evident during the extremely dry years of 1976 and 2003 in Central Europe [25,26]. The European beech trees near their distribution limit have already adapted to extreme conditions, while we have to expect changes in the growth patterns of the beech under mesic conditions [27]. In addition, according to [28], the sensitivity of the European beech to environmental constraints depends on neighborhood identity. Therefore, a systematic formation of mixed stands tends to be an appropriate sylvicultural measure in order to mitigate the effects of global warming and droughts on European beech growth patterns. However, management practices in the area of nature conservation strategies have not been substantially affected by scientific recommendations for forest management, and de [29] cites that 'Pure' science cannot solve the big uncertainties that surround climate change.

In forest ecosystems, the effects of climate change on a regional scale are most evident in the shift of forest vegetation zonation to higher elevations [30,31]. Nevertheless, disturbances, interspecific competition, different phenotype plasticity and various adaptations of dominant species to a particular ecosystem can all introduce considerable uncertainties into regional models of shifts in forest vegetation zones [32]. The growth response to climate change is manifest in long-term horizons [33], and the best recent records are probably in the shifts of the tree line in European mountains [34], although one of the most important drivers for upper tree line shift in the past was the management of alpine areas (e.g., grazing).

An understanding of the ongoing and potential shifts in vegetation zones is important for the development of strategies of sustainable forest management in the context of climate change [35]. The forest vegetation zones represent the regional fundamental frames of climatic conditions for the growth of forest tree species in commercial forests [36], which dominate in the European temperate zone. However, previously published studies of vegetation zone shifts are limited to a few countries dominated by Alpine regions [37,38,39]. Aside from the Alpine region, Svajda et al. [40] studied the altitudinal shift of the Dwarf mountain's pine vegetation at the upper forest limit in the High Tatras (Slovakia)and Kutnar et al. [41] published predictions concerning the changes in the vegetation zones under climate changes in Slovenia. Predictions concerning the influence of the shift of the vegetation zones on tree species in Poland were elaborated on by [42], and the impact of terrain and vegetation structure on the tree line dynamic in the Sudeten region in central Europe was studied by [43].

Regional studies conducted in several countries, with the aim of predicting which forests would be more suitable for future conditions, have come up with similar recommendations concerning reductions in spruce monocultures, and giving preference to mixed forests with a higher proportion of deciduous tree species, particularly the European beech [44,45,46].

Our study aims to apply the forest vegetation zonation for modelling to future changes in climate conditions for the European beech, which is currently one of the most important deciduous tree species in European forests [47] (EUFORGEN 2011). The main aim of this article is to show how predicted changes in climate conditions in forest vegetation zones, as related to regional climate scenarios, can influence the European beech's growth in the Central European landscape. Modelling climate condition changes in relation to the European beech can provide support for the development of strategies of adaptation measures in forest stands dominated by the species. They also provide a science-based rationale for forest management decisions [48] based on the adaptation approach to forest ecosystems [49]. Forest management adaptation strategies [50] aimed at sustainability must take into account a range of economic [51] and social aspects [52,53,54]. There are many uncertainties concerning the impact of climate change on European forests [6], so another goal of this study is to highlight the importance of the somewhat neglected biogeographical models, rooted in the concept of the forest vegetation zones. This article presents the results of the biogeographical model applications for the European beech in the cultural landscape of the Czech Republic for the forecast period 2030–2090.

2. MATERIALS AND METHODS

This biogeographical model of climate conditions in the forest vegetation zones of the Czech Republic (hereinafter, model) is based on the relationship between the present climate and the distribution of vegetation zones in the landscape [55]. The model (Figure 1) is based on fundamental assumptions: (i) that the general ecological relationship between climate conditions and vegetation zonation will be maintained in the future [56,57]; (ii) that predicted climate change will be manifested into modified climate conditions in the present vegetation zones at the regional level; and (iii) that this process can be predicted [58]. It does not imply that the present vegetation zones within the cultural landscape will shift to higher elevations. The model outlines projections of changes in climate conditions to vegetation zonation, not changes in the vegetation zones as such. This fact is an important basic principle of the model. The model uses the forest vegetation zones as a reference framework for climate predictions of particular growing conditions for the European beech, for which the optimal climate conditions in the Czech Republic are unambiguously defined [59]. The current representation of the European beech in the forests of the Czech Republic is only 7.7%, which is considerably less than the expected representation (40.2%) in the potential natural vegetation [60].

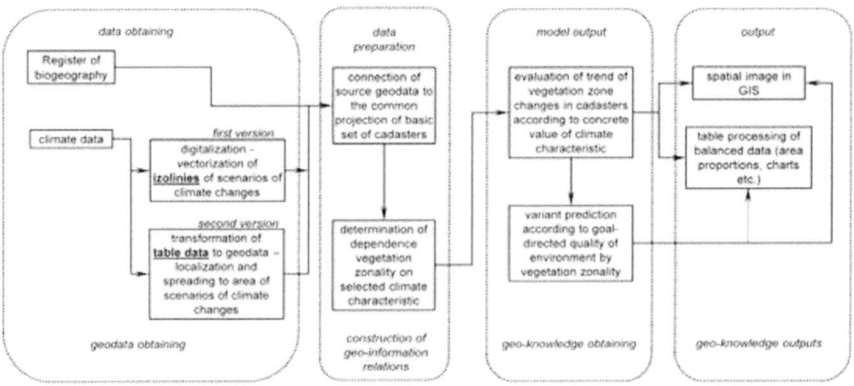

Figure 1. Flowchart of biogeographic envelope for a regional model of climate conditions in vegetation zones of the Czech Republic

The forest vegetation zones of the Czech Republic (Table 1) were defined using the method of bio-indication in the fundamental study by [61], and are widely applied in forest management practice. The detailed characteristics of the natural conditions of forest ecosystems in the forest vegetation zones of the Czech Republic [62] can easily be converted to the European forest, agricultural and protection typology systems related to habitats (e.g., in the NATURA 2000 European Network). Detailed climate characteristics of the vegetation zones are based on Macků [63].

Table 1. Current extent and future areas with projected climatic characteristics corresponding to the respective forest vegetation zones of the Czech Republic.

	Forest Vegetation Zone	Extent of Climate Conditions for Vegetation Zones in				
		2010	2030	2050	2070	2090
		% from the Area of the Czech Republic				
1.	Oak	3.46	3.98	12.78	38.41	40.82
2.	Beech-oak	12.06	14.29	5.49	38.51	38.51
3.	Oak-beech	18.21	20.14	20.14	16.41	16.41
4.a	Beech	39.62	46.22	54.91	5.60	3.98
4.b	Oak-coniferous variety of the fourth vegetation belt	3.62	8.69	0.00	0.00	0.00
5.	Fir-beech	19.40	5.60	5.60	1.07	0.28
6.	Spruce-fir-beech	2.53	0.80	0.80	0.00	0.00
7.	Spruce	1.00	0.28	0.28	0.00	0.00
8.	Dwarf pine subalpine	0.10	0.00	0.00	0.00	0.00

A predictive climate database from the Czech Hydrometeorological Institute (CHMI), known as CLIMDATA, is the source of climate data for the model. It assigns the climate data to the set of 131 points regularly distributed throughout the Czech Republic in the form of the regular trapezia network. It contains a large validated database of climate parameters calculated by the model ALADIN-CLIMATE.CZ for the period 2010–2100 [64], for the scenario SRES A1B [65].

The Register of Biogeography [66] is an ecological database which includes geographical and site information, and it is the source of biogeographical data used in the model. It contains the detailed ecosystem characteristics of the Czech Republic's landscape (i.e., vegetation zonation, soil types and soil conditions the so-called trophic and hydric series, the predominant type of topography) matched with individual cadaster areas. The Czech Republic consists of 13,000 cadasters (polygons with an average area of 6 km^2). The register was created as a component of the Area Integral Information System (AIIS) [67]. The cadaster was selected as a basic spatial unit for the Register because it enables the evaluation of changes in the landscape, and utilizes periodically updated databases from the AIIS which characterize the current state of, and anthropogenic pressure on, the landscape. Census data and the utilization of soil resources are of particular interest, and they both use the cadaster as a basic unit. Cadasters are historic landscape units, originally established to document land tenure and estates, and as such they are not homogeneous in their natural conditions. Nevertheless, on the regional scale (all of the Czech Republic), the cadaster polygons capture the heterogeneity of the natural conditions in the whole country, because the original cadaster system, in existence without many changes since the 18th century, used natural boundaries such as streams, forest edges, and major geomorphological formations in the landscape to delineate its units [68].

The biogeographical model of changes to climatic conditions in the vegetation zones is a suite of specific software applications (FORTRAN programming language, IBM, New York, NY, USA) and a GIS application of Esri products (ArcGIS, Prague, Czech Republic). This is a static model that does not enable prediction of the rate of vegetation change. Climate characteristics were assigned to the points of the Register of Biogeography using an analytical-geometric method, which involved the construction of a network of points with a fine resolution of 250 m. The climatic data from these new points were calculated with a gradient method [69] and taken from the climatic variable of their four closest neighbors in the original climatic database CLIDATA. The predicted climate characteristics of defining points, their corresponding potential vegetation zone and the characteristics of natural climate conditions are expressed in the model algorithms. The algorithms were developed using the method of spatio-temporal analogies, with Lang's Rainfall Factor used as the relationship coefficient (combining total annual precipitation and average annual temperature in one value) [70]. The climate growing conditions for the European beech in the Czech Republic were divided into four categories: "unsuitable climate conditions", "moderately suitable climate conditions", "suitable climate conditions", and "optimal climate conditions" (related to climate conditions for different vegetation zones). These divisions in predicted climate conditions enable a clear visualization of the model outputs. The model outputs for the defined scenario combinations (time period, climate scenario, geographic area, ecological conditions for the European beech) provided a regional scenario of predicted future climate conditions for the European beech in the forecast period 2030–2090.

3. RESULTS

3.1. Regional Scenario of Changes in Climate Conditions of Vegetation Zonation in the Czech Republic

The regional scenario of the changes in climate conditions for vegetation zonation in the Czech Republic for the forecast period 2030–2090 (regional scenario) clearly identifies three main predicted trends (Table 1): (i) a gradual increase in area for regions with climate conditions of the lower forest vegetation zones (1–3); (ii) a relatively stable area with climate conditions of the fourth forest vegetation zone, which remains the vegetation zone with the largest representation (55%) on the landscape in the Czech Republic, in both predicted periods; and (iii) a significant and very fast decrease in area for regions with climate conditions of the higher forest vegetation zones (5–8).

3.2. Application of the Regional Scenario for Sustainable Management of Beech-Dominated Forests

Good and optimal growing conditions for the European beech currently prevail in two thirds of the Czech Republic (Table 2), and cover an area from the second to the sixth forest vegetation zones (Figure 2).

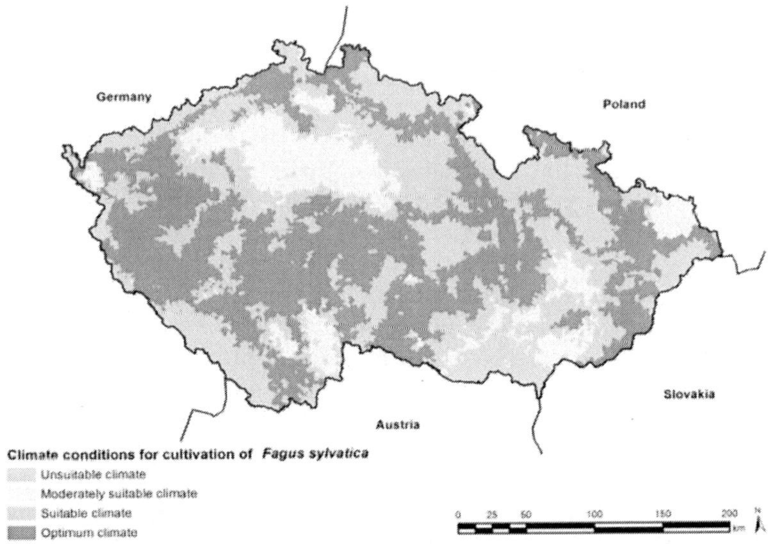

Figure 2. Present status (2010) of climate conditions for European beech in the Czech Republic.

Table 2. Current state and prediction of the climate conditions for European beech in the Czech Republic (forecast period 2030, 2050, 2070, and 2090 year).

Climate Conditions	2010	2030	2050	2070	2090
			km^2		
Unsuitable climate	2144.86	2728.52	2144.86	30,281.86	30,281.86
Moderately suitable climate	10,624.10	18,522.57	12,258.37	23,510.92	24,781.26
Suitable climate	30,883.58	23,772.25	23,027.48	22,092.23	21,281.24
Optimum climate	35,182.71	33,811.91	41,404.54	2950.25	2490.89

The visualization of the regional scenario for the year 2030 (Figure 3) shows a decrease in areas with optimal climate conditions for the European beech in the lower uplands and, instead, a slight enlargement of the area with mediocre climate conditions in this terrain. In the prediction period for 2050, the trend is reversed (Figure 4) and the area with the optimal climate conditions for the growth of European beech is 10% larger than today, according to the regional scenario (Table 2). However, the regional scenario predicts a relatively dramatic deterioration in the growing conditions for the European beech between the years 2070 and 2090. In 2070, in connection with an elevational shift of climate conditions in the forest vegetation zones to higher elevations (Figure S1), the optimal climate for the growth of the European beech will move to the highest mountain areas of the Czech Republic (higher than 1000 m above sea level) and a good climate for growing the European beech will remain in and above the area of the present fifth forest vegetation zone. The proportion of the good and excellent climate conditions will decrease to less than 10% of the whole area of the Czech Republic (Table 2). This trend will remain the same for the year 2090. The predicted trend in the changes in climatic growing conditions for the European beech in the Czech Republic forecasts a significant deterioration of its silviculture in the future. Following the regional scenarios, a strategy of sustainable forest management in the Czech Republic should take into consideration potential future threats likely to affect the European beech, which is now the major economic deciduous tree species in the forest sector in the Czech Republic. The predicted shift in the good and optimal climatic growing conditions for the European beech to the highest mountain areas in the Czech Republic will most likely affect the biodiversity associated with the present vegetation belts. In particular cadasters, according to the regional scenario, from 2070 onwards, the good and optimal climatic growing conditions for European beech dominated forests will only be found in parts of the areas currently occupied by natural European beech, from the fifth to the eighth forest vegetation zones (Figure 5 and Figure 6). Based on the regional scenario results, the national strategy of sustainable forest management in the Czech Republic (National Forest Program) should pay more attention to plans for alternative tree species composition in future planning, with regard to the predicted deterioration of climatic growing conditions for the European beech from the year 2070 (Table 2, Figure 5 and Figure 6).

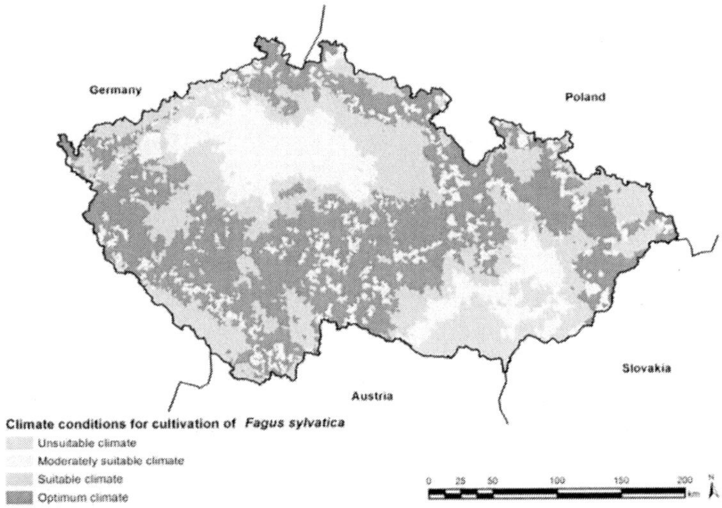

Figure 3. Visualization of predicted distribution of climate conditions for the European beech in the Czech Republic in 2030.

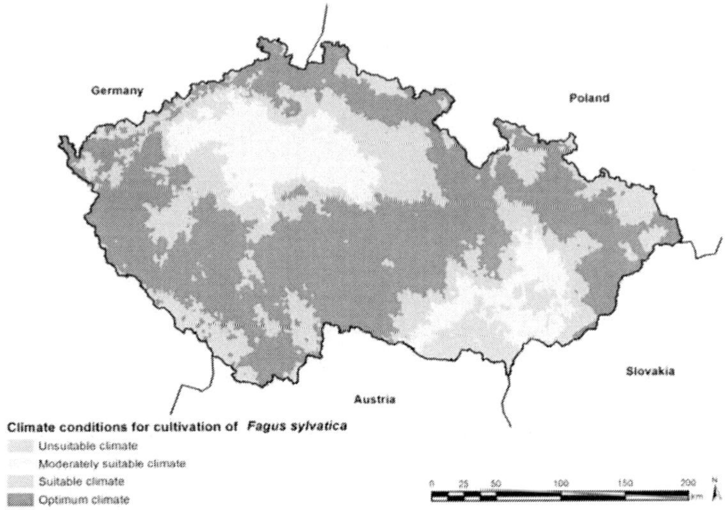

Figure 4. Visualization of predicted distribution of climate conditions for the European beech in the Czech Republic in 2050.

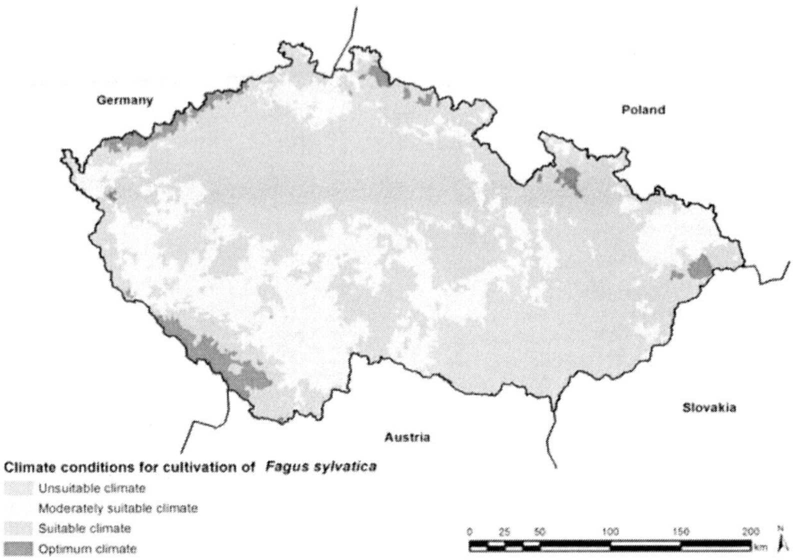

Figure 5. Visualization of predicted distribution of climate conditions for the European beech in the Czech Republic in 2070.

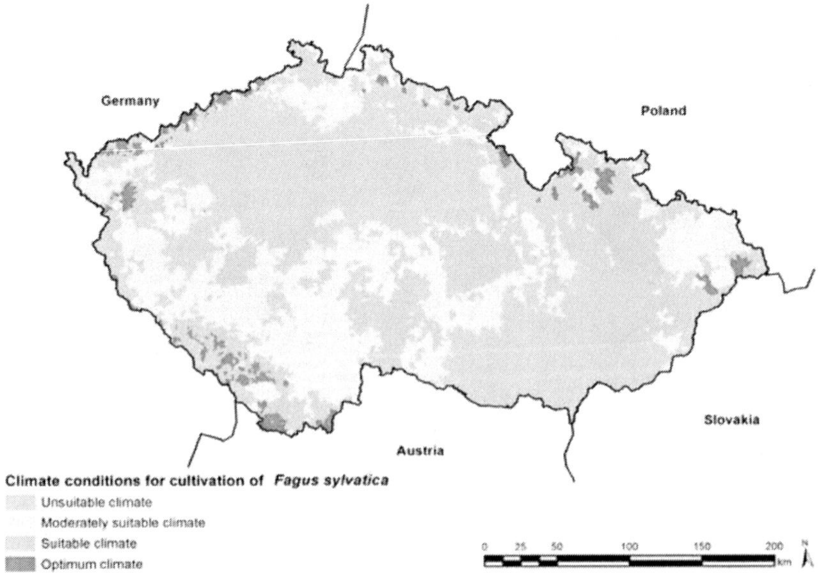

Figure 6. Visualization of predicted distribution of climate conditions for the European beech in the Czech Republic in 2090.

4. DISCUSSION

Global Climate Envelope Models [71], also referred to as Ecological Niche Models or Species Distribution Models, are statistical predictive tools applied in ecological research to estimate the distribution of species, biological communities and habitats [72], and may not always correspond to specific local conditions [73]. This applies to the European beech, which grows in central Europe under fairly specific climate conditions [74], with a natural elevational distribution from 350 m to 990 m.a.s.l. [75], tested the influence of climate and tree-specific traits on inter and intra-site variability in the drought responses of three common European temperate forest tree species in southern Germany and Alpine Austria: Norway spruce, Silver fir and European beech. General patterns of tolerance indicated the high vulnerability of the Norway spruce in comparison to the Silver fir and the European beech, and the strong influence of bioclimatic conditions on the drought responses of all three species. Consequently, drought events led to heterogeneous and variable response patterns in forests stands. These findings support the idea of deliberately using spontaneous selection and adaptation effects as a passive strategy in forest management under climate change conditions. This supports the importance of natural reserves; areas left to natural development without human intervention, where selection processes can take place. Such areas should be an integral part of a broad set of adaptation measures in temperate forests in Europe [76].

The results of the regional modelling of the impact of climate change on European Beech in the Czech Republic are consistent with the main phenomena related to vegetation changes induced by climate change in European beech-dominated forests. These include movement of montane forests dominated by deciduous trees towards a higher elevation, shift of montane beech-fir communities to beech communities and replacement of beech-dominated forests in highland areas by oak-hornbeam forests (especially in the northern Alps) [31]. The results of the biogeography model used in this article suggest that the strategy of sustainable forest management in the lower forest vegetation zones (1st–3rd) of the Czech Republic should focus on adjusting the tree species' composition in commercial forests (plantations) in favour of native broadleaved species, which may be less profitable but more ecologically suitable in these vegetation zones. The area of the fourth vegetation zone (with optimal climate conditions for the growth of the European beech) will continue to dominate the landscape of the Czech Republic in the context of the predicted climate changes. In forest management, the scenario indicates a stable area with optimal climate conditions for European beech (*Fagus sylvatica* L.). Sustainable forest management strategies in the fourth vegetation zone should thus favour such forest systems, which support beech as the dominant tree species. This can be done using the shelter wood system, and with the support of the European beech's natural regeneration, etc. The commercial beech forests located in the fourth vegetation zone, in the northern area of the Bílé Karpaty Mountains in the eastern part of the Czech Republic, are where the European beech currently has its Central European ecological optimum [21]. The conditions suitable for the oak-coniferous variety of the fourth vegetation zone will disappear from the landscape, according to the scenario for 2050 (Figure 4). In comparison to the climate of the zonal variety in the fourth vegetation zone (beech zone), the climate of the oak-coniferous variety is characterized by higher continentality, frequent temperature inversions, late frosts and cooler soil profiles during the growing season which are caused by a higher water content in the soil. Ecosystems associated with these specific climate conditions, possessing the characteristics of the so-called central European taiga [77], will be adversely affected by future climate change, according to the model scenario for 2050 (Figure 3). The area with the conditions of the fifth vegetation zone (fir-beech) will decrease by 14% compared to its present state (Figure 2, Figure 3, Figure 4, Figure 5 and Figure 6). Overall, the shift in the climate conditions of vegetation zones toward higher altitudes will affect forest management strategies in the mountainous areas of the Czech Republic. The scenario for 2070 and 2090 indicate changes towards less favourable climate conditions for European beech, which can enhance a risk of biotic and abiotic disturbances and stressors. Strategies for sustainable forest management should aim to enhance the static stability of the present forest stands and, most importantly, convert the current spruce monoculture stands to mixed forests of European beech and Silver fir (*Abies alba* Mill), where a minor proportion of Norway spruce occur in moist and wet habitats (on pseudogleyed soils).

Projected changes in climate conditions for beech planting linked with forest vegetation zones, correspond to the expected trends in vegetation development in Europe concerning climate changes in the model of European vegetation EUROMOVE [78]. The main trends in the changes are also in line with observed trends in the distribution of wild animals, also explained by climate change, such as certain butterfly species [79] expanding their range to higher elevations. Reference [80] showed that the trend in ongoing changes in bird population sizes and the shifts in nesting areas in the Czech Republic, in the context of climate change, are caused by a decline in species adapted to colder climates in mountainous areas, and an increase in the ranges of lower-elevation species. This confirms the regional scenario of climate characteristics for the beech planting in our study.

Increasing average global temperatures and changes in precipitation over the last 100 years have induced movements in the vegetation zonation of ecosystems around the entire world [81]. The modelling of global vegetation changes using the General Circulation Model [82] clearly shows the effect of global changes on the distribution of vegetation formations. However, assessments of climate parameters vary among global climate models, mainly for the most recent decades [83]. Moreover, global vegetation models cannot take into consideration individual plant species, and the theoretical bases for those models neglect the migration characteristics of individual species and the succession processes occurring at the level of particular ecosystems [84]. The regional scale appears to be the most appropriate for addressing the influences of climate change, species distribution and the succession processes at the level of the ecosystem [85]. In addition, climate change in a particular geographical region often affects certain ecosystems in synergy with other regional influences, such as with landscape fragmentation [86]. Therefore, regional models may provide a fundamental understanding of the significance of climate change for specific target ecosystems [87].

The Fifth Assessment Report for the Intergovernmental Panel on Climate Change (IPCC) defined a set of four new scenarios, known as Representative Concentration Pathways (RCP). Predictive regional climate data for these scenarios are not yet available for the Czech Republic. Therefore, in our model, we had to apply the scenario SRES A1B, based on a validated regional climate database. One of the most important current research projects modelling regional climate change and its effect on organisms is the CORDEX project [88], with its Europe-specific section EURO-CORDEX [89]. The outputs of the regional modelling CORDEX, suitable for regional adaptation studies, should be available in the 500 × 500 m grid in the future.

The modelling of the spatial aspects of occurrence and its development, and the implications for vegetation zonation generate material for further research activities. Refs. [90,91] assessed the impact of elevation and terrain on various species of alpine herbaceous assemblages in the area of Picos de Europa in Spain by modelling on a 15 × 15 m grid. Elevation has the biggest effect on the overall floristic diversity of the community, supporting the close association between elevational climate and vegetation belts. However, the occurrence of individual species was influenced more strongly by slope aspect, topographic index, soil water content and the sun's radiation. The existing models for the future distribution of flora and fauna mostly concentrate on individual target species or groups of species, whereas organisms interact in the ecosystem processes within their ecological niches, and thus the responses of biota to climate change are more likely to be identified at the level of ecosystem diversity [92]. The biogeographical regional models are particularly useful from this point of view [93].

The biogeographical model of changes in climatic conditions for the European beech, as caused by the predicted climate change, is one of a group of process-based biogeographical models used to predict the equilibrium reactions of vegetation to potential climate change on the regional scale [94]. This model type identifies the ecological limitations in the distribution of vegetation formations' (vegetation zones) different equilibrium climate conditions [95]. The majority of proposed models are correlational models based on the mutual dependence among certain environmental bioclimatic variables (average temperature and total precipitation) and the present area of species distribution, or characteristics of species ecological niches [96]. When predictions based on climate scenarios show how the climate can change in the future, the corresponding biological species or their communities with zonal distribution are assigned to the new parameters, whereas, in our study, the vegetation zones were assigned as the new parameters. The biogeographical model used in our study utilizes the dependence of the vegetation belts on the long-term impacts of elevation and aspects characterized by average and extreme temperatures, and their dependence on the amount and distribution of precipitation [97]. The specifications for the present forest vegetation zones in the Czech Republic [98] were updated to a higher precision during the establishment of forest habitat typology [62], which laid the groundwork for silviculture based on ecological conditions [99].

The simulations of vegetation redistribution using biogeographical models are essentially static (equilibrium) views of the analyzed problem, as those models represent modelling based on a certain concentration of atmospheric carbon dioxide in the future. The static models provide a useful "picture" of the terrestrial ecosystems in equilibrium with particular climate conditions at a given time [100]. However, the application of these models is limited by the fact that the models do not simulate all known internal factors of vegetation dynamics, such as interspecies competition, natality and mortality of populations or physiological factors. Dynamic global vegetation models that integrate vegetation dynamics and ecosystem functions have been developed to overcome these limitations [101], but they cannot be used on a regional scale [102]. Some uncertainties in the prediction models arising from insufficient data on the autecology of a particular species can be partly eliminated by using species-specific models. However, these are few and far between and mostly applicable on a continental scale [103], only providing rough guidelines for regional applications. In addition, when interpreting mathematical models of the impact of climate change on biota, we must take into consideration that models do not provide a perfect prognosis of future development [104]. The models based on sophisticated analyses contribute significantly to the prognosis (by providing synthetic scenarios), but their sensitive application must be based on a good knowledge of the biology and ecology of the organisms being modelled [105].

The important benefit of biogeographical models is their potential application for strategies of adaptation and mitigation in the landscape in the context of ecosystem services [106]. The vegetation belts are also important with respect to ecological growing conditions for forest tree species (the European beech currently has optimal growing conditions in the Czech Republic, in the fourth vegetation belt). The first study, which analyzed the continual fluctuations in agro-climatic conditions over the last 200 years, as well as expected shifts in future decades [107], showed an expansion of warmer and drier agro-climatic conditions in the most fertile regions. The development of the climate in Europe can lead to the largest shift in agro-climatic conditions since the onset of agricultural activity long before our historical experience [108]. The same consideration is undoubtedly also valid in forestry [109,110]. Therefore, in forestry, the predictions concerning climate conditions and the analyses of their impact on the forest ecosystem must be considered as crucial initial inputs into the formation of strategies in sustainable forest management [111,112]. In relation to the strategies for the lower uplands and mountain forest vegetation zones in the European temperate zone [113], the European beech has been very intensively researched [114,115,116]. The high stability of old-growth European beech forests under natural conditions can contribute to the overall stability of managed forests in Central Europe [117].

5. CONCLUSIONS

In conclusion, based on the results of the regional models, we recommend that focus is placed on the sustainable forest management strategy of the Czech Republic, and on developing target tree species whose compositions take into consideration the climatic deterioration of growing conditions for the European beech from the year 2070 onwards.

ACKNOWLEDGMENTS

The authors thank the anonymous reviewers for their comments and recommendations. The authors also thank the Czech Hydrometeorological Institute for climate input data to model. This study was supported by the grant "Cultural Landscape of Olomouc Archdiocese—Research, Presentation and Management" DG16P02B014 in the frame of the NAKI II research program funded by the Czech Republic.

AUTHOR CONTRIBUTIONS

Ivo Machar and Antonin Bucek conceived and designed the model; Antonin Bucek contributed to the dendrobiology aspects; Veronika Vlckova, Vit Vozenilek and Lucie Jerabkova analyzed the data by GIS; Lubomir Salek contributed to forest management implications; and Ivo Machar wrote the paper.

CONFLICTS OF INTEREST

The authors declare no conflict of interest.

REFERENCES

1. Kolström, M.; Lindner, M.; Vilén, T.; Maroschek, M.; Seidl, R.; Lexer, M.J.; Netherer, S.; Kremer, A.; Delzon, S.; Barbati, A.; et al. Reviewing the Science and Implementation of Climate Change Adaptation Measures in European Forestry. Forests 2011, 2, 961–982.
2. Nabuurs, G.J.; Pussinen, A.; van Bruselen, J.; Schelhaas, M.J. Future harvesting pressure on European forests. Eur. J. For. Res. 2007, 126, 391–400.
3. Strengbom, J.; Dahlberg, A.; Larsson, A.; Lindelöw, A.; Sandström, J.; Widenfalk, O.; Gustafsson, L. Introducing Intensively Managed Spruce Plantations in Swedish Forest Landscapes will Impair Biodiversity Decline. Forests 2011, 2, 610–630.
4. Seidl, R.; Rammer, W.; Lexer, M.J. Climate change vulnerability of sustainable forest management in the Eastern Alps. Clim. Chang. 2011, 106, 225–254.
5. European Environmental Agency (EEA). Impacts of Europe's Changing Climate; EEA Report No. 2; European Environmental Agency: Copenhagen, Denmark, 2004.
6. Lindner, M.; Fitzgerald, J.B.; Zimmermann, N.E.; Reyer, C.; Delzon, S.; van der Maaten, E.; Schelhaas, M.J.; Lasch, P.; Eggers, J.; van der Maaten-Theunissen, M.; et al. Climate change and European forests: What do we know, what are the uncertainties, and what are the implications for forest management? J. Environ. Manag. 2014, 146, 69–83.

7. Niinemets, Ü. Responses of forest trees to single and multiple environmental stresses from seedlings to mature plants: Past stress history, stress interactions, tolerance and acclimation. For. Ecol. Manag. 2010, 260, 1623–1639.
8. Temperli, C.; Bugman, H.; Elkin, C. Adaptive management for competing forest goods and services under climate change. Ecol. Appl. 2012, 22, 2065–2077.
9. Bohn, U.; Neuhäusl, R.; Gollub, G.; Hettwer, C.; Neuhäuslová, Z.; Schlüter, H.; Weber, H. Map of the Natural Vegetation of Europe. Scale 1:2500000; Landwirtschaftsverlag: Münster, Germany, 2002.
10. Tarp, P.; Helles, F.; Holten-Andersen, P.; Larsen, J.B.; Strange, N. Modelling near-natural silvicultural regimes for beech—An economic sensitivity analysis. For. Ecol. Manag. 2000, 130, 187–198.
11. Silva, D.E.; Rezende Mazzella, P.; Legay, M.; Corcket, E.; Dupouey, J.L. Does natural regeneration determine the limit of European beech distribution under climatic stress? For. Ecol. Manag. 2012, 266, 263–272.
12. Saltré, F.; Duputié, A.; Gaucherel, C.; Chuine, I. How climate, migration ability and habitat fragmentation affect the projected future distribution of European beech. Glob. Chang. Biol. 2015, 21, 897–910.
13. Chiesi, M.; Chirici, G.; Marchetti, M.; Hasenauer, H.; Moreno, A.; Knohl, A.; Matteucci, G.; Pilegaard, K.; Granier, A.; Longdoz, B.; et al. Testing the applicability of BIOME-BGC to simulate beech gross primary production in Europe using a new continental weather dataset. Ann. For. Sci. 2016, 73, 713–727.
14. Saltré, F.; Saint-Amant, R.; Gritti, E.S.; Brewer, S.; Gaucherel, C.; Davis, B.A.S.; Chuine, I. Climate or migration: What limited European beech post-glacial colonization? Glob. Ecol. Biogeogr. 2013, 22, 1217–1227.
15. Di Filippo, A.; Biondi, F.; Maugeri, M.; Schirone, B.; Piovesan, G. Bioclimate and growth history affect beech lifespan in the Italian Alps and Apennines. Glob. Chang. Biol. 2012, 18, 960–972.
16. Seidling, W.; Ziche, D.; Beck, W. Climate responses and interrelations of stem increment and crown transparency in Norway spruce, Scots pine, and common beech. For. Ecol. Manag. 2012, 284, 196–204.
17. Van der Maaten, E. Climate sensitivity of radial growth in European beech (Fagus sylvatica L.) at different aspects in southwestern Germany. Trees 2012, 26, 777–788.
18. Gessler, A.; Keitel, C.; Kreuzwieser, J.; Matyssek, R.; Seiler, W.; Rennenberg, H. Potential risks for European beech (Fagus sylvatica L.) in a changing climate. Trees 2007, 21, 1–11.
19. Latte, N.; Lebourgeois, F.; Claessens, H. Increased tree-growth synchronization of beech (Fagus sylvatica L.) in response to climate change in northwestern Europe. Dendrochronologia 2015, 33, 69–77.
20. Müller-Haubold, H.; Hertel, D.; Seidel, D.; Knutzen, F.; Leuschner, C. Climate Responses of Aboveground Productivity and Allocation in Fagus sylvatica: A Transect Study in Mature Forests. Ecosystems 2013, 16, 1498–1516.
21. Bosela, M.; Štefančík, I.; Petráš, R.; Vacek, S. The effects of climate warming on the growth of European beech forests depend critically on thinning strategy and site productivity. Agric. For. Meteorol. 2016, 222, 21–31.
22. Pötzelsberger, E.; Wolfslehner, B.; Hasenauer, H. Climate change impacts on key forest functions of the Vienna Woods. Eur. J. For. Res. 2015, 134, 481–496.
23. Mette, T.; Dolos, K.; Meinardus, C.; Braüning, A.; Reineking, B.; Blaschke, M.; Pretzsch, H.; Beierkuhnlein, C.; Gohlke, A.; Wellstein, C. Climatic turning point for beech and oak under climate change in Central Europe. Ecosphere 2013, 4, 1–19.
24. Stojanovič, D.B.; Kržič, A.; Matovič, B.; Orlovič, S.; Duputie, A.; Djurdjevič, V.; Galič, Z.; Stojnič, S. Prediction of the European beech (Fagus sylvatica L.) xeric limit using a regional climate model: An example from southeast Europe. Agric. For. Meteorol. 2013, 176, 94–103.
25. Pretzsch, H.; Dieler, J.; Seifert, T.; Rötzer, T. Climate effects on productivity and resource use efficiency of Norway spruce (Picea abies (L.) Karst.) and European

beech (Fagus sylvatica (L.)) in stands with different spatial mixing patterns. Trees 2012, 26, 1343–1360.
26. Pretzsch, H.; Schütze, G.; Uhl, E. Resistance of European tree species to drought stress in mixed versus pure forests: Evidence of stress release by inter-specific facilitation. Plant Biol. 2013, 15, 483–495.
27. Weber, P.; Bugmann, H.; Pluess, A.R.; Walthert, L.; Rigling, A. Drought response and changing mean sensitivity of European beech close to the dry distribution limit. Trees 2013, 27, 171–181.
28. Metz, J.; Annighöfer, P.; Schall, P.; Zimmermann, J.; Kahl, T.; Schultze, E.-D.; Ammer, C. Site-adapted admixed tree species reduce drought susceptibility of mature European beech. Glob. Chang. Biol. 2016, 22, 903–920.
29. De Koning, J.; Turnhout, E.; Winkel, G.; Blondet, M.; Borras, L.; Ferranti, F.; Geitzenauer, M.; Sotirov, M.; Jump, A. Managing climate change in conservation practice: An exploration of the science–management interface in beech forest management. Biodivers. Conserv. 2014, 23, 3657–3671.
30. Bertrand, R.; Lenoir, J.; Piedallu, C.; Riofrío-Dillon, G.; de Ruffray, P.; Vidal, C.; Pierrat, J.C.; Gegout, J.C. Changes in plant community composition lag behind climate warming in lowland forests. Nature 2011, 479, 517–520.
31. Garamvoelgyi, A.; Hufnagel, L. Impacts of climate change on vegetation distribution No. 1. Climate change induced vegetation shifts in the Palearctic region. Appl. Ecol. Environ. Res. 2013, 11, 79–122.
32. Iverson, L.R.; Mckenzie, D. Tree-species range shifts in a changing climate: Detecting, modelling, assisting. Landsc. Ecol. 2013, 28, 879–889.
33. Büntgen, U.; Frank, D.C.; Kaczka, R.J.; Verstege, A.; Zwijacz-Kozica, T.; Esper, J. Growth responses to climate in a multi-species tree-ring network in the Western Carpathian Tatra Mountains. Tree Physiol. 2007, 27, 689–702.
34. Vanoni, M.; Bugmann, H.; Nötzli, M.; Bigler, C. Quantifying the effects of drought on abrupt growth decreases of major tree species in Switzerland. Ecol. Evol. 2016, 6, 3555–3570.
35. Kulhavý, J. A new concept in sustainable forest management—The need for forest ecosystem and landscape research. J. For. Sci. 2004, 50, 520–525.
36. Svobodová, J.; Voženílek, V. Relief for Models of Natural Phenomena. In Landscape Modelling: Geographical Space, Transformation and Future Scenarios (Urban and Landscape Perspectives); Anděl, J., Bičík, I., Dostál, P., Shasneshin, S., Eds.; Springer: Dordrecht, The Netherlands, 2009; Volume 8, pp. 183–196.
37. Elkin, C.; Gutiérrez, A.G.; Leuzinger, S.; Manusch, C.; Temperli, C.; Rasche, L.; Bugmann, H. A 2 °C warmer world is not safe for ecosystem services in the European Alps. Glob. Chang. Biol. 2013, 19, 1827–1840.
38. Egli, H.R. Landscape change in the Bavarian Alpine Region and political approaches to management. Erde 2010, 141, 383–384.
39. Rosbakh, S.; Bernhardt-Römermann, M.; Poschlod, P. Elevation matters: Contrasting effects of climate change on the vegetation development at different elevations in the Bavarian Alps. Alp Bot. 2014, 124, 143–154.
40. Švajda, J.; Solar, J.; Janiga, M.; Buliak, M. Dwarf Pine (Pinus mugo) and selected abiotic habitat conditions in the Western Tatra Mountains. Mt. Res. Dev. 2011, 31, 220–228.
41. Kutnar, L.; Kobler, A. Prediction of forest vegetation shift due to different climate-change scenarios in Slovenia. Sumar. List 2011, 135, 113–126.
42. Zajaczkowski, J.; Brzeziecki, B.; Perzanowski, K.; Kozak, I. Impact of potential climate changes on competitive ability of main forest species in Poland. Sylwan 2013, 157, 253–261. (In Polish)
43. Treml, V.; Chuman, T. Ecotonal Dynamics of the Altitudinal Forest Limit are Affected by Terrain and Vegetation Structure Variables: An Example from the Sudetes Mountains in Central Europe. Arct. Antarct. Alp. Res. 2015, 47, 133–146.

44. Griess, V.C.; Acevedo, R.; Härtl, F.; Staupendahl, K.; Knoke, T. Does mixing tree species enhance stand resistance against natural hazards? A case study for spruce. For. Ecol. Manag. 2012, 267, 284–296.
45. Lasch-Born, P.; Suckow, F.; Gutsch, M.; Reyer, C.; Hauf, Y.; Murawski, A.; Pilz, T. Forests under climate change: Potential risks and Opportunities. Meteorol. Z. 2015, 24, 157–172.
46. Matthies, B.D.; Valsta, L.T. Optimal forest species mixture with carbon storage and albedo effect for climate change mitigation. Ecol. Econ. 2016, 123, 95–105.
47. EUFORGEN 2011. European Forest Genetic Resources Programme. Distribution Maps. Available online: http://www.euforgen.org/distribution-maps (accessed on 6 August 2016).
48. Parviainen, J.; Frank, G. Protected forests in Europe approaches-harmonizing the definitions for international comparison and forest policy making. J. Environ. Manag. 2003, 67, 27–36.
49. Idle, E.T.; Bines, T.J.H. Management Planning for Protected Areas. A Guide for Practitioners and Their Bosses; Eurosite, England Nature: Peterborough, UK, 2005; pp. 86–144.
50. Reyer, C.; Bugmann, H.; Nabuurs, G.J.; Hanewinkel, M. Models for adaptive forest management. Reg. Environ. Chang.2015, 15, 1483–1487.
51. Hanewinkel, M.; Hummel, S.; Cullmann, D. Modelling and economic evaluation of forest biome shifts under climate change in southwest Germany. For. Ecol. Manag. 2010, 259, 710–719.
52. Keskitalo, E.C.H. How can forest management adapt to climate change? Possibilities in different forestry systems. Forests 2011, 2, 415–430.
53. Klenk, N.L.; Adams, B.W.; Bull, G.Q.; Innes, J.L.; Cohen, S.J.; Larson, B.C. Climate change adaptation and sustainable forest management: A proposed reflexive research agenda. For. Chron. 2011, 87, 351–357.
54. Mermet, L.; Farcy, C. Contexts and concepts of forest planning in a diverse and contradictory world. For. Policy Econ.2011, 13, 361–365.
55. Vlčková, V.; Buček, A.; Machar, I.; Daněk, T.; Pechanec, V.; Brus, J.; Kilianová, H. The application of geobiocoenological landscape typology in the modelling of climate change implications. J. Landsc. Ecol. 2015, 8, 69–81.
56. Kirilenko, A.P.; Solomon, A.M. Modelling dynamic vegetation response to rapid climate change using bioclimatic classification. Clim. Chang. 1998, 38, 15–49.
57. Yee, T.W.; Mitchell, N.D. Generalized additive models in plant ecology. J. Veg. Sci. 1991, 2, 587–602.
58. Woodward, F.I.; Lomas, M.R.; Betts, R.A. Vegetation-climate feedback in a greenhouse world. Philos. Trans. R. Soc. Lond. B Biol. Sci. 1998, 353, 38–39.
59. Úradníček, L.; Maděra, P.; Kolibáčová, S.; Koblížek, J. Woody Plant Species in Czech Republic; Matice Lesnická: Písek, Czech Republic, 2001; p. 333. (In Czech)
60. Moravec, J.; Husová, M.; Chytrý, M.; Neuhäuslová, Z. Vegetation Survey of the Czech Republic. Volume 2. Hygrophilous, Mesophilous and Xerophilous Deciduous Forests; Academia: Praha, Czech Republic, 2000; p. 319. (In Czech)
61. Zlatník, A. Overview of geobiocoene groups originally forest and shrub type in ČSR. Zprávy Geografického ČSAV 1976, 13, 55–64. (In Czech)
62. Viewegh, J.; Kusbach, A.; Mikeska, M. Czech forest ecosystem classification. J. For. Sci. 2003, 49, 85–93.
63. Macků, J. Climatic characteristics of forest vegetation zones of the Czech Republic. J. Landsc. Ecol. 2015, 7, 39–48.
64. Pretel, J. Current Climate Development and its Outlook. Ochr. Prir. 2009, 46, 2–7. (In Czech)
65. Nakićenović, N.; Swart, R. Special Report on Emissions Scenarios; A Special Report of Working Group III of the IPCC; IPCC: New York, NY, USA, 2000; p. 612.
66. Machar, I. Applying of the Biogeography Register to Predicting the Consequences of Global Climate Changes on the Landscape in the Czech Republic. In Proceedings of

the 11th International Conference on Environment, Ecosystems and Development, Brasov, Romania, 1–3 June 2013; pp. 15–18.
67. Buček, A.; Vlčková, V. Collection of Maps with Prognosis of Global Climate Changes Consequences for Nature in the Czech Republic. Acta Pruhon. 2011, 98, 83–87.
68. Skaloš, J.; Engstová, B. Methodology for mapping non-forest wood elements using historic cadastral maps and aerial photographs as a basis for management. J. Environ. Manag. 2010, 91, 831–843.
69. Vlčková, V. System Nature of Modelling of Possible Trends in the Effects of Climate Change through the Technology of Geographic Information Systems Tools. Acta Inform. Prag. 2014, 3, 70–88. (In Czech)
70. Kupka, I. Is the Lang's rain factor usable for assessing the microclimate influence on growth height of forest culture? Rep. For. Res. 2006, 51, 153–156.
71. Botkin, D.B.; Saxe, H.; Araújo, M.B.; Betts, R.; Bradshaw, R.H.W.; Cedhagen, T.; Chesson, P.; Dawson, T.P.; Etterson, J.R.; Faith, D.P.; et al. Forecasting the effects of global warming on biodiversity. BioScience 2007, 57, 227–236.
72. Elith, J.; Leathwick, J.R. Species distribution models: Ecological explanation and prediction across space and time. Annu. Rev. Ecol. Evol. Syst. 2009, 40, 677–697.
73. Bedia, J.; Herrera, S.; Gutiérrez, J.M. Dangers of using global bioclimatic datasets for ecological niche modeling. Limitations for future climate projections. Glob. Planet. Chang. 2013, 107, 1–12.
74. Bublinec, E.; Luptáková, A.; Kúdelová, D.; Macko, J. Selected characteristics of climate in beech ecosystems. Ecology2011, 30, 282–287.
75. Zang, C.; Hartl-Meier, C.; Dittmar, C.; Rothe, A.; Menzel, A. Patterns of drought tolerance in major European temperate forest trees: Climatic drivers and levels of variability. Glob. Chang. Biol. 2014, 20, 3767–3779.
76. Plesník, J.; Pelc, F. Nature and landscape in the Czech Republic and their adaptation to climate change. Ochr. Prír.2009, 64, 30–34. (In Czech)
77. Divisek, J.; Chytry, M.; Grulich, V.; Polakova, L. Landscape classification of the Czech Republic based on the distribution of natural habitats. Preslia 2014, 86, 209–231.
78. Bakkenes, M.; Alkemade, J.R.M.; Ihle, F.; Leemans, R.; Latour, J.B. Assessing effects of forecasted climate change on the diversity and distribution of European higher plants for 2050. Glob. Chang. Biol. 2002, 8, 390–407.
79. Konvička, M.; Maradová, M.; Beneš, J.; Fric, Z.; Kepka, P. Uphill shifts in distribution of butterflies in the Czech Republic: Effects of changing climate detected on a regional scale. Glob. Ecol. Biogeogr. 2003, 12, 403–410.
80. Reif, J.; Storch, D.; Voříšek, P.; Šťastný, K.; Bejček, V. Bird-habitat associations predict population trends in central European forest and farmland birds. Biodivers. Conserv. 2008, 17, 3307–3319.
81. Gonzales, P.; Neilson, R.P.; Lenihan, J.M.; Drapek, R.J. Global patterns in the vulnerability of ecosystems to vegetation shift due to climate change. Glob. Ecol. Biogeogr. 2010, 19, 755–768.
82. Grassl, H. Status and improvements of coupled general circulation models. Science 2000, 288, 1991–1997.
83. Dubrovský, M.; Hayes, M.; Pierpaolo, D.; Trnka, M.; Svoboda, M.; Pierpaolo, Z. Multi-GCM projections of future drought and climate variability indicators for the Mediterranean region. Reg. Environ. Chang. 2014, 14, 1907–1919.
84. Neilson, R.P.; Pitelka, L.F.; Solomon, A.M.; Nathan, R.; Midgley, G.F.; Fragoso, J.M.V.; Lishke, H.; Thompson, K. Forecasting Regional to Global Plant Migration in Response to Climate Change. Bioscience 2005, 55, 749–759.
85. Drégelyi-Kiss, Á.; Drégelyi-Kiss, G.; Hufnagel, L. Ecosystems as climate controllers-biotic feedbacks. Appl. Ecol. Environ. Res. 2008, 6, 111–134.
86. Opdam, P.; Wascher, D. Climate change meets habitat fragmentation: Linking landscape and biogeographical scale levels in research and conservation. Biol. Conserv. 2004, 117, 285–297.
87. Walther, G.R. Community and ecosystem responses to recent climate change. Philos. Trans. R. Soc. Lond. B Biol. Sci.2010, 365, 2019–2024.

88. World Climate Research Programme. Available online: http://wcrp.ipsl.jussieu.fr/SF_RCD_CORDEX.html (accessed on 15 March 2017).
89. EURO-CORDEX. Available online: http://www.euro-cordex.net/060378/index.php.en (accessed on 15 March 2017).
90. Tuček, P.; Caha, J.; Janoška, Z.; Vondráková, A.; Samec, P.; Voženílek, V.; Bojko, J. Forest vulnerability zones in the Czech Republic. J. Maps 2014, 10, 179–182.
91. Jiménez-Alfaro, B.; Marcenó, C.; Bueno, A.; Gavilán, R.; Obeso, J.R. Biogeographic deconstruction of alpine plant communities along altitudinal and topographic gradients. J. Veg. Sci. 2014, 25, 160–171.
92. Parmesan, C. Biotic Response: Range and Abundance Changes. In Climate Change and Biodiversity, 1st ed.; Lovejoy, T.E., Hannah, L., Eds.; Yale University Press: New Haven, CT, USA; London, UK, 2005; pp. 41–55.
93. Lomolino, M.V.; Riddle, B.R.; Brown, J.H. Biogeography, 3rd ed.; Sinauer Associates Inc.: Sunderland, UK, 2009; p. 752.
94. Peterson, A.T.; Tian, H.; Martínez-Meyer, E.; Soberón, J.; Sánchez-Cordero, V.; Huntley, B. Modelling Distributional Shifts of Individual Species and Biomes. In Climate Change and Biodiversity, 1st ed.; Lovejoy, T.E., Hannah, L., Eds.; Yale University Press: New Haven, CT, USA; London, UK, 2005; pp. 211–228.
95. Giorgi, F.; Hewitson, B.; Christensen, J.; Hulme, M.; von Storch, H.; Whetton, P.; Jones, R.; Merns, I.; Fu, C. Regional climate information—Evaluation and projection. In Climate Change 2001: The Scientific Basis; Houghton, J.T., Ding, Y., Griggs, D.J., Noguer, M., van der Linden, P., Dai, X., Maskell, K., Johnson, C.I., Eds.; Cambridge University Press: New York, NY, USA, 2002; pp. 583–638.
96. Peterson, A.T.; Soberon, T.J.; Sanchez-Cordero, V. Conservatism of ecological niches in evolutionary time. Science1999, 285, 1265–1267.
97. Vahalík, P.; Mikita, T. Possibilities of forest altitudinal vegetation zones modelling by geoinformatic analysis. J. Landsc. Ecol. 2012, 4, 49–61.
98. Mackovčin, P. A multi-level ecological network in the Czech Republic: Implementing the territorial system of ecological stability. GeoJournal 2000, 51, 211–220.
99. Průša, E. Silviculture on Typological Foundations, 1st ed.; Lesnická Práce: Kostelec nad Cernymi lesy, Czech Republic, 2001; p. 593. (In Czech)
100. Neilson, R.P.; Prentice, I.C.; Smith, B.; Kittel, T.; Viner, D. Simulated changes in vegetation distribution under global warning. In The Regional Impacts of Climate Change: An assessment of Vulnerability; Special Report of IPCC Working Group II; Watson, R.T., Zinyowera, M.C., Moss, R.H., Dokken, D.J., Eds.; Cambridge University Press: Cambridge, UK, 1998; pp. 439–456.
101. Prentice, I.C.; Webb, R.S.; Ter-Mikhaelian, M.T.; Solomon, A.M.; Smith, T.M.; Pitovranov, S.E.; Nikolov, N.T.; Minin, A.A.; Leemans, R.; Lavorel, S.; et al. Developing a Global Vegetation Dynamics Model. Results of an IIASA Summer Workshop; IIASA Research Report, RR-89-007; Institute for Applied Systems Analysis: Laxenburg, Austria, 1989; p. 48.
102. Bachelet, D.R.; Neilson, R.P.; Lenihan, J.M.; Drapek, R.J. Climate Change Effects on Vegetation Distribution and Carbon Budget in the U.S. Ecosystems 2001, 4, 164–185.
103. Morin, X.; Thuiler, W. Comparing niche- and process-based models to reduce prediction uncertainty in species range shifts under climate change. Ecology 2009, 90, 1301–1313.
104. Ackerman, F.; DeCanio, S.J.; Howarth, R.B.; Sheeran, K. Limitations of integrated assessment models of climate change. Clim. Chang. 2009, 95, 297–315.
105. Walther, G.R.; Post, E.; Convey, P.; Menzel, A.; Parmesan, C.; Beebee, T.J.C.; Fromentin, J.M.; Hoegh-Guldberg, O.; Fairlein, F. Ecological responses to recent climate change. Nature 2002, 416, 389–395.
106. Schröter, D.; Cramer, W.; Leemans, R.; Prentice, I.C.; Araújo, M.B. Ecosystem service supply and vulnerability to global change in Europe. Science 2005, 310, 1333–1337.
107. Trnka, M.; Brázdil, R.; Dubrovský, M.; Semerádová, D.; Štěpánek, P.; Dobrovolný, P.; Možný, M.; Eitzinger, J.; Málek, J.; Formayer, H.; et al. A 200-year climate record in Central Europe: Implications for agriculture. Agron. Sustain. Dev.2011, 31, 631–641.

108. Spulerova, J.; Dobrovodska, M.; Izakovicova, Z.; Kenderessy, P.; Petrovic, F.; Stefunkova, D. Developing a strategy for the protection of traditional agricultural landscapes based on a complex landscape-ecological evaluation (the case study of a mountain landscape in Slovakia). Morav. Geogr. Rep. 2013, 21, 15–26.
109. Kongsager, R.; Locatelli, B.; Chazarin, F. Addressing Climate Change Mitigation and Adptation Together: A Global Assessment of Agriculture and Forestry Projects. Environ. Manag. 2016, 57, 271–282.
110. Pretel, J. Specification of Existing Estimates of the Climate Change Impacts in the Sectors of Water Management, Agriculture and Forestry Sectors, and Proposals for Adaptation; Final Report 2007–2011, Project VaV—SP/1a6/108/07; ČHMÚ: Praha, Czech Republic, 2011; p. 126. (In Czech)
111. Campioli, M.; Vincke, C.; Jonard, M.; Kint, V.; Demarée, G.; Ponette, G. Current status and predicted impact of climate change on forest production and biogeochemistry in the temperate oceanic European zone: Review and prospects for Belgium as a case study. J. For. Res. 2012, 17, 1–18.
112. Jönsson, A.M.; Lagergren, F.; Smith, B. Forest management facing climate change—An ecosystem model analysis of adaptation strategies. Mitig. Adapt. Strateg. Glob. Chang. 2015, 20, 201–220.
113. Torres-Rojo, J.M.; Vilcko, F.; von Gadow, K. Evaluating management regimes for European beech forests using dynamic programming. For. Syst. 2014, 23, 470–482.
114. Bilek, L.; Remes, J.; Zahradnik, D. Managed vs. unmanaged. Structure of beech forest stands (Fagus sylvatica L.) after 50 years of development, Central Bohemia. For. Syst. 2011, 20, 122–138.
115. Pretzsch, H.; Biber, P.; Uhl, E.; Dauber, E. Long-term stand dynamics of managed spruce-fir-beech mountain forests in Central Europe: Structure, productivity and regeneration success. Forestry 2015, 88, 407–428.
116. Vacek, Z.; Vacek, S.; Bilek, L.; Kral, J.; Remes, J.; Bulusek, D.; Kralicek, I. Ungulate Impact on Natural Regeneration in Spruce-Beech-Fir Stands in Cerny Dul Nature Reserve in the Orlicke Hory Mountains, Case Study from Central Sudetes. Forests 2014, 5, 2929–2946.
117. Kucbel, S.; Saniga, M.; Jaloviar, P.; Vencurik, J. Stand structure and temporal variability in old-growth beech-dominated forests of the northwestern Carpathians: A 40-years perspective. For. Ecol. Manag. 2012, 264, 125–133.

CHAPTER 17

System Dynamics Modeling for Agricultural and Natural Resource Management Issues: Review of Some Past Cases and Forecasting Future Roles

Benjamin L. Turner [1,*], Hector M. Menendez III [2], Roger Gates [3], Luis O. Tedeschi [4] and Alberto S. Atzori [5]

[1]*Department of Agriculture, Agribusiness, and Environmental Science, Texas A&M University-Kingsville, Kingsville, TX 78363, USA*

[2]*Department of Natural Resource Management, South Dakota State University, Brookings, SD 57006, USA*

[3]*Department of Natural Resource Management and West River Ag Center, South Dakota State University, Rapid City, SD 57702, USA*

[4]*Department of Animal Science, Texas A&M University, College Station, TX 77843, USA*

[5]*Sezione di Scienze Zootecniche, Dipartimento di Agraria, University of Sassari, Sassaari 07100, Italy*

ABSTRACT

Contemporary issues in agriculture and natural resource management (AGNR) span a wide spectrum of challenges and scales—from global climate change to resiliency in national and regional food systems to the sustainability of livelihoods of small-holder farmers—all of which may be characterized as complex problems. With rapid development of tools and technologies over the previous half century (e.g., computer simulation), a plethora of disciplines have developed methods to address individual components of these multifaceted, complex problems, oftentimes neglecting unintended consequences to other systems. A systems thinking approach is needed to (1) address these contemporary AGNR issues given their multi- and interdisciplinary aspects; (2) utilize a holistic perspective to accommodate all of the elements of the problem; and (3) include qualitative and quantitative techniques to incorporate "soft" and "hard" elements into the analyses. System dynamics (SD) methodology is uniquely suited to investigate AGNR given their inherently complex behaviors. In this paper, we review applications of SD to AGNR and discuss the potential contributions and roles of SD in addressing emergent problems of the 21st century. We identified numerous SD cases applied to water, soil, food systems, and smallholder issues. More importantly, several case studies are shown illustrating the tradeoffs between short-term and long-term strategies and the pitfalls of relying on quick fixes to AGNR problems (known as "fixes that backfire" and "shifting the burden", well-known, commonly occurring, systemic structures—or archetypes—observed across numerous management situations [Senge, P.M. *The Fifth Discipline*, 1st ed.; Doubleday: New York, NY, USA, 1990.]). We conclude that common attempts to alleviate AGNR problems, across continents and regardless of the type of resources involved, have suffered from reliance on short-term management strategies. To effectively address AGNR problems, longer-term thinking and strategies aimed at fundamental solutions will be needed to better identify and minimize the often delayed, and unintended, consequences arising from feedback between management interventions and AGNR systems.

Keywords: complex systems; system dynamics; computer simulation; interdisciplinary; systems analysis; natural resources; agriculture; management; unintended consequences

1. INTRODUCTION

Contemporary agriculture and natural resource (AGNR) problems are becoming increasingly evident and affect the livelihoods of people and resources globally, such as local or global changes in climate (e.g., drought frequency and intensity; [1,2,3,4]), hydrological or water resource management issues (e.g., water security; agricultural water management; [5,6,7,8]), land resource issues (e.g., land transformation, soil quality and soil erosion, urbanization; [9,10,11,12]), biodiversity resource conservation [13,14,15], agriculture and food system challenges (e.g., food security, human health; [16,17,18,19,20]), and/or rural economic conditions and small-holder development. In many circumstances, these problems could be characterized as complex, as they have many interacting and overlapping feedbacks, where a systems approach to problem solving has been increasingly promoted, including strategies for enhancing ecosystem services [21], agricultural intensification [22], manipulation of multiple leverage points [23], and improving systems and resources integration [24].

Complex problems differ from simple or complicated problems in that they exhibit several key system properties: (a) components are tightly coupled and organized ("everything influences almost everything else"); (b) observed behaviors are dynamic ("change occurs at many time scales"); (c) interventions are most often policy resistant ("obvious solutions fail or make things worse"); (d) causal relationships are counterintuitive ("causes and effects are distant in time and space"); and (e) tradeoffs in preferred system pathways are presented ("long-term and short-term solutions are often at odds") [25,26]. In the real-world, resources and systems often overlap and interact through complex feedback processes, which involve numerous variables, can operate at multiple temporal and spatial scales, and involve human decision making that can exacerbate perturbations or create new and unintended problems [26]. Because of the importance of human decision making, mental models of system stakeholders must be accounted for. Mental models "are deeply ingrained assumptions, generalizations, or even pictures or images that influence how we understand the world and how we take action" [25]. Mental models are dynamic in the sense that they can change as the stakeholder learns new or forgets old information, adopts or discards systems of belief, can change with changing perceptions about the system or problem of interest, and are always incomplete [25,27,28].

Because AGNR systems are complex and difficult to comprehend, previous efforts to address AGNR challenges have historically used traditional methods that are familiar and easy to accept [29], and that were promoted within disciplinary silos. In many of these methods, scientists employ a *linear* mental model of problems, which assumes simple cause-and-effect relationships between system components and focuses on progressively narrower model boundaries of investigative efforts to isolate components [30,31]. Such isolation exposes any analysis to the risk of not adequately recognizing or diagnosing root causes of issues or not incorporating all of the pertinent factors at work [30,31,32], which could lead to flawed or unsustainable recommendations regarding strategy or policy implementation as well as perpetuating the symptoms of the original problem (i.e., not adequately addressing the root problem) or making the problem even worse. For many of the problems described above, the problem symptoms continue to persist or are amplified (as shown in the above case descriptions [1,2,3,4,5,6,7,8,9,10,11, 12,13,14,15,16,17,18,19,20]), despite the massive attempts to curtail the problems (e.g., Kyoto and Paris climate agreements; the "Green Revolution" of the 20th century).

System dynamics (SD) is a scientific framework for addressing complex, nonlinear feedback systems [33]. As a methodology, SD draws upon both qualitative (e.g., survey and interview methods) and quantitative techniques (e.g., computer programming and simulation), emphasizes stakeholder involvement (to define mental models within the system), and encourages the researchers themselves to adopt a *nonlinear* mental model (to seek and describe the feedback processes of a problem's dynamics). Specifically, SD modeling tools have proven to be useful in addressing AGNR problems. Our objective is to review the use of SD methodology, primarily the use of simulation modeling, for AGNR problems and provide a discussion on the role of SD for future applications in addressing 21st-century resource challenges. We begin with an overview of what SD modeling is and the general tenets of the methodology. Then, we overview successful examples of SD applications to a variety of AGNR issues. From those successful examples, we discuss the potential contribution and role of SD in addressing contemporary natural resource issues.

2. WHAT IS SYSTEM DYNAMICS (SD) MODELING METHODOLOGY?

The SD method was developed to enhance learning in complex systems, is fundamentally an interdisciplinary science (i.e., pertaining to more than one branch of knowledge; where two or more scientific disciplines are involved in a coordinated scientific investigation), is grounded in the theory of nonlinear dynamics and feedback control, and draws on cognitive and social psychology, economics, and other social sciences to incorporate human dimensions and decision making [33].

There are five general steps (similar to many other modeling approaches) used in applying the SD modeling process: (1) problem articulation; (2) development of a dynamic hypothesis; (3) formulation of a simulation model; (4) testing the simulation model; and (5) policy or strategy design, experimentation, and analysis (Table 1) [33]. The first step describes the researchers' intentional effort to "admire the problem" rather than jumping to conclusions about the underlying mechanisms perpetuating an issue [34]. This may be achieved through stakeholder interviews or surveys, focus groups, eliciting mental models of the problem from key personnel, as well as collecting or aggregating all the relevant data that can describe the behavior of the problem over time (i.e., a reference mode). The second step aims to synthesize all that is known about the problem into an endogenous (i.e., feedback-based) theory upon which to evaluate the quantitative model (step 3). These first two steps are often associated or compared to "soft systems" methodology due to the emphasis placed on stakeholder engagement, defining decision-making criteria and mental models of the system actors/stakeholders, and conceptual modeling of the root causes of the problem of interest [33,34,35].

Table 1. An overview of the steps employed in the system dynamics (SD) methodology.

Step of the Process [1]	Purpose/Objective [1]	Description of Activities [2]	Advice of Practitioners
1. Problem articulation	Determine boundaries, variables, time horizons, and data sources	1. Interviews/surveys 2. Describing mental models 3. Collecting/aggregating reference mode data	"Admire the problem" [34]
2. Development of dynamic hypothesis	Initial explanation of the endogenous dynamics of the problem at hand	1. Identify current theories of the problem 2. Causal loop diagramming 3. Stock-and-flow mapping	
3. Formulation of a simulation model	Move from qualitative to quantitative understanding of the problem	1. Specifying model structure, decision rules 2. Parameter estimation and setting initial conditions 3. Checking model consistency with dynamic hypothesis	"Challenge the clouds" [33,36]
4. Testing the simulation model	Building confidence in the quantitative model	1. Reference mode comparisons 2. Extreme condition testing 3. Sensitivity analyses	"Try to 'break' the model" [33,37]
5. Policy/strategy design and analysis	Identifying leverage or tipping points	1. Scenario design and analysis 2. Stakeholder outreach	"Ask, 'What if?'" [33]

[1] From [33]; [2] A non-exhaustive list of activities.

The third step is the construction of the quantitative model. Construction is aided by icon-based programming (consisting of stocks, flows, auxiliaries, information links, and clouds; Figure 1) used to conceptualize the primary feedback mechanisms and describe them using coupled partial differential equations. Practitioners advise modelers just setting out to "challenge the clouds" (clouds representing the boundaries of the quantitative model) by expanding their own mental and conceptual models about the problem at hand and resisting temptations to reduce the number of components included in the model for the sake of simplicity alone [36]. The fourth step attempts to "break" the model (i.e., test the model with extreme conditions and/or parameter values far outside the calibrated values which closely correspond to values in the real world) to investigate if assumed parameter values are realistic, if the direction of model responses correspond to expected feedback polarity to check model consistency, and to identify variables that could break the system or improve system function (e.g., potential leverage points) [33,37]. The third and fourth steps are often compared to "hard systems" methodology used in other types of systems analyses due to the emphasis placed on specifying system equations, objectives, constraints, ensuring basic natural laws are respected, and testing the model to understand its change in behavior quantitatively [33,37,38]. Interdisciplinary science is built into the SD methodology by the integration of both "soft" and "hard" components of a system or problem. The final step involves asking and applying "What if?" questions to the model based on proposed strategy or policy interventions (e.g., "what if government subsidies are raised or lowered?"; "what if different management practices are implemented?") to identify places of management leverage or potential, and future tipping points. This summarizes the basic SD process behind the real-world applications described below. For methodological considerations of SD applied to different disciplines, see [39,40,41,42,43,44,45].

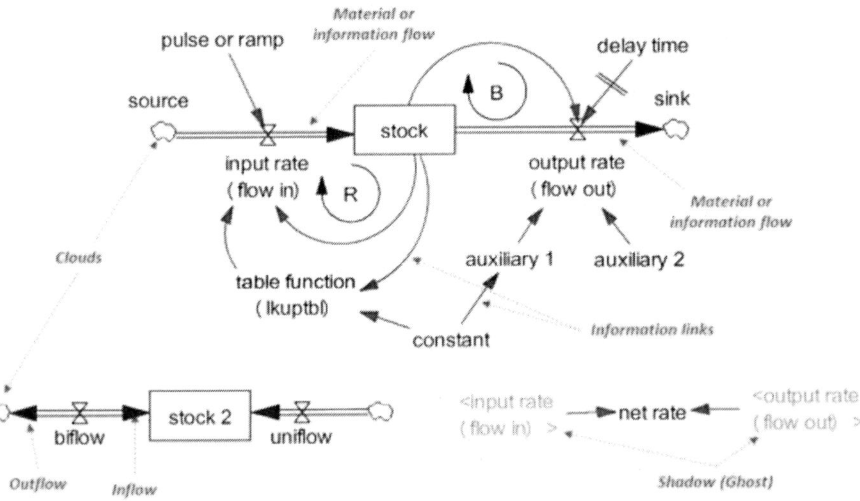

Figure 1. Icons used in system dynamics programming. Stocks represent accumulations (expressed mathematically as integrals). Inflows and outflows change the level of the stock over the given time-step and are influenced by current system stock levels, auxiliary functions (which can take on any large number of potential mathematical functions; e.g., pulses; ramps; graphical or table; etc.), and delays, each connected through an information link. Clouds represent the model boundaries (i.e., sources and sinks), while shadows represent variables used in one location that have been formulated in another. The "R" symbol represents reinforcing (or positive) feedback (also denoted "+") while

the "B" symbol represents balancing (or negative) feedback processes (also denoted "–"). Here the Vensim PLE program (Ventana Systems, Inc., Harvard, MA, USA) is used, although there are some different SD programs available (e.g., iThink; STELLA; Powersim Studio; AnyLogic; etc.).

3. APPLICATION OF SYSTEM DYNAMICS TO AGRICULTURE AND NATURAL RESOURCE PROBLEMS

Before proceeding to the contemporary cases we have reviewed, it is important to acknowledge the first SD model incorporating AGNR relationships, which was the pioneering work of the World3 model published in *The Limits to Growth* in 1972 ([46]; updated in 1991 and 2004 [47,48]). The World3 model explored the upper limits of human developmental capacity by modeling the interacting feedback loops among five factors: population growth, food per capita production, nonrenewable resource depletion, industrial output, and pollution generation. Food production was considered one of the main drivers of population growth and was limited by reduced land fertility caused by pollution, which in turn was driven by population growth. The model assumed that higher food demand would cause higher land (arable) allocation to food production, thereby representing a limit to human population growth, and food production increases would cease due to land degradation when the arable land reaches its maximum level. Scenarios run with the World3 model resulted in overshoot and collapse of the human population in the mid-21st century due to the depletion of natural resources. The authors were largely criticized both for the novelty and uncertainty of the model and were labeled "pessimistic." However, a recent comparison of the observed world data from 1970 to 2000 with the original estimates performed by [46] showed that historical data match favorably with the modeled trends [49,50,51].

Having summarized some persistent agriculture and natural resource (AGNR) problems and given an overview of the system dynamics (SD) modeling methodology (including the seminal work in *The Limits to Growth* study), this section reviews the application of SD to some contemporary AGNR problems, including hydrology and water resources; agriculture, land and soil resources; food system resiliency; and small holder development. Of the case studies presented, many involved stakeholder outreach and participation. Such activities were outside the scope of this review. Here we focus primarily on the integration of multiple scientific disciplines through the use of SD modeling. Although brief, each section directs the reader to multiple relevant resources that provide more detailed descriptions of the work described.

3.1. Hydrology and Water Resource Management

Hydrology and water resource management issues are inherently complex due to local climate characteristics (including variability in rainfall distribution patterns), surface water-groundwater connectivity, natural and man-made reservoir storage, growing populations and demand, among other factors. These characteristics make hydrologic and water resource management problems well suited to study by SD methodology. Scientists have used SD to study such problems beginning as the early as the 1980s with applications on small-scale hydropower analyses [51]. Various SD applications in hydrologic and water resource management problems can generally be categorized into two types: (1) water resource or watershed planning problems (where the research is used to better understand the current situation and/or to inform stakeholders regarding the current state of the system); and (2) scenario analyses of the impacts economics or policies have on water resources (i.e., to explore the behavior space of the current system given changed conditions or alternative strategies). In real-world situations, planning and analysis activities are often performed in tandem (i.e., one activity informs the other—a feedback loop in decisions and results) since they are each a respective step in the resource management process. In the literature, however, articles often focus on one activity or the other. Below, we provide some citations of each use (planning or analysis) with some illustrative applications.

Because of the integrative abilities of the SD method to connect physical and social system components, the emphasis on stakeholder participation, as well as the visual attractiveness of SD models, SD has been an effective tool for water resource planning problems throughout the world. There are many examples, both peer-reviewed [52,53,54,55,56,57,58] or presented at conferences [59,60,61]. Two illustrative cases dealing with groundwater management planning are shown, one in the Tenggeli Desert region, China, using qualitative causal loop diagramming [62], and another in the Palouse region, USA, which created a quantitative model to estimate groundwater drawdown rates useful for regional water planning and management [63].

Water resource planning case 1 Yaoba Oasis, China: Yaoba Oasis is an artificial oasis developed in the 1960s to resettle displaced herdsmen ("ecological refugees") and relieve pressure from stressed grasslands

[62]. Due to the arid environment, groundwater was pumped to support irrigation of newly cultivated land (Figure 2, loop 1). By doing so, aquifer levels were lowered, allowing salt water from the underground Taosuhu Lake to flow into the aquifer as water tables were lowered. Increased water salinity has degraded soils and limited the expansion of cultivated land (Figure 2, loop 2) as well as usable irrigation water (loop 3). As water wells are discarded due to salinity issues, water tables can recover and reduce salinity issues (loop 4). However, due to the delay in management perceptions of water availability, cultivated land continues to grow despite ongoing salinity concerns and a depleting water supply (loop 5). This behavior mimics both "fixes that backfire" (Figure 2) and "tragedy of the commons" archetypes [25,64]. Several contributing policies (subsidized water resource fees to reduce costs of crop production) as well as interventions (encouraging water-saving crops; investments in irrigation technology and well management, and institutional strengthening; loops 6 and 7) were identified. The addition of such strategies into the feedback loop structure highlighted potentially important leverage points for stakeholders regarding mechanisms to *balance* water consumption with water availability; similar situations of water resource constraints and strategies have been occurring globally.

Water resource planning case 2 The Palouse, USA: The Palouse region in Washington and Idaho, USA, has a growing rural population that is dependent on groundwater, relying on two basalt aquifers for potable water [63]. By synthesizing existing water resources data, researchers developed an SD model to simulate the Palouse hydrologic cycle and project trends in aquifer characteristics assuming that the current infrastructure does not change. By integrating population dynamics and hydrological processes at the surface with the geologic characteristics of the region, they found, in the near future, groundwater withdrawals will likely exceed recharge rates, making maintenance of groundwater resources unsustainable [63]. Another important result was identification of uncertainty in key system components (e.g., storativity and recharge) which are not readily affected by groundwater management efforts. This was a major contribution to the water resource planning effort since a better understanding of the behaviors of inflows and outflows of the aquifer helped inform stakeholders about managing the water resource. For additional material regarding systems thinking applied to water resource planning, see [65].

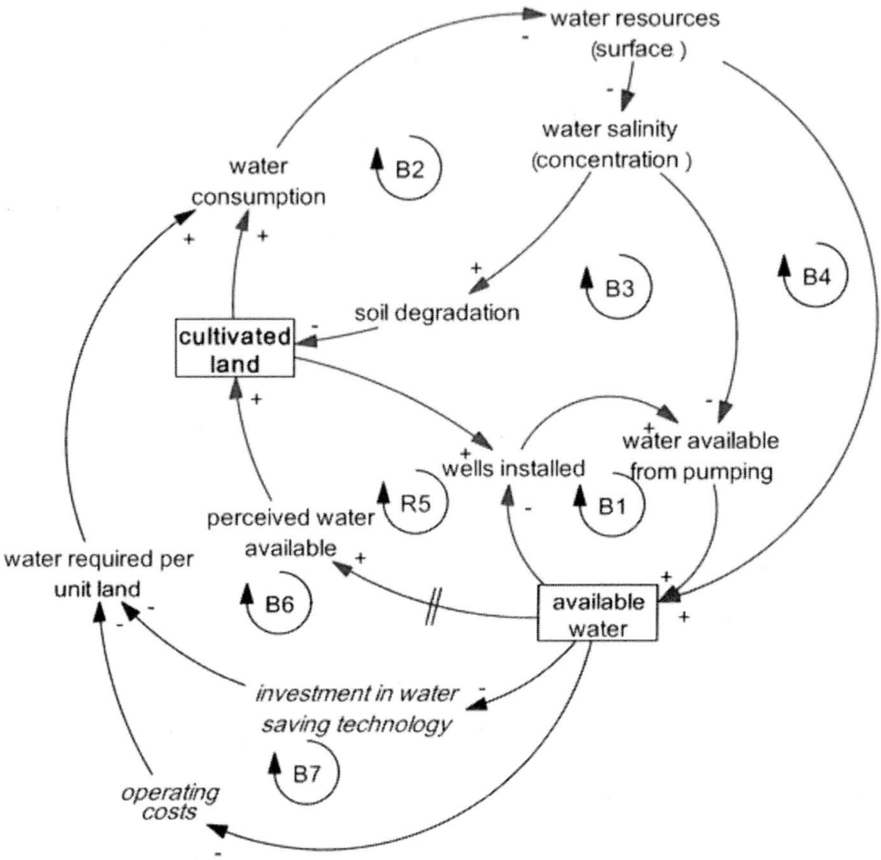

Figure 2. Feedback loop structure of cultivated land in the Yaoba oasis region, constrained by water availability, salinity, and well discarding (negative/balancing feedback loops B1–B4) and perpetuated by delayed perceptions of water availability (positive/reinforcing feedback loop R5), with potential leverage points (negative/balancing feedback loops B6 and B7); adapted and modified based on [56]. The common "fix" (installing more wells to pump water) eventually "backfires" as the perceived water availability outpaces actual water available, further developing more land in cultivation.

Similar to SD applications in water resource planning, water resource problem analyses (i.e., testing changing conditions or alternative policies) have had widespread application in the literature [55,66,67,68,69,70,71,72,73,74,75] as well as in the System Dynamics

Society [76,77,78,79,80,81] across a broad range of issues from energy production to agriculture to municipal water management. We highlight two cases, one dealing with sensitivity analyses of uncertain socioeconomic parameters in a traditional irrigation community in northern New Mexico, USA [75], and another on scenario analyses of alternative water transfer policies in the Zayandeh-Rud River Basin of Iran [72].

Water resource analysis case 1 New Mexico, USA: Historic irrigation communities in northern New Mexico, USA (known as "acequias") provide an array of ecosystem and socioeconomic goods and services based on the mechanisms of water distribution and management along hand-dug canals supplied by mountain snowmelt runoff. An SD model was developed to integrate information regarding community leadership, local hydrology, and agricultural economics. The researchers then tested (via scenario and sensitivity analyses) the extent to which community resource management practices centering on shared resources (e.g., water for floodplain irrigation) and community mutualism (i.e., shared responsibility of residents to maintain irrigation policies and cultural traditions) influenced the irrigation system function and ecosystem goods and services [75]. Similar to [63], the model revealed uncertainty in numerous system components, which the researchers explored through comprehensive sensitivity analyses. Sensitivity analyses revealed that agricultural profitability, community make-up (i.e., percentage of residents with historical familial ties), and land use (residential vs. cultivated) were key determinants of irrigation system response. This was due to their influence on feedback processes responding to input parameter perturbations [75]. Their results correspond well with other types of system analysis efforts in the field of socio-hydrology [82].

Water resource analysis case 2 Zayandeh-Rud River Basin, Iran: The Zayandeh-Rud River Basin has traditionally used a supply-chain oriented approach to deal with water stress in the past 60 years [72]. Researchers integrated the basin hydrologic, socioeconomic, and agricultural sub-systems (e.g., the hydrology sub-system, Figure 3) and tested alternative policy options for managing water stress, including business as usual (BAU; basin transfers assumed constant; ag water use efficiency = 45%), agricultural water demand management I, II, and III (AWDM; basin transfers similar to BAU but with different cropping patterns of 80% or 45% water use efficiency), and inter-basin water transfer with and without demand management (IBWT; representing transfers with or without groundwater pumping and improvements in agricultural water use efficiency). Forecasts for each of the policy scenario tests revealed divergent dynamics between the BAU scenario (increasing water shortages) and the AWD and IBWT strategies (consistent or decreasing water shortages). Their results indicated that common policy options for managing water shortages (primarily inter-basin transfers) promoted the growth of the system, and therefore, water demand, further perpetuating water shortage problems. Without considering the dynamics of the interrelated problems (i.e., the connections between hydrology, socioeconomics, and agriculture), water managers are bound to succumb to recurring and more severe water shortage problems. Management "quick-fix" strategies that are often employed (e.g., water transfers) are diagnosed by a common SD archetype (a commonly occurring systemic structure observed across numerous management situations) called "fixes that backfire" [25]. Overcoming this phenomenon requires a shift from short-term to long-term thinking (Figure 4) [25].

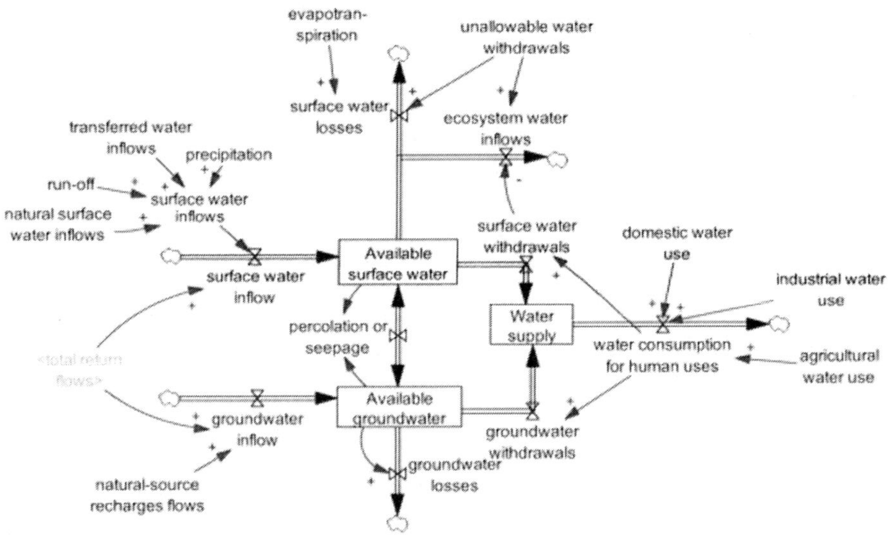

Figure 3. Hydrology sub-system similarly adapted and simplified from [65], displaying natural and anthropogenic-driven water flows, including domestic, industrial, and agricultural water demand that drives inter-basin transfers from surface water, further developing the watershed and ultimately total water demand.

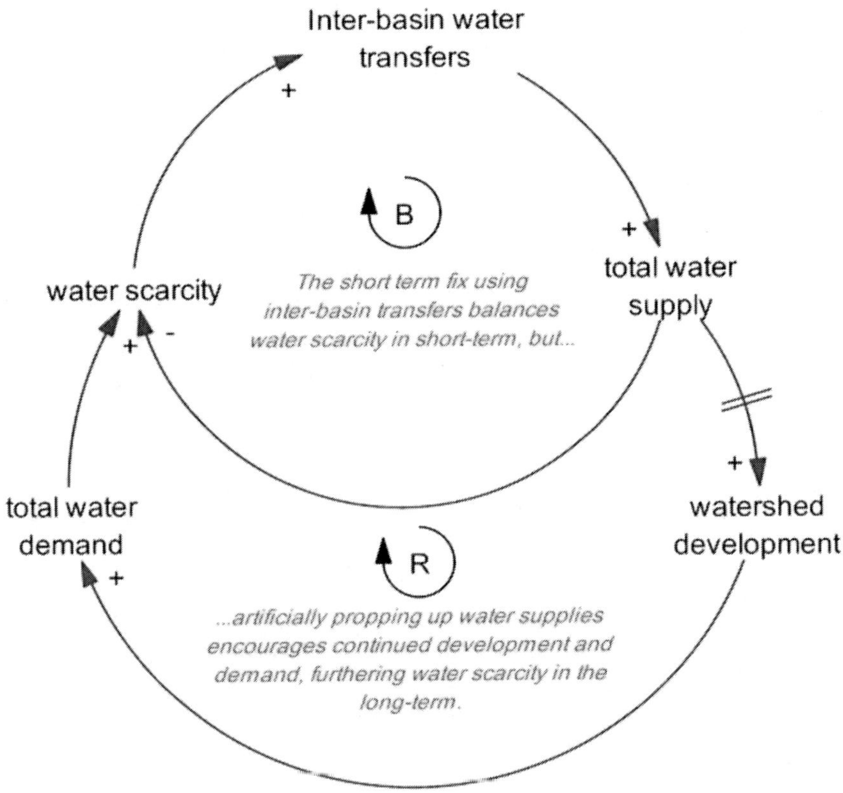

Figure 4. "Fixes that backfire" archetype identified by [65]. The "quick fix" of delivering inter-basin transfers alleviates (or balances) water scarcity in the short-term, but makes it worse in the long-term by encouraging (or reinforcing) watershed development. Adapted and expanded based on [65].

3.2. Agricultural Land and Soil Resources

As global populations continue to grow and more land becomes devoted to agriculture, agricultural production will have to be sustained on lands and soils traditionally not rated for intensive or long-term cultivation [11,16,17,19], creating elevated risk of soil erosion and other environmental externalities resulting from agricultural production [11,22,83]. Land use and agricultural practices may vary widely and can change based on modifications in public policies, pressure from urban expansion, changing economic conditions and profitability, and personal and cultural characteristics, among other factors. Because of the variety of influences, SD provides a useful framework for investigating how these socioeconomic characteristics influence the sustainability of land and soil resources, including both their productivity and ecosystem goods and services [84,85,86]. Land and soil models identified by our review may be categorized as: (1) soil modeling at the field scale via horizons; and (2) land management and erosion prevention at the watershed scale. Soil models at the field scale have focused on nutrient management [87], infiltration and water-holding capacity useful for reclamation decisions [88], or soil-water interactions useful for managing irrigation application for improved water use efficiency, reduced salinity, and reduced costs [89]. Likewise, several modeling efforts have documented the development, evaluation, and application of SD models representing reconstructed watersheds [90,91,92], which have been used to test and corroborate the implemented reclamation strategies within certain ranges of hydrologic conditions as well as compare different vegetation alternative for future reclaimed covers. Here, we highlight three illustrative SD applications, one at the soil layer scale, and two at the watershed scale.

Soil management at the soil layer scale with groundwater interactions: Due to the complex nature of water behavior in soils, managing irrigation water can be challenging due to common unintended consequences, including soil salinity issues as well as water table reductions due to reliance on pumping. In order to better understand the complex (nonlinear) interdependent relationships throughout the soil profile affecting soil water storage and the agricultural system, researchers developed a two-part SD model of soil-water-plant relationships with a dynamic link for interactions with the shallow groundwater table to account for natural recharge as well as pumping [89]. The surface layer model (Figure 5a) consisted of two balancing loops (ET → soil water content drawdown → reduction in ET; water content increase → percolation → reduction in soil water at surface) and one reinforcing loop (ET → capillary rise → soil water content → ET) around the soil water stock, while the surface-ground water interaction incorporated seepage, percolation, and lateral flows (Figure 5b). The model was then used to understand supplemental irrigation in aerobic rice systems, and demonstrated various water table drawdowns depending on irrigation application and groundwater abstraction rates. Because maintaining a sustainable groundwater resource is critical to sustainability, increasing irrigation applications can result in a "fix that backfires," since continued irrigation applications may draw down the water table to the point that maintaining irrigation is too costly for the agricultural system.

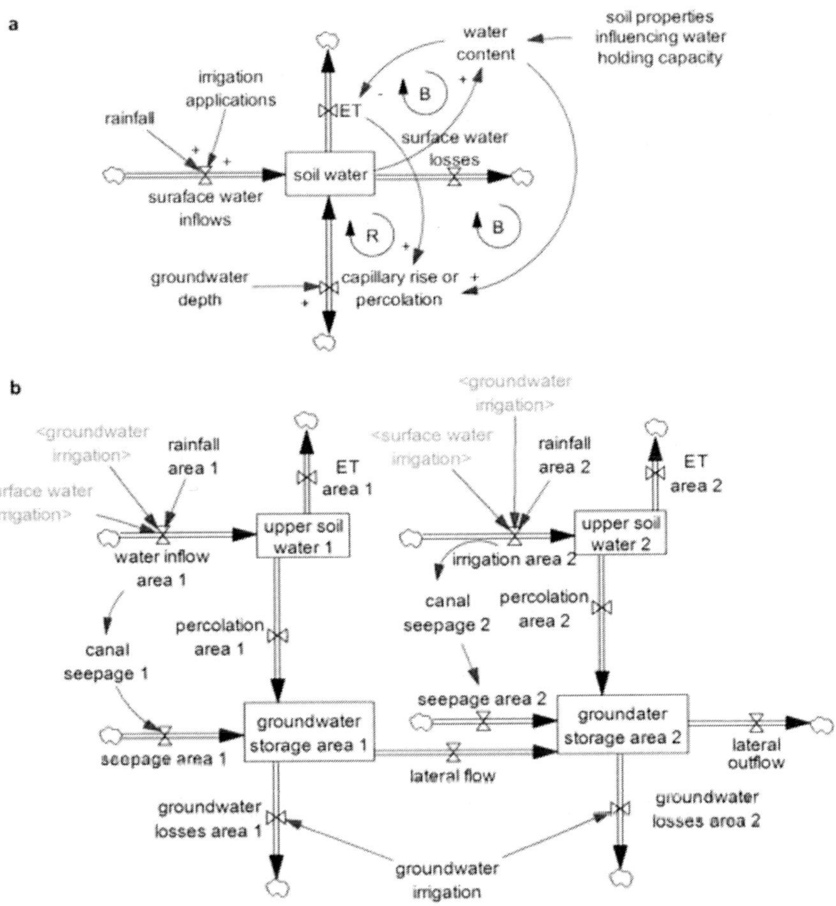

Figure 5. Soil water model used to simulate (**a**) soil water in the surface layer used for estimating soil-water (irrigation)-plant interactions; and (**b**) the surface water-groundwater interactions driven by irrigation applications and groundwater abstraction between fields. Adapted and simplified from [90].

Land management and soil erosion case 1 Alqueva, Portugal: The first case studied erosion and sedimentation in the Alqueva dam watershed, an agriculturally dominated area in southern Portugal [93]. The model used the Revised Universal Soil Loss Equation (RUSLE; [94]), which involves complex equations and large quantities of data. SD enabled the authors to circumvent this challenge by using the structural concepts of RUSLE and account for their interconnections while still generating meaningful predictions of soil erosion and sediment deposition at the watershed level (Figure 6). Their findings indicated erosion rates of 2.5–249.4 tons/ha. To aid in the improvement of agriculture practices utilized by farmers in the watershed, the model also identified the most influential factors on soil erosion given the particular soils in the watershed, aiding efforts to prevent soil loss and maximize capital for soil erosion prevention and remediation efforts.

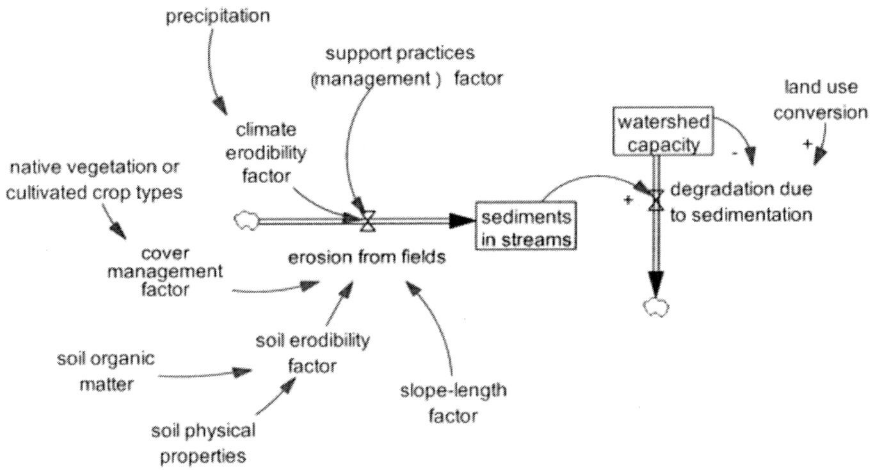

Figure 6. Soil erosion model based on the RUSLE framework used to simulate erosion dynamics at the field scale, based on local soil, climate, and management practices, with land use change in the watershed. Adapted and simplified from [94].

Land management and soil erosion case 2 Keelung River, Taiwan: The second soil erosion case evaluated soil erosion and nutrient impact using an SD model in the Keelung watershed, one of Taiwan's largest rivers which runs through the capital city Taipei [95]. For two decades prior to the study, Taipei's urban and agriculture expansion had encroached on areas with steep slopes making the soils increasingly susceptible to erosion. Using a modified version of the generalized Watershed Loading Function [96], they estimated daily erosion up to 18,000 tons/day. Perhaps more importantly, the model captured the complexity of the problem by accounting for feedback between erosion, runoff, sediment, and economic (policy) sub-models. The economic component included subsidies to mitigate environmental risks, including afforestation subsidies, which reduced soil-related damage compared to the status quo. The approach the researchers developed offered a unique interface and data integration to conduct policy analyses for complex AGNR problems (Figure 7), allowing integration of other common computing platforms with SD. It is likely that advancements in Python™ and Statistical Program *R* [97] will continue to improve GIS integration into SD models.

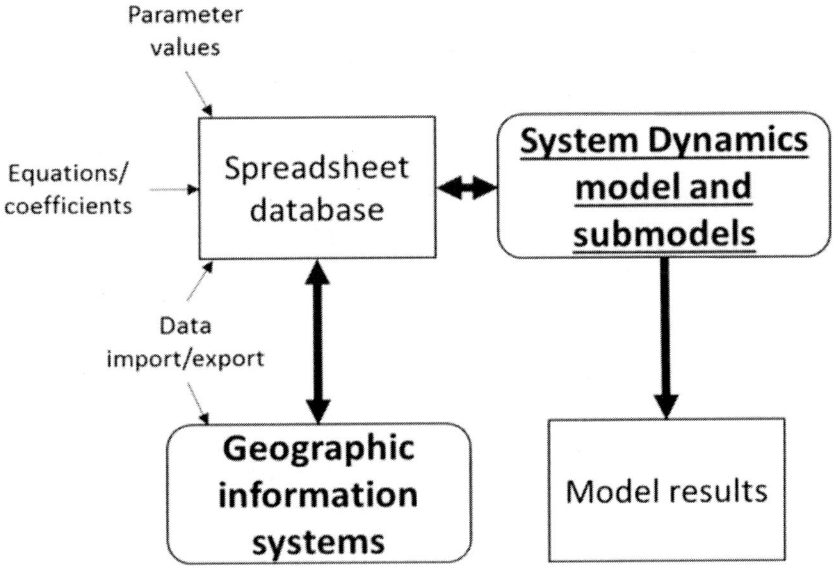

Figure 7. Conceptual diagram displaying the interface capabilities of SD with other programs used in land use and soil erosion analyses; modified from [96].

3.3. Food System Resiliency

From food to food systems: World population increase in the 20th century and ongoing challenges of under- or malnourished populations has increased global food insecurity trends and therefore pressure on food security (i.e., all people at all times have physical and economic access to sufficient and nutritious food to meet their needs for an active life [98]) provided by food systems (i.e., the set of activities from production to consumption, including geographic, biophysical, human and socioeconomic features, and environmental constraints [99]). Since then, models of single food-related variables have been used to predict future outcomes and support territorial food security policies. In particular, food production and consumption and partial indicators of food security are often used to deduce implications of future trends [100] or by performing descriptive flow analysis [101]. On the other hand, focus on food-related boundaries continue to change from those in the last century as society moves from food products to food systems. Others have compared the main features of modern food systems to traditional food systems showing that the modern transformation may be attributed to: increases in urban populations relative to rural populations; increased number of national and global stakeholders of food production relative to local agents; enhancement of the processing phases of the food system relative to production phases; and affirmation of the long-supply chains relative to shorter or more local supply chains [102].

The main factors limiting the achievement of food security were traditionally linked to production shocks due to natural factors (i.e., climate trends such as poor rainfall). However, in modern food systems, food security has been more associated with variation in international prices, foreign trade problems, and income shocks causing food poverty [102]. It is now evident that food systems are affected by AGNR drivers but also produce outcomes in social capital (welfare, employment, wealth, etc.) and natural capital (environmental security, ecosystem services, stock variation, etc.) [93]. In response to these issues, food policies enacted by national ministries or departments of agriculture have switched from agricultural technology and production improvements to regulation around industrial competition, health and food safety, and waste management related to the food chains [99,102].

Other papers have reported insights to support food security policy at territorial and local levels. A non-exhaustive list of published models focusing on regional or local food systems was presented by [103]. SD models have focused on closing the food sufficiency gap. For example, in Colombia, food sufficiency is highly dependent on the land use and food demand, requiring a sustainable goal-seeking behavior of increasing producer training, service infrastructure, as well as adjustments in irrigation and drainage. Models have corroborated that increased productivity and efficiency will be required rather than agricultural land expansion [104]. Other uses of SD in food system research include food system distribution optimization for developing and transition countries [105] showing how food dynamics are deeply embedded in urban and rural dynamics and therefore the food systems cannot be managed as separate parts (e.g., urban versus rural; production versus retailing) [106,107,108].

To be sustainable, food policies have to be directed toward efficient AGNR management, including the supporting infrastructure of the food system (e.g., technology, organizational quality, roads, urban planning, etc.) and will have to be based on minimizing food gaps and maximizing food resilience. The pressure or temptation to lower food security and food system resiliency goals (e.g., 100% to 90% of food gaps closed by 2050) will be immense [107], leading to "drifting goals" behaviors among policymakers likely to support food resiliency for some albeit to the exclusion of others through the persistence of the food availability gap (Figure 8). The food availability gap varies due to two primary drivers: food demand (B1; which is the desired goal) and current food supplies provided by the system (B2; representing urban and rural resources) [106]. Rash policies directed towards infrastructure development in order to facilitate exchanges and food distribution (e.g., road building to generate greater accessibility) might be effective in the short-term but will incentivize continued urbanization, exacerbating congestion of the infrastructure and thereby eroding the ability to deliver food supplies, creating a "fix that backfires" (loops B2 and R3). Policies on resource use should account for natural limits to the system and respect the "limits to growth" (i.e., natural resources capacity; B2 and R4). In this context, it should be identified and emphasized that policies that degrade natural resource capacity (the carrying capacity of the system) might negatively affect food system production and increase the food gap. It indicates that food policy design should be oriented to maintain the system's capacity to produce and distribute food, taking into account the current socioeconomic structures and the nation-state as a whole in order to support their human developmental capacity while respecting the natural constraints enforced by the local environmental and biophysical limits [108].

From food systems to food resilience: Systems perspectives of food systems that utilize holistic approaches such as SD (which focus on the relationships among the parts rather than the parts only [25,32]) can account for whole-system interactions and improve sustainability outcomes of food systems [109]. Internal and external drivers of change that appear as long-term pressures to the system can increase the system's insecurity and exacerbate the problem of low food system resilience.

As described by [109], resilience thinking may assume different definitions depending on the area of application. The same authors reported that, in general, resilience is a dynamic concept that is broadly defined as the capacity of a system to continue to achieve goals despite disturbances and shocks [109]. This is an essential part of sustainability and might be measured by the system's performance in the long run. Resilience is complementary to the concept of sustainability (e.g., the measure of the system performance in the long run) by measuring the

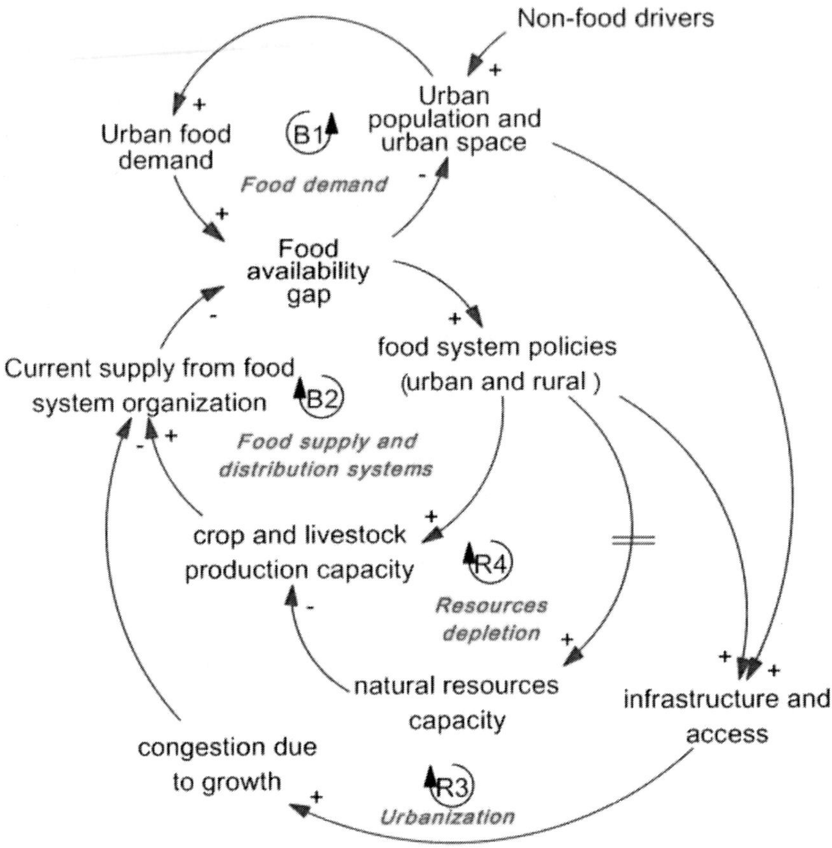

Figure 8. Systems analysis of food supply and distribution system (FSDS) adapted from [106]. Each labeled loop indicates an individual flow of information and/or materials. Well-known systemic structures, or archetypes, can be identified considering synergistic actions of the loops: drifting goals (B1 and B2); fixes that backfire (B2 and R3), and limits to growth (B2 and R4), where natural resources capacity represents the system-carrying capacity.

capacity of the system to face disturbances. Therefore, food system resiliency was defined as "the capacity over time of a food system and its units at multiple levels, to provide sufficient, appropriate and accessible food to all, in the face of various and even unforeseen disturbances" [109]. A more resilient system can absorb a shock with a more rapid recovery than a less resilient one (Figure 9). Reduced resilience can cause incomplete system recovery resulting in sustainability albeit at an overall lower system capacity (similar behaviors are observed in the system archetype of drifting goals, where original standards are abandoned or forgotten in favor of short-term success or new, albeit lower, standards of resilience). Following the resilience action cycle approach, food resilience problems have been studied and analyzed with SD to develop viable solutions for sustainable economies, especially in developing countries [110,111,112,113]. To build resilience in food systems with aid from quantitative models, three different boundary-scale levels are recommended [109]: (1) national and regional on which territorial policies will be considered; (2) supply chain on which special products and value-chains are considered from local to global level, and (3) individual perspectives regarding smallholders and their livelihoods.

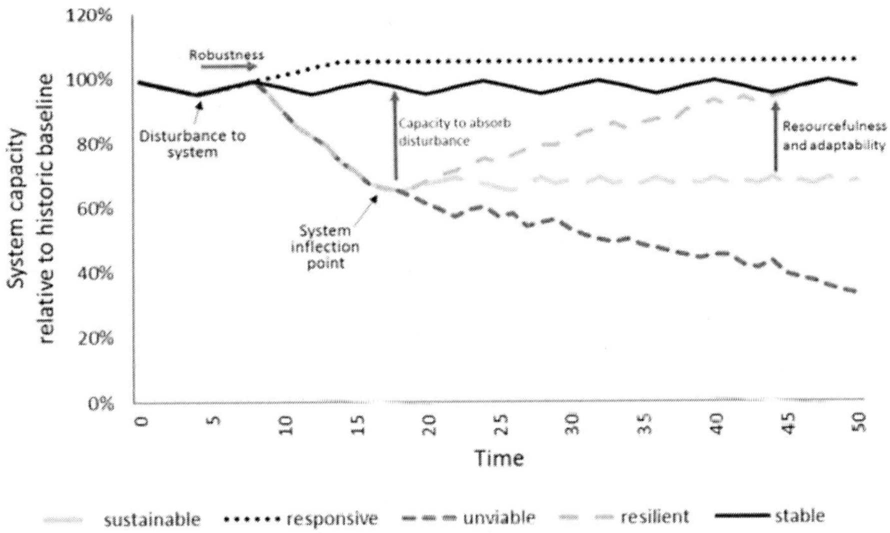

Figure 9. The food system resilience concept, adapted after both [109,114], displaying how a disturbance event acts on the function of food security and the possible responses from the food system depending on the system robustness (ability to withstand force), redundancy (ability to absorb force), flexibility (ability to recover rapidly), and resourcefulness and adaptability (ability to return the system back to previous condition). Possible system responses include proactive responses to improve system capacity, stable systems able to resist disturbances, sustainable equilibrium with low redundancy and high capacity losses, resilient recovery back to previous conditions, or unviable system degradation. Both system degradation and stable equilibria at a reduced level are both characteristic of the drifting goals archetype. Time is relative to the system in question, potentially days to years.

3.4. Integration of Smallholder Crop-Livestock Production and Smallholder Development

Sustainable intensification is a topic recently disseminated to solve social and economic inequalities around the world, and livestock is an intrinsic part of this sustainability enigma [114]. However, because of the complexity of the social-economic-environmental aspects of sustainability, system-oriented approaches for decision making are needed to succeed [24,115,116]. A holistic approach for integrating the crop-livestock-social components of sustainable intensification is required for effective deployment of technologies and assessment of failures. Many attempts in using decision support systems have been documented [114], but the application of SD in livestock is incipient [116]. Smallholder animal farmers are part of the sustainable livestock intensification program; mainly those in tropical regions that are mostly vulnerable to climate change [114]. The social aspect of smallholder crop-livestock production systems is extremely important because it entails subsistence farmers and herders, and small communities that are adapted to specific regions. SD models can be used to link social, economic, production, and environmental aspects because they rely on a big-picture perspective rather than being concerned with small details. For example, researchers have used an SD model to understand the functioning of mixed farming systems in the tropics, specifically in the Yucatán peninsula, including different components such as nutrition and management of sheep; partitioning of nutrients; flock dynamics; local and regional sheep production, marketing and economics; and labor [117]. A multi-objective modeling approach was adopted by combining different computer programs: the Agriculture Production Systems Simulator (APSIM) for crop modeling [118] and the Small Ruminant Nutrition System (SRNS) for animal modeling [119] with an SD model interfacing both programs (Figure 10). Applying their model to compare specialized systems versus mixed farming, they concluded the mixed farming scenario provided more income than specialized enterprises [120]. They suggested that for smallholders in that region, specialization was not sustainable, and therefore, mixed crop-livestock was the best production option.

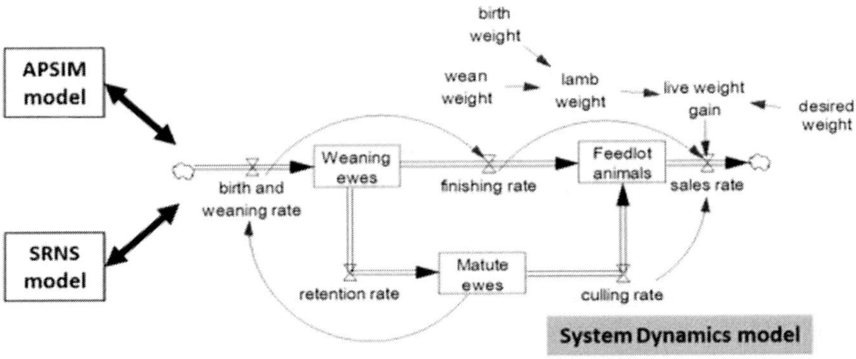

Figure 10. A graphical representation of an SD model interfacing with two non-SD models (APSIM and SRNS). Adapted and simplified from [119].

Smallholders are considered the core of rural growth, especial in developing countries. Smallholders are often subsidence farmers that consider food security as the main objective and profit as the second objective [121]. SD models developed in the last decade were applied to describe rural communities and smallholder farm dynamics to support technical choices [121,122], promote rural growth and improve an efficient use of the AGNR [113,123,124], increase the technical training of farmers in rural communities of developing countries [125], and integrate smallholders in their socioeconomic and environmental context [126].

4. DISCUSSION ON THE ROLES OF SYSTEM DYNAMICS FOR EMERGENT AGRICULTURE AND NATURAL RESOURCE MANAGEMENT CHALLENGES

Based on the above cases, it is clear that SD has a role in addressing AGNR management problems. The role of SD in contemporary AGNR problems dates back to the earliest SD models, including World3 [46], and over the subsequent decades, SD has branched out into varying disciplines in order to account for greater specificity and complexity of the AGNR problems at hand. It is now evident that food systems are primarily affected by AGNR drivers. However, they also produce outcomes in social capital (welfare, employment, wealth, etc.) and natural capital (environmental security, ecosystem services, etc.) through feedback mechanisms between food consumer and producer sectors [99]. SD models have increased their focus on natural resource constraints that potentially limit the behavior of socioecological systems (e.g., food system infrastructure; growth of urban centers) (see [108]). We have highlighted several noteworthy cases of SD applied to water resources, land and soil, food systems, or smallholder livelihood issues. In each of these cases, the SD modeling method facilitated identification of key insights not previously recognized by researchers or practitioners in their respective disciplines. It follows from these cases that AGNR features and dynamics need to be included in future model boundaries of investigations around social and political stability, especially those where natural resources may be a component of the problem or conflict. Incorporating agricultural system resiliency measures into SD models will be useful if the aim is to find appropriate adaptation strategies to changing conditions in the long term [126].

Interestingly, several system archetypes were identified in a number of the case studies (e.g., fixes that backfire [62,65,89,106], Figure 2, Figure 4, and Figure 8; drifting goals [106,109,114], Figure 8 and Figure 9; limits to growth [106]; and tragedy of the commons [62]). The two most commonly identified archetypes were fixes that backfire and drifting goals. The generic fix that backfires situation is characterized by a "quick fix" to alleviate a problem (which works in the short term), but employing the fix creates unintended consequences that reinforce, or perpetuate, the original problem (i.e., the problem persists in the long term). The feedback loop structure is a balancing loop (the "quick fix") embedded within a reinforcing loop (the unintended consequence). On the other hand, the drifting goals situation is characterized by the person, organization, or system that strives to meet a certain goal (e.g., a quality standard; profitability or production goal), but while waiting to see the results of invested efforts, it becomes easier to be satisfied with less and therefore lower the original goal [25]. The feedback loop structure of the drifting goals archetype is two connected balancing loops (one long-term loop capturing the invested effort to meet the original goal; one short-term loop that captures the pressure to lower the goal to a level that is easier to achieve).

Within SD modeling, archetypes are not unrelated [127,128]. We find that AGNR problems, based on the above cases, are on a path to falling into the common trap known as shifting the burden (Figure 11). The shifting the burden archetype is characterized by two balancing loops (similar to drifting goals) embedded within a reinforcing loop (similar to fixes that backfire). One loop captures a long-term solution to alleviate a chronic problem (e.g., meet societal needs for food and fiber production through regenerative solutions that protect biodiversity and enhance soil and water resources; B1 in Figure 11). Another loop captures the "quick-fix" solution (e.g., reliance on capitalizing easily attainable resources without regard for any social or environmental externalities; "take-make-waste" within B2 of Figure 11). The "quick fix" also represents lowering the overall sustainability goal (i.e., content with "take-make-waste" alternatives since they are generally faster and cheaper). The trap of shifting the burden lies in the unintended consequences (i.e., R loop in Figure 11), where reliance on short-term solutions are reinforced through unintended consequences on social systems (e.g., sunk costs which reduce management adaptability) and environmental systems (e.g., degraded ecosystems that lengthen restoration times and overall productivity). These short-term actions are often based on erroneous assumptions (e.g., infinite and cheap energy; non-limiting waste disposal capacity; non-limiting water and soil; stronger environmental resilience; etc.) [129]. In order to formulate effective policies that support fundamental solutions in the long term, supporting systemic perspectives and testing proposed strategies with tested models and scenarios of future behaviors will require collaborative approaches, where system researchers and stakeholders work together to identify and implement sustainable strategies [129,130]. Although our review has focused on examples of discipline integration through SD modeling, readers interested in stakeholder outreach and participation and how SD can provide aid in the decision-making process are encouraged to see [130].

Whether or not researchers fully adopt SD as their chosen methodology (and certainly not all scientists will become SD *modelers* in the sense they formulate models and test different policies), future approaches will increasingly become interdisciplinary in order to deal with the multifaceted and complex nature of AGNR problems. In such cases, the recognition and adoption of SD can hold great value, given it provides a proven modeling methodology to identify, describe, and quantitatively test the feedback loops interacting within a given system [33]. Even in most circumstances where not all of the interdisciplinary team members are SD modelers, they will still be integral to the research investigation due to their disciplinary expertise, which can enlighten researchers from other disciplines about the complexity of each respective component of the problem as well as inform the model formulation through the provision of expert knowledge (either personally or through guidance in the scientific literature) to the model developers.

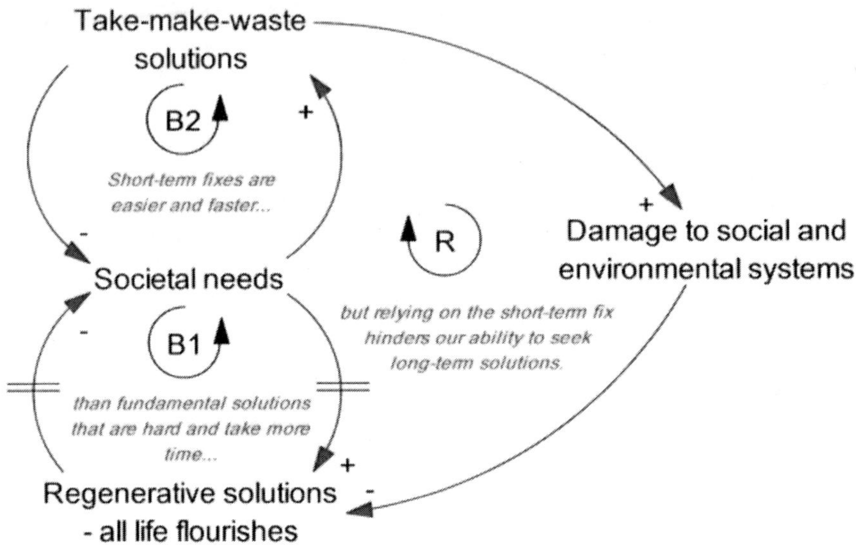

Figure 11. Shifting the burden archetype seen across many agricultural and natural resource problems, where societal needs can be met through long-term (regenerative) solutions or short-term (wasteful) solutions. Often short-term solutions are employed because they are faster and less expensive, but create unintended consequences that hinder society's ability to invest in long-term solutions through more degraded systems and sunk costs associated with relying on short-term strategies.

Another unique capability that was observed in multiple case studies is that SD has the ability to connect and interface with other computer programs in a user-friendly manner. This makes SD a useful and necessary tool that should be integrated with different modeling platforms. Those platforms might be less adept at capturing feedback loop-driven processes and could offer different useful information such as data specialization in the case of GIS, agent behavior with agent-based modeling platforms, or input values to set up initial conditions in the case of empirical models specially developed for technical purposes. SD aids in "seeing the big picture" and with the array of SD programs, researchers now have a suite of model-building platforms that can integrate other types of models that otherwise could not be used in tandem.

A major limitation of the SD approach is the risk to formulating erroneous policies by trusting simulations of poor (i.e., unvalidated) models that produce inaccurate results (due to poor technical and evaluation considerations). The wide-ranging availability of modeling tools and software has made model development easier (particularly for novices without deep knowledge of biophysical or socioeconomic processes) and tempts modelers to quickly develop and test models without fully understanding the endogenous dynamics. Model critique and evaluation is critical before policy strategy recommendations are made. In order to build confidence in model results and recommendations, good modeling practices require (1) use of local knowledge and/or historical data to calibrate model predictions to reality; (2) sensitivity analyses of key variables (e.g., Monte Carlo simulation) and model boundary and extreme condition testing to gauge the model's respect of physical and economic laws; and (3) analysis of scenarios compared to expert opinions (for model calibration and evaluation techniques, see [33,131,132]). Other limitations or criticisms of SD modeling include applications to the work type of problems, incorrectly applying the SD modeling method and its frameworks, and/or the tendency to build unnecessarily large models for "big" problems (for robust descriptions of these and other limitations as well as their handling strategies, see [133,134,135]). In this sense, SD modeling requires equal collaboration across boundaries [129]. For the scientific community, it will require increasing model building and model evaluation capacities, especially enhancing interdisciplinary approaches in modeling, as well as robust protocols of policy evaluation on the effects of applied actions on AGNR systems. Since all models are simplifications of reality and systems continually evolve, it is important that models are updated and improved as we learn more about the complex interrelationships contributing to the problems we face.

Finally, adoption of SD to a wider range of practitioners will remain difficult due to learning barriers observed in complex systems (e.g., limited information; confounding variables and ambiguity; misperceptions of feedback; flawed cognitive maps or (linear) mental models; defensive routines [136,137]). Overcoming these barriers requires improving the learning process (including primary and secondary education offerings in systems thinking), developing learning opportunities through management flight simulators (interactive computer models allowing participants to practice and learn dynamic decision making in real-time), reorganizing organizations to incentivize individuals to pursue systemic cause-and-effect solutions [136,137], and perhaps most importantly, overcoming deeply ingrained preconceptions by pausing to "admiring the problem" to assist in the shift from short-term to long-term thinking (Figure 11; [129]).

5. CONCLUSIONS

In this paper, we have described a variety of agriculture and natural resource management (AGNR) problems occurring globally. System dynamics (SD) provides a valuable framework for investigating these complex AGNR issues, specifically through the use of computer simulation modeling. After briefly describing the SD methodology, we provided an extensive review of SD applications to water, soil, food systems, and smallholder issues and illustrated in these cases several system archetype behaviors (e.g., fixes that backfire; drifting goals). Continuing down these paths are likely to lead to reliance upon short-term solutions to AGNR problems (i.e., "quick fixes" or "take-make-waste"), making it more difficult to employ fundamental solutions (e.g., regenerative solutions). If AGNR goals continue to "drift" (i.e., settle for lower sustainability goals regarding resource use and externalities), AGNR systems are likely to fall into the trap of "shifting the burden," where reliance on "quick fixes" become the only feasible alternative. We conclude that common attempts to alleviate AGNR problems, across continents and regardless of the types of resources involved, have suffered from reliance on short-term management strategies. To effectively address AGNR problems, longer-term thinking and strategies aimed at fundamental solutions will be needed to better identify and minimize the often delayed, and unintended, consequences arising from feedback between management interventions and AGNR systems.

The ability of SD to aid in recognition of complex interacting factors and scientifically test different management or policy strategies was also of key importance. Several cases described ways in which SD was used to integrate with different types of models (e.g., RUSLE or APSIM) or computer applications (e.g., Microsoft Excel or ArcGIS), highlighting the potential that SD programs have to address complex AGNR issues by integrating different types of scientists and models into collaborative interdisciplinary investigations. However, adoption of SD into other disciplines remains slow due to a number of barriers, hindering more transformative or transdisciplinary research. Overcoming these barriers is possible and it will be essential to integrate concepts and models across a wide array of sciences in order to adequately address the emerging AGNR challenges. Due to the strengths embedded in SD methodology, SD provides a valuable framework to investigate AGNR problems, both independently and in tandem with other types of models and disciplines. SD should be a central tool for conducting transdisciplinary research capable of addressing our most pressing AGNR issues.

ACKNOWLEDGMENTS

We'd like to thank two anonymous reviewers for their very helpful comments provided during the review process—both of which greatly improved the paper.

AUTHOR CONTRIBUTIONS

B.T. organized the outline of the manuscript and Section 1 and Section 2. H.M., L.T., and A.A. constructed Section 3 and Section 4 and provided editorial comments throughout the entire manuscript. R.G. constructed the conclusions and abstract and provided editorial comments and direction throughout the development of the manuscript.

CONFLICTS OF INTEREST

The authors declare no conflict of interest.

REFERENCES

1. Trenberth, K.E.; Dai, A.; van der Schrier, G.; Jones, P.D.; Barinchivich, J.; Briffa, K.R.; Sheffield, J. Global warming and changes in drought. Nat. Clim. Chang. 2013, 4, 17–22.
2. Akerlof, K.; Maiback, E.W.; Fitzgerald, D.; Cedeno, A.Y.; Neuman, A. Do people "personally experience" global warming, and if so how, and does it matter? Glob. Environ. Chang. 2012.
3. Corlett, R.T.; Westcott, D.A. Will plant movements keep up with climate change? Trends Ecol. Evol. 2013.
4. Sheffield, J.; Wood, E.F.; Roderick, M.I. Little change in global drought over the past 60 years. Nature 2012, 491, 435–438.
5. Taylor, R.G.; Scanlan, B.; Doll, P.; Rodell, M.; van Beek, R.; Wada, Y.; Longuevergne, L.; Leblanc, M.; Famiglietti, J.S.; Edmunds, M.; et al. Groundwater and climate change. Nat. Clim. Chang. 2012.
6. Haddeland, I.; Heinke, J.; Biemans, H.; Eisner, S.; Florke, M.; Hanasaki, N.; Konzmann, M.; Ludwig, F.; Masald, Y.; Schewe, J.; et al. Global water resources affected by human interventions and climate change. PNAS 2014, 111, 3251–3256.
7. Walsh, C.L.; Blenkinsop, S.; Fowler, H.J.; Burton, A.; Dawson, R.J.; Glenis, V.; Manning, L.J.; Kilsby, C.G. Adaptation of water resource systems to an uncertain future. Hydrol. Earth Syst. Sci. Discuss. 2015, 12, 8853–8889.
8. Savenige, H.H.G.; Hoekstra, A.Y.; van der Zaag, P. Evolving water science in the Anthropocene. Hydrol. Earth Syst. Sci. 2014, 18, 319–332.

9. Mahmood, R.; Pielke, R.A., Sr.; Hubbard, K.G.; Niyogi, D.; Dirmeyer, P.A.; McAlpine, C.; Carleton, A.M.; Hale, R.; Gameda, S.; Beltran-Przekurat, A.; et al. Land cover changes and their biogeophysical effects on climate. Int. J. Climatol. 2014, 34, 929–953.
10. Seto, K.C.; Gueralp, B.; Hutyra, L.R. Global forecasts of urban expansion to 2030 and direct impacts on biodiversity and carbon pools. PNAS 2012, 109, 16083–16088.
11. Van den Bergh, J.C.J.M.; Grazi, F. Ecological footprint policy? Land use as an environmental indicator. J. Ind. Ecol.2013.
12. Nepstad, D.C.; Boyd, W.; Stickler, C.M.; Bezerra, T.; Azevedo, A. Responding to climate change and the global land crisis: REDD+, market transformation and low emissions rural development. Philos. Trans. R. Soc. B 2013.
13. Bellard, C.; Bertelsmeier, C.; Leadley, P.; Thuiller, W.; Courchamp, F. Impacts of climate change on the future of biodiversity. Ecol. Lett. 2012, 15, 365–377.
14. Cheung, W.W.L.; Sarmiento, J.L.; Dunne, J.; Frolicher, T.L.; Lam, V.W.Y.; Deng Palomares, M.L.; Watson, R.; Pauly, D. Shrinking of fishes exacerbates impacts of global ocean changes on marine ecosystems. Nat. Clim. Chang. 2012.
15. Pauls, S.U.; Nowak, C.; Balint, M.; Pfenninger, M. The impact of global climate change on genetic diversity within populations and species. Mol. Ecol. 2013, 22, 925–946.
16. Wheeler, T.; von Braun, J. Climate Change Impacts on Global Food Security. Science 2013, 341, 508–513.
17. Van Ittersum, M.K.; Cassman, K.G.; Grassini, P.; Wolf, J.; Tittonell, P.; Hochman, Z. Yield gap analysis with local to global relevance—A review. Field Crops Res. 2013, 143, 4–17.
18. Vermeulen, S.J.; Campbell, B.M.; Ingram, J.S.I. Climate Change and Food Systems. Annu. Rev. Environ. Resour. 2012, 37, 195–222.
19. Teixeira, E.I.; Fischer, G.; van Velthuizen, H.; Walter, C.; Ewert, F. Global hot-spots of heat stress on agricultural crops due to climate change. Agric. For. Meteorol. 2013, 170, 206–215.
20. Shindell, D.; Kuylenstierna, J.C.I.; Vignati, E.; van Dingene, R.; Amann, M.; Klimont, Z.; Anenberg, S.C.; Muller, N.; Janssens-Maenhout, G.; Raes, F.; et al. Simultaneously mitigating near-term climate change and improving human health and food security. Science 2012, 335.
21. Bommarco, R.; Kleijn, D.; Potts, S.G. Ecological intensification: Harnessing ecosystem services for food security. Trends Ecol. Evol. 2012.
22. Garnett, T.; Appleby, M.C.; Balmford, A.; Bateman, I.J.; Benton, T.G.; Bloomer, P.; Burlingame, B.; Dawkins, M.; Dolan, L.; Fraser, D.; et al. Sustainable intensification in agriculture: Premises and policies. Science 2013, 341, 33–34.
23. West, P.C.; Gerber, J.S.; Engstrom, P.M.; Mueller, N.D.; Brauman, K.A.; Carlson, K.M.; Cassidy, E.S.; Johnston, M.; MacDonald, G.K.; Ray, D.K.; et al. Leverage points for improving global food security and the environment. Science2014, 345, 325–328.
24. Liu, J.; Mooney, H.; Hull, V.; Davis, S.J.; Gaskell, J.; Hertel, T.; Lubchenco, J.; Seto, K.C.; Gleick, P.; Kremen, C.; et al. Systems integration and global sustainability. Science 2015, 347.
25. Senge, P.M. The Fifth Discipline, 1st ed.; Doubleday: New York, NY, USA, 1990.
26. Cilliers, P.; Biggs, H.C.; Blignaut, S.; Choles, A.G.; Hofmeyr, J.-H.S.; Jewitt, G.P.W.; Roux, D.J. Complexity, modeling, and natural resource management. Ecol. Soc. 2013, 18, 1.
27. Doyle, J.K.; Ford, D.N. Mental Models Concepts for system dynamics research. Syst. Dyn. Rev. 1998, 14, 3–29.
28. Forrester, J.W. The "model versus a modeling process". Syst. Dyn. Rev. 1985, 1, 133–134.
29. Beautement, P.; Broenner, C. Complexity Demystified: A Guide for Practitioners; Triarchy Press: Devon, UK, 2011.
30. Bawden, R.J. Systems Thinking and Practice in Agriculture. J. Dairy Sci. 1991, 74, 2362–2373.
31. Dahlberg, K. Beyond the Green Revolution, 1st ed.; Plenum Press: New York, NY, USA, 1979.

32. Stirzaker, R.; Biggs, H.; Roux, D.; Cilliers, P. Requisite simplicities to help negotiate complex problems. Ambio 2010, 39, 600–607.
33. Sterman, J.D. Business Dynamics, 1st ed.; Irwin McGraw-Hill: New York, NY, USA, 2000.
34. Goodman, M. "Everyone's Problem to Solve: Systems Thinking Cross-Functionally". The Systems Thinker Newsletter, June/July 2006. Available online: https://thesystemsthinker.com/everyones-problem-to-solve-systems-thinking-cross-functionally/ (accessed on 14 September 2016).
35. Lane, D.C. Should System Dynamics be Described as a 'Hard' or 'Deterministic' Systems Approach? Syst. Res. Behav. Sci. 2000, 17, 3–22.
36. Richmond, B. An Introduction to Systems Thinking, STELLA, 3rd ed.; High Performance Systems, Inc.: Lebanon, NH, USA, 2001.
37. Miller, J. Active nonlinear tests (ANTs) of complex simulation models. Manag. Sci. 1998, 44, 820–830.
38. Checkland, P.; Holwell, S. "Classic" OR and "soft" OR—An asymmetric complementarity. In Systems Modeling: Theory and Practice; Pidd, M., Ed.; John Wiley & Sons, Inc.: Chichester, UK, 2004.
39. Ford, A. Modeling the Environment, 1st ed.; Island Press: Washington, DC, USA, 1999.
40. Deaton, M.I.; Winebrake, J.J. Dynamic Modeling of Environmental Systems, 1st ed.; Springer: New York, NY, USA, 2000.
41. Grant, W.E.; Pedersen, E.K.; Marin, S.L. Ecology and Natural Resource Management Systems Analysis and Simulation, 1st ed.; John Wiley and Sons, Inc.: New York, NY, USA, 1997.
42. McGarvey, B.; Hannon, B. Dynamic Modeling for Business Management, 1st ed.; Springer: New York, NY, USA, 2004.
43. Ruth, M.; Hannon, B. Model Dynamic Economic Systems; Springer: New York, NY, USA, 1997.
44. Hannon, B.; Ruth, M. Modeling Dynamic Biological Systems; Springer: New York, NY, USA, 1997.
45. Costanza, R.; Voinov, A. Landscape Simulation Modeling; Springer: New York, NY, USA, 2004.
46. Meadows, D.; Randers, J.; Meadows, D.; Behrens, W.W. The Limits to Growth; Universe Books: New York, NY, USA, 1972.
47. Meadows, D.H.; Meadows, D.L.; Randers, J. Beyond the Limits; Chelsey Green: Post Mills, VT, USA, 1992.
48. Meadows, D.H.; Randers, J.; Meadows, D.L. Limits to Growth—The 30-Year Update; Chelsea Green: White River Junction, VT, USA, 2004.
49. Turner, G. A comparison of the Limits to Growth with 30 years of reality. Glob. Environ. Chang. 2008, 18, 397–411.
50. Meadows, D.L. Evaluating past forecasts: Reflections on one critique of The Limits to Growth. In Sustainability or Collapse? An Integrated History and Future of People on Earth; Costanza, R., Grqumlich, L., Steffen, W., Eds.; MIT Press: Cambridge, MA, USA, 2007; pp. 399–415.
51. Paqualino, R.; Monasterolo, I.; Jones, A.W.; Philips, A. Understanding Global Systems Today—A Calibration of the World3-03 Model between 1995 and 2012. Sustainability 2015, 7, 9864–9889.
52. Kirshen, P.H. Computer Model for Small-Scale Hydropower Policy Analysis. J. Water Resour. Plan. Manag. 1981, 107, 61–76.
53. Simonovic, S.P.; Fahmy, H.; El-Shorbagy, A. The Use of Object-Oriented Modeling for Water Resources Planning in Egypt. Water Resour. Manag. 1997, 11, 243–261.
54. Hoekema, D.J.; Sridhar, V. A System Dynamics Model for Conjunctive Management of Water Resources in the Snake River Basin. J. Am. Water Resour. Assoc. 2013, 49, 1327–1350.

55. Tidwell, V.C.; Passell, H.D.; Conrad, S.H.; Thomas, R.P. System dynamics modeling for community-based water planning: Application to the Middle Rio Grande. Aquat. Sci. 2004, 66, 357–372.
56. Sehlke, G.; Jacobson, J. System dynamics modeling of transboundary systems: The Bear River basin model. Ground Water 2005, 43, 722–730.
57. Beall, A.; Fiedler, F.; Boll, J.; Cosens, B. Sustainable Water Resource Management and Participatory System Dynamics. Case Study: Developing the Palouse Basin Participatory Model. Sustainability 2011, 3, 720–742.
58. Ryu, J.H.; Contor, B.; Johnson, G.; Allen, R.; Tracy, J. System Dynamics to Sustainable Water Resources Management in the Eastern Snake Plain Aquifer Under Water Supply Uncertainty. J. Am. Water Resour. Assoc. 2012, 48, 1204–1220.
59. Bender, M.J.; Simonovic, S.P. A Systems Approach for Collaborative Decision Support in Water Resources Planning. Int. J. Technol. Manag. 2000, 19, 546–556.
60. Mojtahedzadeh, M.T. A Dynamic Model for Development Planning in an Arid Area. In Proceedings of the 1992 International System Dynamics Conference, Utrecht, The Netherlands, 14–17 July 1992.
61. Ho, C.C.; Yang, C.C.; Chang, L.C.; Chen, T.W. The Application of System Dynamics Modeling to Study Impact of Water Resources Planning and Management in Taiwan. In Proceedings of the 23rd International Conference of the System Dynamics Society, Boston, MA, USA, 17–21 July 2005.
62. Xu, H.G. Exploring effective policies for underground water management in artificial oasis: A system dynamics analysis of a case study of Yaoba Oasis. J. Environ. Sci. 2001, 13, 476–480.
63. Dhungel, R.; Fiedler, F. Water Balance to Recharge Calculation: Implications for Watershed Management Using Systems Dynamics Approach. Hydrology 2016, 3, 13.
64. Hardin, G. The tradegy of the commons. Science 1968, 13, 1243–1248.
65. Mirchi, A.; Madani, K.; Watkins, D.; Ahmad, S. Synthesis of System Dynamics Tools for Holistic Conceptualization of Water Resources Problems. Water Resour. Manag. 2012, 26, 2421–2442.
66. Xu, Z.X.; Takeuchi, K.; Ishidaira, H.; Zhang, X.W. Sustainability Analysis for Yellow River Water Resources Using the System Dynamics Approach. Water Resour. Manag. 2002, 16, 239–261.
67. Davies, E.G.R.; Simonovic, S.P. Global water resources modeling with an integrated model of the social-economic-environmental system. Adv. Water Resour. 2011, 34, 684–700.
68. Tidwell, V.; Kobos, P.; Malczynski, L.; Klise, G.; Castillo, C. Exploring the Water-Thermoelectric Power Nexus. J. Water Resour. Plan. Manag. 2011, 138, 491–501.
69. Roach, J.; Tidwell, V. A Compartmental-Spatial System Dynamics Approach to Ground Water Modeling. Ground Water 2009, 47, 686–689.
70. Bassi, A.M.; Tan, Z.; Goss, S. An Integrated Assessment of Investments towards Global Water Sustainability. Water2010, 2, 726–741.
71. Balali, H.; Viaggi, D. Applying a System Dynamics Approach for Modeling Groundwater Dynamics to Depletion under Different Economical and Climate Change Scenarios. Water 2015, 7, 5258–5271.
72. Gohari, A.; Eslamian, S.; Mirchi, A.; Abedi-Koupaei, J.; Bavani, A.M.; Madani, K. Water transfer as a solution to water shortage: A fix that can backfire. J. Hydrol. 2013, 491, 23–39.
73. Stave, K.A. A system dynamics model to facilitate public understanding of water management options in Las Vegas, Nevada. J. Environ. Manag. 2003, 67, 303–313.
74. Ahmad, S.; Simonovic, S.P. System Dynamics Modeling of Reservoir Operations for Flood Management. J. Comput. Civ. Eng. 2000, 14, 190–198.
75. Turner, B.L.; Tidwell, V.; Fernald, A.; Rivera, J.; Rodriguez, S.; Guldan, S.; Ochoa, C.; Hurd, B.; Boykin, K.; Cibils, A. Modeling acequia irrigation systems using system dynamics: Model development, evaluation, and sensitivity analyses to investigate effects of socio-economic and biophysical feedbacks. Sustainability 2016, 8, 1019.

76. Keith, B.; Enos, J.; Garlick, C.B.; Simmons, G.; Copeland, D.; Cortizo, M. Limits to Population Growth and Water Resource Adequacy in the Nile River Basin, 1994–2100. In Proceedings of the 31st International Conference of the System Dynamics Society, Boston, MA, USA, 21–25 July 2013.
77. Keith, B.; Horton, R.; Bower, E.; Lee, J.; Pinelli, A.; Dittrick, D. Estimating the Effect of Climate Change on the Hydrology of the Nile River in the 21st Century. In Proceedings of the 32nd International Conference of the System Dynamics Society, Delft, The Netherlands, 20–24 July 2014.
78. Kwakkel, J.H.; Slinger, J.S. A System Dynamics Model-Based Exploratory Analysis of Salt Water Intrusion in Coastal Aquifers. In Proceedings of the 30th International Conference of the System Dynamics Society, St. Gallen, Switzerland, 22–26 July 2012.
79. Shanshan, D.; Lanhai, L.; Honggang, X. The system dynamic study of regional development of Manas Basin Under the constraints of water resources. In Proceedings of the 28th International Conference of the System Dynamics Society, Seoul, Korea, 25–29 July 2010.
80. Hansen, J.K. Estimating Impacts of Water Scarcity Pricing. In Proceedings of the 27th International Conference of the System Dynamics Society, Albuquerque, NM, USA, 26–30 July 2009.
81. Luo, Y.; Khan, S.; Cui, Y. A System Dynamics Model for Sustainable Irrigation Water Management in the Lower Yellow River Basin. In Proceedings of the 23rd International Conference of the System Dynamics Society, Boston, MA, USA, 25–29 July 2005.
82. Elshafei, Y.; Tonts, M.; Sivapalan, M.; Hipsey, M.R. Sensitivity of emergent sociohydrologic dynamics to internal system properties and external sociopolitical factors: Implications for water management. Water Resour. Res. 2016, 52.
83. Turner, B.L.; Wuellner, M.; Nichols, T.; Gates, R.; Tedeschi, L.O.; Dunn, B.H. Development and Evaluation of a System Dynamics model for Investigating Agriculturally Driven Land Transformation in the North Central United States. Nat. Resour. Model. 2016, 29, 179–228.
84. Rozman, C.; Pažek, K.; Škraba, A.; Turk, J.; Kljajić, M. Determination of Effective Policies for Ecological Agriculture Development with System Dynamics—Case Study in Slovenia. In Proceedings of the 30th International Conference of the System Dynamics Society, St. Gallen, Switzerland, 22–26 July 2012.
85. Amelia, D.F.; Kopainsky, B.; Nyanga, P.H. Exploratory Model of Conservation Agriculture Adoption and Diffusion in Zambia: A Dynamic Perspective. In Proceedings of the 32nd International Conference of the System Dynamics Society, Delft, The Netherlands, 20–24 July 2014.
86. Dent, J.B.; Edward-Jones, G.; McGregor, M.J. Simulation of Ecological, Social and Economic Factors in Agricultural Systems. Agric. Syst. 1995, 49, 337–351.
87. Saysel, A.K. Analyzing Soil Nitrogen Management with Dynamic Simulation Experiments. In Proceedings of the 16th International Conference of the System Dynamics Society, Delft, The Netherlands, 20–24 July 2014.
88. Huang, M.; Elshorbagy, A.; Barbour, S.L.; Zettl, J.D.; Si, B.C. System dynamics modeling of infiltration and drainage in layered coarse soil. Can. J. Soil Sci. 2011, 91, 185–197.
89. Khan, S.; Yufeng, L.; Ahmad, A. Analysing complex behavior of hydrological systems through a system dynamics approach. Environ. Model. Softw. 2009, 24, 1363–1372.
90. Elshorbagy, A.; Jutla, A.; Barbour, L.; Kells, J. System dynamics approach to assess the sustainability of reclamation of disturbed watersheds. Can. J. Civ. Eng. 2005, 32, 144–158.
91. Elshorbagy, A.; Jutla, A.; Kells, J. Simulation of the hydrological processes on reconstructed watersheds using system dynamics. Hydrol. Sci. J. 2007, 52, 538–562.
92. Keshta, N.; Elshorbagy, A.; Carey, S. A generic system dynamics model for simulating and evaluating the hydrological performance of reconstructed watersheds. Hydrol. Earth Syst. Sci. 2009, 13, 865–881.

93. Cakula, A.; Ferreira, V.; Panagopoulos, T. Dynamic Model of Soil Erosion and Sediment Deposit in Watersheds. In Recent Researches in Environment, Energy Systems and Sustainability; Ramos, R.A.R., Straupe, I., Panagopoulos, T., Eds.; WSEAS Press: Faro, Portugal, 2012; pp. 33–38.
94. Widman, N. RUSLE2-Instructions and Users Guide. United States Department of Agriculture; Natural Resources Conservation Service: Washington, DC, USA, 2004.
95. Yeh, S.C.; Wang, C.A.; Yu, H.C. Simulation of soil erosion and nutrient impact using an integrated system dynamics model in a watershed in Taiwan. Environ. Model. Softw. 2006, 21, 937–948.
96. Haith, D.A.; Shoemaker, L.L. Generalized watershed loading function for stream flow nutrients. Water Resour. Bull.1987, 12, 471–478.
97. Duggen, J. System Dynamics Modeling with R; Springer International Publishing: Cham, Switzerland, 2016.
98. United Nations Food and Agriculture Organization. Rome Declaration. 1996. Available online: http://www.fao.org/docrep/003/w3613e/w3613e00.htm (accessed on 13 September 2016).
99. Ericksen, P.J. Conceptualizing food systems for global environmental change research. Glob. Environ. Chang. 2008, 18, 234–245.
100. Gomez, M.I.; Ricketts, K.D. Food value chain transformations in developing countries: Selected hypotheses on nutritional implications. Food Policy 2013, 42, 139–150.
101. Pina, W.H.A.; Pardo Martinez, C.I. Urban material flows analysis: An approach for Bogotà, Colombia. Ecol. Indic.2014, 42, 32–42.
102. Maxwell, L.; Slater, R. Food Policy old and New. Dev. Policy Rev. 2003, 21, 531–553.
103. Giraldo, D.P.; Betancour, M.; Arango, S. Food Security in Developing countries, a systemic perspective. In Proceedings of the 26th International Conference of the System Dynamics Society, Athens, Greece, 20–24 July 2008.
104. Giraldo, D.P.; Betancour, M.; Arango, S. Effects of Food Availability Policies on National Food Security: Colombian case. In Proceedings of the 32nd International Conference of the System Dynamics Society, Washington, DC, USA, 24–28 July 2011.
105. Aragrande, M.; Argenti, O. Studying Food Supply and Distribution Systems to Cities in Developing Countries and Countries in Transition. Methodological and Operational Guide; "Food into Cities" Collection, DT/36-01E; FAO: Rome, Italy, 2001.
106. Armendariz, V.; Atzori, A.; Armenia, A.; Romano, A. Analyzing Food Supply and Distribution Systems using complex systems methodologies. In Proceedings of the 9th Igls-Forum on System Dynamics and Innovation in Food Networks, Innsbruck, Austria, 9–13 February 2015.
107. Armendariz, V.; Atzori, A.; Armenia, A. Understanding the Dynamics of Food Supply and Distribution Systems (FSDS); "Complex-Systems Dynamics Principles Applied to Food Systems" Initiative from FAO "Meeting Urban Food Needs" Project; FAO: Rome, Italy, 2015.
108. Armendariz, V.; Armenia, S.; Atzori, A. SD Updates of FAO Methodological Guide to manage the Food Supply and Distribution Systems (FSDS). In Proceedings of the 33rd International Conference of the System Dynamics Society, Cambridge, MA, USA, 19–23 July 2015.
109. Tendall, D.M.; Jörin, J.; Kopainsky, B.; Edwards, P.; Shreck, A.; Le Quang, B.; Grant, M.; Six, J. Food system resilience: Defining the concept. Glob. Food Secur. 2015, 6, 17–23.
110. Stave, K.; Kopainsky, B. A system dynamics approach for examining mechanisms and pathways of food supply vulnerability. J. Environ. Stud. Sci. 2015, 5, 321–336.
111. Kopainsky, B.; Huber, R.; Pedercini, M. Food provision and environmental goals in the Swiss agri-food system: System dynamics and the social-ecological systems framework. Syst. Res. Behav. Sci. 2015, 32, 414–432.
112. Kopainsky, B.; Nicholson, C.F. System dynamics and sustainable intensification of food systems: Complementarities and challenges. In Proceedings of the 33rd International Conference of the System Dynamics Society, Cambridge, MA, USA, 19–23 July 2015.

113. Stephens, E.; Nicholson, C.; Brown, D.; Parsons, D.; Barrett, C.; Lehmann, J.; Mbugua, D.; Ngoze, S.; Pell, A.; Riha, S. Modeling the Impacts of Natural-Resource Based Poverty Traps on Food Security in Kenya: The Crops, Livestock and Soils in Smallholder Economic Systems (CLASSES) Model. Food Secur. 2012, 4, 423–439.

114. Tedeschi, L.O.; Muir, J.P.; Riley, D.G.; Fox, D.G. The role of ruminant animals in sustainable livestock intensification programs. Int. J. Sustain. Dev. World Ecol. 2015, 22, 452–465.

115. Garnett, T.; Godfray, C. Sustainable Intensification in Agriculture: Navigating a Course through Competing Food System Priorities; Food Climate Research Network and the Oxford Martin Programme on the Future of Food, University of Oxford: Oxford, UK, 2012; 51p. Available online: http://www.futureoffood.ox.ac.uk/sustainable-intensification (accessed on 13 February 2015).

116. Tedeschi, L.O.; Nicholson, C.F.; Rich, E. Using System Dynamics modelling approach to develop management tools for animal production with emphasis on small ruminants. Small Rumin. Res. 2011, 98, 102–110.

117. Parsons, D.; Nicholson, C.F.; Blake, R.W.; Ketterings, Q.M.; Ramírez-Avilés, L.; Fox, D.G.; Tedeschi, L.O.; Cherney, J.H. Development and evaluation of an integrated simulation model for assessing smallholder crop-livestock production in Yucatán, Mexico. Agric. Syst. 2010, 104, 1–12.

118. Keating, B.A.; Carberry, P.S.; Hammer, G.L.; Probert, M.E.; Robertson, M.J.; Holzworth, D.; Huth, N.I.; Hargreaves, J.N.G.; Meinke, H.; Hochman, Z.; et al. An overview of APSIM, a model designed for farming systems simulation. Eur. J. Agron. 2003, 18, 267–288.

119. Tedeschi, L.O.; Cannas, A.; Fox, D.G. A nutrition mathematical model to account for dietary supply and requirements of energy and nutrients for domesticated small ruminants: The development and evaluation of the Small Ruminant Nutrition System. Small Rumin. Res. 2010, 89, 174–184.

120. Parsons, D.; Nicholson, C.F.; Blake, R.W.; Ketterings, Q.M.; Ramírez-Avilés, L.; Cherney, J.H.; Fox, D.G. Application of a simulation model for assessing integration of smallholder shifting cultivation and sheep production in Yucatán, Mexico. Agric. Syst. 2010, 104, 13–19.

121. Gerber, A. Agricultural theory in system dynamics: A case study from Zambia. In Proceedings of the 33rd International Conference of the System Dynamics Society, Cambridge, MA, USA, 19–23 July 2015.

122. Derwisch, S.; Morone, P.; Tröger, K.; Kopainsky, B. Investigating the drivers of innovation diffusion in a low income country context. The case of adoption of improved maize seed in Malawi. Futures 2016, 81, 161–175.

123. Pedercini, M.; Barney, G.O. Dynamic analysis of interventions designed to achieve Millennium Development Goals (MDG). The Case of Ghana. Socio-Econ. Plan. Sci. 2010, 44, 89–99.

124. Molina, R.; Atzori, A.S.; Campos, R.; Sanchez, H. Using System Thinking to study sustainability of Colombian dairy system. Bus. Syst. Rev. 2014, 3, 123–141.

125. Hager, G.; Kopainsky, B.; Nyanga, P. Learning as conceptual change during community based group interventions. A case study with smallholder farmers in Zambia. In Proceedings of the 33rd International Conference of the System Dynamics Society, Cambridge, MA, USA, 19–23 July 2015.

126. Herrera, H.; Kopainsky, B. Rethinking agriculture in a shrinking world: Operationalization of resilience with a System Dynamics perspective. In Proceedings of the 33rd International Conference of the System Dynamics Society, Cambridge, MA, USA, 19–23 July 2015.

127. Senge, P.M. The Fifth Discipline Fieldbook: Strategies and Tools for Building a Learning Organization; Doubleday: New York, NY, USA, 1994.

128. Goodman, M. Using the Archetype Family Tree as a Diagnostic Tool. The Systems Thinker. Available online: https://thesystemsthinker.com/using-the-archetype-family-tree-as-a-diagnostic-tool/ (accessed on 20 October 2016).

129. Senge, P.M. The Necessary Revolution: How Individuals and Organizations Are Working Together to Create a Sustainable World; Nicholas Brealey: London, UK, 2010.
130. Bourget, L. (Ed.) Converging Waters Integrating Collaborative Modeling with Participatory Processes to Make Water Resources Decisions; Institute for Water Resources, U.S. Army Corps of Engineers: Alexandria, VA, USA, 2011; Available online: http://www.iwr.usace.army.mil/Portals/70/docs/ maasswhite/ Converging_Waters.pdf (accessed on 19 October 2016).
131. Tedeschi, L.O. Assessment of the adequacy of mathematical models. Agric. Syst. 2005, 89, 225–247.
132. Oliva, R. Model calibration as a testing strategy for system dynamics models. Eur. J. Oper. Res. 2003, 151, 552–568.
133. Featherson, C.R.; Doolan, M. A Critical Review of the Criticisms of System Dynamics. In Proceedings of the 30th International Conference of the System Dynamics Society, St. Gallen, Switzerland, 22–26 July 2012.
134. Barlas, Y. Leverage points to march "upward from the aimless plateau". Syst. Dyn. Rev. 2007, 23, 469–473.
135. Forrester, J.W. System dynamics-the next fifty years. Syst. Dyn. Rev. 2007, 23, 359–370.
136. Richmond, B. Systems Thinking: Four Key Questions; High Performance Systems, Inc.: Lebanon, NH, USA, 1991.
137. Sterman, J.D. Learning in and about complex systems. Syst. Dyn. Rev. 1994, 10, 291–330.

CHAPTER 18

Resolving Conflicts between Agriculture and the Natural Environment

Andrew J. Tanentzap[1] *, Anthony Lamb[2] , Susan Walker[3] , Andrew Farmer[4]

[1] *Ecosystems and Global Change Group, Department of Plant Sciences, University of Cambridge, Cambridge, United Kingdom,*

[2] *Conservation Science Group, Department of Zoology, University of Cambridge, Cambridge, United Kingdom,*

[3] *Landcare Research, Dunedin, New Zealand,*

[4] *Institute for European Environmental Policy, London, United Kingdom*

ABSTRACT

Agriculture dominates the planet. Yet it has many environmental costs that are unsustainable, especially as global food demand rises. Here, we evaluate ways in which different parts of the world are succeeding in their attempts to resolve conflict between agriculture and wild nature. We envision that coordinated global action in conserving land most sensitive to agricultural activities and policies that internalise the environmental costs of agriculture are needed to deliver a more sustainable future.

INTRODUCTION

Agriculture is the most dominant land use on Earth and will remain so as the world's population and global food demand rise [1] (Figure 1). This poses a fundamental challenge to the natural environment, which requires access to many of the same resources as agricultural producers and is harmed by agriculture's external impacts [2,3]. Using land resources for agriculture prioritises crop and livestock production over the many other benefits that natural ecosystems provide and that agriculture displaces, including biodiversity, carbon storage, and water purification [4]. Governments also heavily support agricultural production in ways that are often bad for the environment, such as rewarding output or lowering input prices [5,6]. Very little is spent on actively mitigating the environmental impacts (Figure 1).

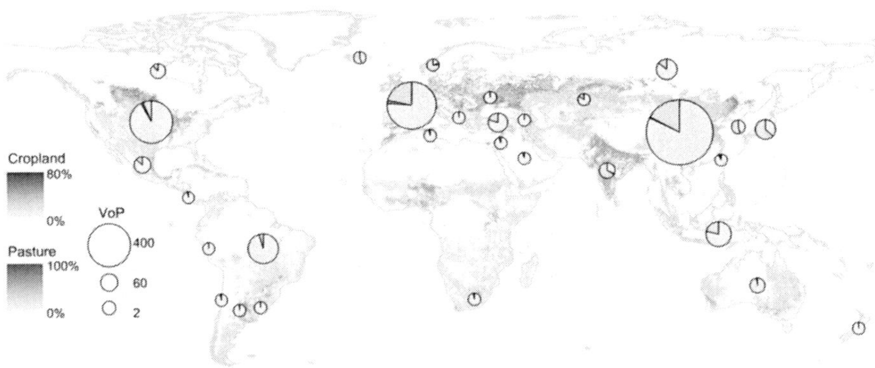

Figure 1. Financial support to farmers from taxpayers and consumers associated with agricultural policies as a proportion of the total value of agricultural production (VoP) at the farm gate.

We distinguished between policies aiming to achieve specific environmental objectives beyond those required by regulation (purple segments) and all other types of support irrespective of their influence on farm production or income (orange segments), e.g., market price support, monetary transfers based on output, input subsidies. Sizes of symbols were scaled to VoP in hundreds of millions of United States dollars. Data are from the most recent year available (2008–2013) and described in S1 Text. Green and brown shading in the background is the percent of cropland and pasture at 5' resolution in the year 2000 [7]. Financial support data were preferentially sourced from the Organisation for Economic Cooperation and Development (OECD) (n = 21 countries), supplemented from the World Trade Organization (WTO) where possible for additional countries (n = 12). Data used to make this figure are provided in S1 Data.

Here, we summarise how agriculture impacts the natural environment and then scrutinise the success of government agricultural policies that we believe offer the most promise for safeguarding nature while feeding a growing human population. We use economic and political theory to help explain why some measures are more prone to succeed than others, recognising that scientific understanding of a problem does not mean it will be solved [8]. Throughout, we touch upon examples from centres of agricultural production.

A CULTIVATED PLANET

Agriculture's demand for land drives conversion of natural habitats, and this is arguably its largest environmental cost. Approximately 40% of ice-free land (4,300 million ha) is already covered by crops or used to raise livestock [7], and an additional 2.7–4.9 million ha of cropland annually, the size of a small European nation, may be required to feed the world's population by 2030 [1,9]. Conversion of land for agriculture is estimated to account for 80% of global deforestation [10], and ca. 53% of terrestrial species assessed as threatened by the International Union for Conservation of Nature (IUCN) [11] are negatively impacted by agriculture (Figure 2). Land conversion also reduces the size of the terrestrial carbon sink. Global simulation models predict that 24% and 10% less carbon is stored in vegetation and soil, respectively, than if present-day landscapes retained their natural vegetation [12].

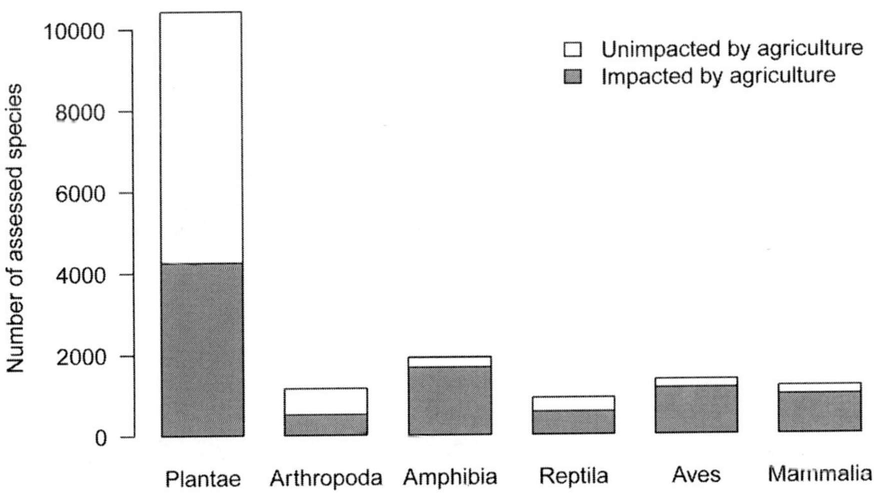

Figure 2. Estimated numbers of threatened species negatively impacted by agriculture.

We counted the total number of species classified as either critically endangered, endangered, or vulnerable by the IUCN Red List in different taxonomic units and the number that were assessed to be threatened by at least one agricultural activity [11]. Agricultural activities were annual and perennial nontimber crops, wood and pulp plantations, livestock farming and ranching, logging and wood harvesting, abstracting of surface water (agricultural use), abstraction of ground water (agricultural use), and agricultural and forestry effluents. Many other species may become threatened with agricultural expansion (S2 Fig).

Agriculture also contributes more to other forms of environmental degradation than any other economic sector. Between 30%–35% of global greenhouse gas (GHG) emissions come from agriculture [3], and crop irrigation accounts for 70% of the world's freshwater withdrawals [13]. The use of synthetic nitrogen fertiliser has increased nearly 21-fold since 1950, and more nitrogen is now added to agricultural soils than from natural processes [14]. Virtually all human-derived nitrogen is lost to the atmosphere or receiving waters [14,15]. This eutrophication suffocates aquatic ecosystems [16]. Phosphorus from fertiliser and livestock manure [14,17], pesticides [18–20], and nanoparticles [21,22] are also exacerbating environmental pollution. Additionally, poor agricultural management causes soil degradation, reducing agricultural productivity and creating further demand for nutrients, water, and land conversion [23,24].

Solutions for limiting agriculture's environmental costs are increasingly well understood. These include reducing land clearance by improving yields of existing crops, minimizing excess nutrient and water use, and shifting production from livestock towards crops [3]. Implementing these practices for only a handful of commodities in a subset of the world can have a large environmental benefit [25]. The success of these practices, however, depends on how well policy puts them into practice.

CHALLENGES TO LIMITING AGRICULTURE'S ENVIRONMENTAL COSTS

Designing and implementing policies to reduce agriculture's environmental costs is difficult. Like other industries, farmers and the businesses reliant upon them, such as input suppliers, processors, and retailers, have a rational self-interest in maximising economic returns and oppose policies that threaten their viability [26,27]. In developed economies, this has historically allowed them to organise into focused lobby groups with considerable advantage over diffuse public interests seeking to safeguard the environment [28,29]. Governments rarely benefit from challenging agri-food lobbies over environmental issues that lack immediate returns for the public good. These lobbies reward government for less demanding environmental policies [30,31], and tailoring policies appropriate for the range of physical and social conditions that farms occupy is costly [26,27]. Concerns about foodborne illness (including genetically modified organisms), environmental impacts (e.g., pesticide use), and animal welfare have ceded some power in the policy arena of developed countries from producers to consumers [32], and increasingly to large retailers, but existing governance structures remain highly susceptible to farmer interests despite spatiotemporal variation in power relations [29]. For example, corporate lobbying efforts in the agricultural sector were deemed effective by 52% of respondents in a 2013 survey of nearly 600 European Union (EU) decision makers, and >80% effectiveness was reported in Denmark, Latvia, and the United Kingdom [33]. Policymakers therefore face major challenges in aligning the private interests of the agri-food sector with public interest in environmental goods across the diversity and extent of farm entities.

In addition to political challenges, measures that do succeed in reducing the environmental harms of agriculture in one place can perversely increase them elsewhere. This is because, in a global market, local policies or technologies that cause agricultural production to carry a greater opportunity cost than using land for environmental goods will displace production to regions where these costs are lower [34]. For example, regulations limiting fertilisers can reduce local water pollution and GHG emissions. But because such measures may reduce yields (i.e., tonnage of production per unit area), food demand will be met elsewhere, leading to intensification and land clearance that potentially creates a net environmental loss [35]. Incentivising low-intensity systems, seen as beneficial to biodiversity conservation in regions such as Europe, may similarly limit production and displace environmental impacts while also lacking the full benefits of natural ecosystems [36]. Efforts to limit the environmental costs of agriculture must therefore consider the environmental impacts beyond farm boundaries as well as the consequences for food production.

POTENTIAL RESOLUTIONS TO THE AGRICULTURE–NATURE CONFLICT

Agricultural policy with explicit environmental objectives is generally organized around three approaches (Table 1 and expanded upon in S1 Table; see Box 1 for an exception):

1. regulations, such as limits on pesticide use or abstracting water, which can be enforced through penalties and conditions placed on financial support to farmers;
2. community-based approaches, which support farmers and local stakeholders to work collectively in addressing environmental impacts; and
3. economic instruments, which pay farmers directly or create markets for adopting practices that minimise environmental impacts and provide noncommodity outputs beyond those required by existing regulation; tariffs can also be used to internalise environmental costs.

Table 1. Successes associated with different policy approaches and the local[1] obstacles to their implementation.

Approach	Success story	Obstacles
Regulations	Critical Counties programme launched in 2008 by Brazilian government suspended access to credit for farmers in counties with the highest deforestation rates. This stimulated collective actions to conserve native vegetation, resulting in deforestation rates dropping to <20% of previous 10-yr average [40].	Complexity, enforceability, cost of implementation. Where compliance is linked to payments, as in EU's Common Agricultural Policy, clear environmental standards and operational guidance required at farm level [41]. Payments must also incorporate regional and/or sectorial variation in the cost of compliance.
Community-based	Landcare Australia is among the best known partnerships between communities, government, and organisations. Established by a $360 million programme in 1989 and engaging about 30% of farming community at its height [42]. Notable successes improving water quality, reducing soil degradation, restoring habitat [43].	Delivering social and/or short-term economic benefits to individual farmers; lack of these benefits has limited uptake of Landcare practices in Australia [44,45].
Economic instruments	Conservation Reserve Program (CRP) established in 1985 by US government provides farmers with annual rental payments for removing land most sensitive to agricultural impacts from production for ≥10 years. Currently, >10 million ha enrolled at annual cost of ca. US$2 billion [46]. CRP has eliminated millions of tonnes of GHG emissions and fertiliser additions and established new habitat supporting tens of millions of birds and other wildlife [47]. Gains arisen despite conversion of non-cropland into production predicted over area equal to 20% of CRP land [48].	Can perform poorly if farmers paid for actions rather than actual environmental outcomes associated with their land [49]. For example, temporary retirement schemes emerged initially in EU to reduce overproduction. As no environmental outcome was explicitly targeted, farmers tended to retire least profitable land that was often the poorest quality, limiting environmental benefits [50,51].

[1]Globally, the major obstacle for all approaches is to avoid displacing production and causing environmental degradation elsewhere.

Box 1. Abandoning the Agricultural Sector to Market Forces

New Zealand radically altered its agricultural policy in 1984 to embrace minimal government intervention. It now relies exclusively on regulation to minimise the environmental impacts of agriculture while providing among the least support to domestic producers out of all developed countries (ca. 0.5% of the value of production; S1 Text). Consequently, the natural environment has been left vulnerable to the pressures of market forces, which are rapidly accelerating production and land conversion with increasing demand for dairy products in particular. The rate of conversion of indigenous grassland to exotic pasture in the South Island has increased by 67% from the period between 1990–2001 to 2001–2008 [52], and New Zealand leads the world in livestock emissions per capita, exceeding the global total across developed countries by more than 13-fold [53].

The nonprescriptive Resource Management Act (RMA) is the primary national legislation intended to oversee natural and physical resources, with land regulation devolved to elected district councils. Although the RMA tasks councils with "maintaining" biodiversity in balance with resource development, they generally expedite the latter. Fewer than 5% of agricultural sector consents requiring biodiversity or ecosystem services maintenance were recently found to comply with conditions [54]. Limited resources are often blamed for poor enforcement, but political interference also occurs [55]. Other regulations, such as the Wildlife Act and Native Plant Protection Act, are rarely invoked, and the latter does not apply to landowners [55].

New Zealand also has a dedicated judicial system to contest decisions made under the RMA. While this system has delivered several notable "wins" for the environment (e.g., [56]), many local actions escape scrutiny. In practice, biodiversity protection by rural landowners occurs mainly on small, voluntary, set-asides of residual unproductive land (known as covenants) and through patchy predator-trapping and stock fencing activities, supported by limited local and national funds. It appears that the laissez-faire approach to the agricultural sector does little to safeguard the environment.

Farmers will be motivated to accept one or more of these approaches where they incur small socioeconomic costs relative to the benefits that they receive from achieving environmental objectives [37,38]. Eliminating financial support for farmers to maximise agricultural production is important in helping to reduce the socioeconomic costs of adopting environmental practices. Reforms in developed economies since the 1980s are now supporting producer incomes more through direct payments than by subsidising production or protecting domestic prices [2,39] (Box 2).

Even with farmer acceptance, success of any policy approach will ultimately depend on whether farmers maintain or boost yields on existing agricultural lands rather than convert new natural lands to production. Halting local land conversion can be achieved with strict enforced regulations (Table 1), but in a globalized market, changes in the local supply of a given cash crop will lead to changes in demand and supply in other distant locations [59]. Trends in cereal production in 19 countries between 1990 and 2005 clearly show this link between local actions and global consequences. These countries managed to boost cereal yields on existing lands while removing cropland from production, but without fully meeting local demand. Per capita cereal imports consequently increased by 12-fold in these countries compared with those lacking land retirement policies [60], clearly suggesting that production was displaced elsewhere. Where subsistence agriculture is important, limiting its local production may similarly boost reliance on cash crops produced abroad [61]. National governments implementing any of the three main policy approaches need to begin working in a coordinated way to combat land use displacement.

Box 2. Environmental Harms of Agricultural Price Support

Price support policies, such as import tariffs or minimum commodity prices, encourage overproduction in the absence of supply management. This strains natural resources and makes marginal land artificially profitable [57]. In the US, the federal government historically managed the dairy industry by guaranteeing to purchase milk if its market price fell below a set value. Modelling revealed that this would drive abandonment and reversion to natural forest across thousands of hectares of land in the state of Wisconsin if milk prices were decreased by only 5%–15%, with millions saved in water treatment and government expenditure [58]. The US has now adopted a strategy of paying producers should their margins fall beneath national averages, likely with similar environmental consequences.

Finding the Right Mixture of Solutions

Maintaining the status quo with agricultural policy is unsustainable for the environment [3,35,62]. On their own, regulations and community approaches will offer little help if damaging the environment remains financially lucrative because the associated penalties are trivial, governments continue to subsidise production (Figure 1, S1 Fig), and farmers are not accountable for the environmental costs of their actions [63–65] (Box 3). Economic instruments, such as payments for public goods, are also contentious, especially for biodiversity conservation [66–68]. Positive outcomes often arise only from certain actions, such as enhancing specific food resources or nesting habitat [69], and catch-all interventions may avoid targeting rare species [70,71]. Removing production subsidies and making farmers financially responsible for their environmental impacts, such as with markets for public goods, can help local outcomes but must avoid curbing intensification because this will encourage land conversion elsewhere to meet demand for food [35].

Box 3. Balancing Large-Scale Incentives for the Rural Poor with Strict Penalties for Agri-business

China—the world's largest agricultural producer—offers a stark contrast in government policy to that implemented by New Zealand described in Box 2. The country's economy is heavily controlled by centralised government and relatively free from the political influence wielded by farmers lobbies. Government can therefore act largely in its own self-interest, which may align more often with the public good. This has allowed the Chinese government to implement environmental policies on scales that would be virtually impossible elsewhere in the world. The Grain to Green Project is one example, having converted nearly 9.1 million hectares of cropland into forest whilst incentivising farmers to participate with grain and cash subsidies [72]. This permanent land retirement scheme has reduced soil erosion and increased carbon sequestration at landscape scales [73]. China is also adopting a stricter approach to large industrial agriculture. In 2014, the country's top legislature revised the national environmental protection law for the first time in 25 years. The revisions included harsher penalties to deter companies from accepting fines rather than investing in pollution control [74]. Whether enforcement follow suits remains to be seen.

Ruhl [26] argued that successful agricultural policies will target large industrial operations that can be readily regulated while incentivising changes in production practices by small farmers without compromising their livelihoods. Recent developments in China are therefore promising, but the country's agricultural sector still causes major environmental harm, including poor water quality in almost all rivers and lakes [75], and nearly 40% more GHG emissions per tonne of nitrogen fertiliser than in Europe [76]. Subsidies on inputs such as fertilisers and pesticides are largely to blame as they promote overuse. Fertilisers are estimated to be overused by 30%–60% [77], yet nearly CMY¥108,000 million (US$17,000 million) was spent in 2012 on just one scheme to help farmers purchase more of these inputs [78]. Removing harmful financial subsidies would help greatly in aligning the country's agricultural and environmental aims.

We assert that conflict between agriculture and the environment will be best resolved by policies dedicating high-quality habitat towards nature conservation, while encouraging intensive production on existing farmland with stringent limits on environmental impacts [35,36,79–81]. Measures that make farmland itself more benign—so called "land-sharing" approaches—also deliver local environmental and social benefits [82,83], but can reduce agricultural yields and cause land conversion elsewhere, effectively displacing environmental burdens [80]. Intensifying production on existing farmland may thus be a better option because it avoids displacing impacts and can allow other land to be freed for nature conservation and restoration, with no net loss in agricultural production.

Our consideration of policy instruments approaches and the obstacles to their success (Table 1) suggests that our proposed resolution would practically entail a combination of punitive measures, such as taxes that reduce existing overuse of farm inputs [84], with positive incentives linked to measurable outcomes [49], such as for permanently removing land from production. Taxes on farm inputs can improve upon existing payments to limit input use because they do not explicitly cap nutrient, pesticide, and water use. Without input limits, some farmers may find it profitable to intensify existing production, despite the added taxes, while others might profit more from reducing resource consumption. Across the landscape, yields will be maintained, and potentially increased without additional land conversion. For example, water surcharges in Mexico are used to pay landowners to retain forested watersheds. Displaced land conversion from exploitation of communal properties and changing market pressures reduced the environmental benefits of this programme by only ca. 4% [85]. By enabling farms to respond differently to the same policy, input taxes carry an additional advantage over traditional regulations that hold all farmers to the same standard and ignore sectorial diversity [86]. An aligned but more audacious approach would be to replace taxes on income with taxes on environmental consumption. The idea is to shift from taxing the production of private wealth toward taxing the consumption of public environmental goods, thus promoting economic growth from sustainable use rather than depletion of natural capital. Brown et al. [55] have proposed that such a system would most heavily tax intensive land uses that generate the most negative environmental impacts, while areas of intact ecological function would entitle the owners to a rebate. This approach might encourage farmers to work intensively with the most productive land but set aside and restore less productive areas with net gains for the environment.

Evidence at national levels suggests that investments in land retirement and input-limiting schemes, consistent with our suggested policy approach, do deliver local environmental benefits. To illustrate this, we examined trends in two environmental responses, farmland bird populations and GHG emissions from synthetic fertilisers, across countries with similar agricultural patterns, i.e., dominated by livestock and arable crops (see S1 Text for methods). We calculated the annual, per hectare monetary investment in a specific environmental outcome beyond that required by regulation from a detailed inventory of government policies compiled by the OECD (annotated in S1 Data). As expected, because habitat loss is a major threat to farmland and grassland biota, we found that farmland birds benefited as more money was invested in retiring agricultural land (Figure 2). We also found that total emissions of N_2O declined with increasing investment in reducing fertiliser use (Figure 3). While cursory in nature, these results suggest that targeted policies have the potential to deliver positive environmental outcomes.

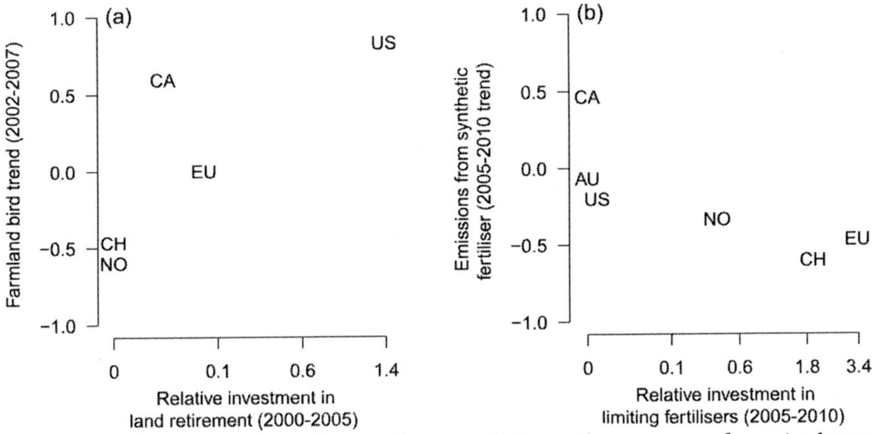

Figure 3. Improvements in environmental performance of agriculture correlate with investments in agri-environment schemes (AES).

Temporal trend in (**A**) farmland bird populations and (**B**) GHG emissions from synthetic fertiliser were calculated over time for each country using Kendall's τ (see S1 Text for full methods). Investments in AES were scaled by the ratio of investments in AES relative to financial transfers to agricultural producers for all other purposes so as to account for the fact that environmental improvements can be offset by policies that maximise production. Values were then summed across 5-y periods with a pre-emptive 2-y lag for land retirement. Habitat quality will take at least one growing season to improve following land retirement and any changes in bird populations from improved breeding success will thus largely be measurable the following year, i.e., two years later [87,88]. Spearman's rank correlation: $\rho = 0.95, p = 0.005$ and $\rho = -0.93, p = 0.008$ for (**A**) and (**B**), respectively. Points are responses in individual countries: AU = Australia [data available for (**B**) only], CA = Canada, EU (aggregated together), NO = Norway, CH = Switzerland, and US. Data used to make this figure are provided in S1 Data.

Measures delivering environmental benefits must also be coupled with mechanisms that promote intensification to avoid displacing production elsewhere, and they must do so with fewer external costs. Research suggests increased yields and better environmental practices are possible by improving nutrient and water management and breeding more efficient crops [3]. Reductions in food waste and human diets with less livestock and dairy consumption can also help meet rising food demands without increasing resource use [89,90]. Dietary change is, however, contentious. Governments may need to consider policies that tax livestock products and/or subsidize the development of livestock-free protein sources, e.g., [91], in order to dislodge entrenched norms.

Here, we have laid out a vision for limiting the environmental impacts of agriculture, yet at least two major challenges remain. First, winning political support for policy change remains a fundamental requirement to any of our proposed reforms. Ultimately, democratically elected governments, while not immune to the lobbying power of agribusiness, will generally respond to electoral pressures [31]. Electorates may express little desire for policy change if they are poorly informed, unmotivated, or misled by industry about limiting the environmental costs of agriculture. Second, implementation of our policy mixture will vary spatially. For example, successful land retirement depends on features of the surrounding landscape, such as habitat connectivity [92], and can reduce yields in regions where the most productive land also has the most valuable environmental assets. Nonetheless, improvements in the management of land resources in developing countries [93], new technologies that optimally allocate farm inputs to reduce overuse [94], advances in crop breeding [95], and tools for better conservation planning [96] give us hope that innovations exist for meeting the need for more sustainable and intensive agriculture.

ACKNOWLEDGMENTS

We thank Andrew Balmford, David Coomes, and staff at the OECD for discussion.

REFERENCES

1. Tilman D, Balzer C, Hill J, Befort BL (2011) Global food demand and the sustainable intensification of agriculture. Proc Natl Acad Sci USA 108: 20260–20264. pmid:22106295
2. Warren J, Lawson C, Belcher K (2007) The agri-environment. Cambridge: Cambridge University Press. 232 p.
3. Foley JA, Ramankutty N, Brauman KA, Cassidy ES, Gerber JS, et al. (2011) Solutions for a cultivated planet. Nature 478: 337–342. pmid:21993620
4. Millennium Ecosystem Assessment. Ecosystems and human well-being: synthesis. Washington, DC: World Resources Institute; 2005. 155 p.
5. James AN, Gaston KJ, Balmford A (1999) Balancing the Earth's accounts. Nature 401: 323–324. pmid:16862091
6. Myers N, Kent J (2001) Perverse subsidies: how tax dollars can undercut the environment and the economy. Washington, DC: Island Press. 277 p.
7. Ramankutty N, Evan AT, Monfreda C, Foley JA. (2008) Farming the planet: 1. Geographic distribution of global agricultural lands in the year 2000. Global Biogeochem Cycles 22: GB1003.

8. Lawton JH (2007) Ecology, politics and policy. J Appl Ecol 44: 465–474.
9. Lambin EF, Meyfroidt P (2011) Global land use change, economic globalization, and the looming land scarcity. Proc Natl Acad Sci USA 108: 3465–3472. pmid:21321211
10. Kissinger G, Herold M, de Sy V (2012) Drivers of deforestation and forest degradation: a synthesis report for REDD+ policymakers. Lexeme Consulting, Vancouver. 46 p.
11. International Union for Conservation of Nature [IUCN] (2014) The IUCN Red List of Threatened Species. Version 2014.3. http://www.iucnredlist.org via the Internet. Accessed 2015 March 14.
12. Bondeau A, Smith PC, Zaehle S, Schaphoff S, Lucht W, et al. (2007) Modelling the role of agriculture for the 20th century global terrestrial carbon balance. Glob Chang Biol 13: 679–706.
13. Food and Agriculture Organization of the United Nations [FAO]. (2015) AQUASTAT database. Available: http://www.fao.org/nr/water/aquastat/data/ via the Internet. Accessed 14 March 2015.
14. Bouwman L, Goldewijk KK, Hoek KWVD, Beusen AHW, Van Vuuren DP, et al. (2013) Exploring global changes in nitrogen and phosphorus cycles in agriculture induced by livestock production over the 1900–2050 period. Proc Natl Acad Sci USA 110: 20882–20887. pmid:21576477
15. Galloway JN, Aber JD, Erisman JW, Seitzinger SP, Howarth RW, et al. (2003) The nitrogen cascade. BioScience 53: 341–356.
16. Diaz RJ, Rosenberg R (2008) Spreading dead zones and consequences for marine ecosystems. Science 321: 926–929. pmid:18703733
17. MacDonald GK, Bennett EM, Potter PA, Ramankutty N (2011) Agronomic phosphorus imbalances across the world's croplands. Proc Natl Acad Sci USA 108: 3086–3091. pmid:21282605
18. Rohr JR, Schotthoefer AM, Raffel TR, Carrick HJ, Halstead N, et al. (2008) Agrochemicals increase trematode infections in a declining amphibian species. Nature 455: 1235–1239 pmid:18972018
19. Henry M, Béguin M, Requier F, Rollin O, Odoux J-F, et al. (2012) A common pesticide decreases foraging success and survival in honey bees. Science 336: 348–350. pmid:22461498
20. Hallmann CA, Foppen RPB, van Turnhout CAM, de Kroon H, Jongejans E (2014) Declines in insectivorous birds are associated with high neonicotinoid concentrations. Nature 511: 341–343. pmid:25030173
21. Priester JH, Ge Y, Mielke RE, Horst AM, Moritz SC, et al. (2012) Soybean susceptibility to manufactured nanomaterials with evidence for food quality and soil fertility interruption. Proc Natl Acad Sci USA 109: E2451–E2456. pmid:22908279
22. Santiago-Martín A, Constantin B, Guesdon G, Kagambega N, Raymond S, et al. (2015) Bioavailability of engineered nanoparticles in soil systems. J. Hazard. Toxic Radioact. Waste in press.
23. Pimentel D, Harvey C, Resosudarmo P, Sinclair K, Kurz D, et al (1995) Environmental and economic costs of soil erosion and conservation benefits. Science 267: 1117–1123. pmid:17789193
24. Lal R (1997) Degradation and resilience of soils. Phil Trans R Soc Lond B 352: 997–1010.
25. West PC, Gerber JS, Engstrom PM, Mueller ND, Brauman KA, Carlson KM, et al. (2014) Leverage points for improving global food security and the environment. Science 345: 325–328. pmid:25035492
26. Ruhl JB (2000) Farms, their environmental harms, and environmental law. Ecol Law Q 27: 263–349.
27. Walker S, Brower AL, Stephens RTT, Lee WG (2009) Why bartering biodiversity fails. Conserv Lett 2: 149–157.
28. Olson M (1965) The logic of collective action. Cambridge: Harvard University Press. 208 p.

29. Daugbjerg C, Swinbank A (2012) An introduction to the "new" politics of agriculture and food. Policy Soc 31: 259–270.
30. Eskridge WN Jr (1988) Politics without romance: implications of public choice theory for statutory interpretation. Va Law Rev 74: 275–338.
31. Bellemare MF, Carnes N (2015) Why do members of congress support agricultural protection? Food Policy 50: 20–34.
32. Grant WP (2012) Can political science contribute to agricultural policy? Policy Soc 31: 271–279.
33. .Burson-Marsteller (2013) A guide to effective lobbying in Europe: the view of policy-makers. Available: http://burson-marsteller.nl/wp-content/uploads/2013/06/Lobbying-survey-Report.pdf. Accessed 1 June 2015.
34. Meyfroidt P, Rudel TK, Lambin EF (2010) Forest transitions, trade, and the global displacement of land use. Proc Natl Acad Sci USA 107: 20917–20922. pmid:21078977
35. Balmford A, Green R, Phalan B (2012) What conservationists need to know about farming. Proc Roy Soc B 279: 2714–2724.
36. Merckx T, Pereira HM (2015) Reshaping agri-environmental subsidies: From marginal farming to large-scale rewilding. Basic and Applied Ecology 16: 95–103.
37. Lastra-Bravo XB, Hubbard C, Garrod G, Tolón-Becerra A (2015) What drives farmers' participation in EU agri-environmental schemes? Results from a qualitative meta-analysis. Environmental Science & Policy 54: 1–9.
38. Wilson GA, Hart K (2000) Financial imperative or conservation concern? EU farmers' motivations for participation in voluntary agri-environmental schemes. Environment and Planning A 32: 2161–2185.
39. Organisation for Economic Co-operation and Development [OECD] (2004) Agriculture and the environment: lessons learned from a decade of OECD work. Paris: OECD. 36 p.
40. Nepstad D, McGrath D, Stickler C, Alencar A, Azevedo A, et al. (2014) Slowing Amazon deforestation through public policy and interventions in beef and soy supply chains. Science 344: 1118–1123. pmid:24904156
41. European Court of Auditors (2008) Is cross compliance an effective policy? Special report no 8/2008. Luxembourg: Office for Official Publications of the European Communities. 57 p.
42. Curtis A, Lockwood M (2000) Landcare and catchment management in Australia: lessons for state-sponsored community participation. Soc Nat Resour 13: 61–73.
43. OECD (2013) Providing agri-environmental public goods through collective action. Paris: OECD. 306 p.
44. Curtis A, Lacy TD (1998) Landcare, stewardship and sustainable agriculture in Australia. Environ Values 7: 59–78.
45. Pannell DJ, Marshall GR, Barr N, Curtis A, Vanclay F, et al. (2006) Understanding and promoting adoption of conservation practices by rural landholders. Aust J Exp Agric 46: 1407–1424.
46. Stubbs M (2014) Conservation Reserve Program (CRP): status and issues. Washington, DC: Congressional Research Service. 24 p.
47. United States Department of Agriculture (2011) CRP benefits. http://www.fsa.usda.gov/Internet/FSA_File/united_states.pdf via the Internet. Accessed 14 March 2015.
48. Wu J (2000) Slippage effects of the conservation reserve program. Am J Agric Econ 82: 979–992.
49. Franks JR, McGloin A (2007) Joint submissions, output related payments and environmental co-operatives: Can the Dutch experience innovate UK agri-environment policy? Journal of Environmental Planning and Management 50: 233–256.
50. Sotherton NW (1998) Land use changes and the decline of farmland wildlife: An appraisal of the set-aside approach. Biol Conserv 83: 259–268.

51. Robinson GM, Lind M (1999) Set-Aside and environment: a case study in southern England. Tijdschr Econ Soc Geogr 90: 296–311.
52. Weeks ES, Walker SF, Dymond JR, Shepherd JD, Clarkson BD (2012) Patterns of past and recent conversion of indigenous grasslands in the South Island, New Zealand. N Z J Ecol 37: 127–138.
53. Caro D, Davis SJ, Bastianoni S, Caldeira K (2014) Global and regional trends in greenhouse gas emissions from livestock. Climatic Change 126: 203–216.
54. Brown MA, Clarkson BD, Stephens RT, Barton BJ (2014) Compensating for ecological harm–the state of play in New Zealand. N Z J Ecol 38: 139–146.
55. Brown MA, Stephens RTT, Peart R, Fedder B (2015) Vanishing nature: facing New Zealand's biodiversity crisis. Auckland: Environmental Defence Society. 208 p.
56. Forest & Bird (2013) 2013 Annual Report. http://www.forestandbird.org.nz /files/file/FB%20Annual%20Report%202013%20V4%20low%20res.pdf via the Internet. Accessed 14 March 2015.
57. Abler D (2004) Multifunctionality, agricultural policy, and environmental policy. Agricultural and Resource Economics Review 33: 8–17.
58. Plantinga AJ (1996) the effect of agricultural policies on land use and environmental quality. Am J Agr Econ 78: 1082–1091.
59. MacDonald GK, Brauman KA, Sun S, Carlson KM, Cassidy ES, Gerber JS, et al. (2015) Rethinking agricultural trade relationships in an era of globalization. BioScience 65: 275–289.
60. Rudel TK, Schneider L, Uriarte M, Turner BL, DeFries R, Lawrence D, et al. (2009) Agricultural intensification and changes in cultivated areas, 1970–2005. Proc Natl Acad Sci USA 106: 20675–20680 pmid:19955435
61. Murray BC (2008) Leakage from an avoided deforestation compensation policy: concepts, empirical evidence, and corrective policy options. Raleigh: Nicholas Institution for Environmental Policy Solutions. 32 p.
62. Pe'er G, Dicks LV, Visconti P, Arlettaz R, Báldi A, et al. (2014) EU agricultural reform fails on biodiversity. Science 344:1090–1092. pmid:24904142
63. Pretty JN, Brett C, Gee D, Hine RE, Mason CF, et al. (2000) An assessment of the total external costs of UK agriculture. Agric Syst 65: 113–136.
64. Pretty J, Brett C, Gee D, Hine RE, Mason CF, et al. (2001) Policy challenges and priorities for internalizing the externalities of modern agriculture. Journal of Environmental Planning and Management 44: 263–283.
65. Balmford A, Bruner A, Cooper P, Costanza R, Farber S, et al. (2002) Economic reasons for conserving wild nature. Science 297: 950–953. pmid:12169718
66. Kleijn D, Sutherland WJ (2003) How effective are European agri-environment schemes in conserving and promoting biodiversity? J Appl Ecol 40: 947–969.
67. Kleijn D, Baquero RA, Clough Y, Díaz M, De Esteban J, et al. (2006) Mixed biodiversity benefits of agri-environment schemes in five European countries. Ecol Lett 9: 243–254. pmid:16958888
68. Whittingham MJ (2007) Will agri-environment schemes deliver substantial biodiversity gain, and if not why not? J Appl Ecol 44: 1–5.
69. Baker DJ, Freeman SN, Grice PV, Siriwardena GM (2012) Landscape-scale responses of birds to agri-environment management: a test of the English Environmental Stewardship scheme. J Appl Ecol 49: 871–882.
70. Kleijn D, Berendse F, Smit R, Gilissen N (2001) Agri-environment schemes do not effectively protect biodiversity in Dutch agricultural landscapes. Nature 413: 723–725. pmid:11607029
71. 71.71.Vickery JA, Bradbury RB, Henderson IG, Eaton MA, Grice PV (2004) The role of agri-environment schemes and farm management practices in reversing the decline of farmland birds in England. Biol Conserv 119: 19–39.
72. State Forestry Administration of China (2014) Forestry in China. http://english.forestry.gov.cn/uploads/Information_Services/Latest_Publication/Forestry_in_China.pdf via the Internet. Accessed 14 March 2015.

73. Liu J, Li S, Ouyang Z, Tam C, Chen X (2008) Ecological and socioeconomic effects of China's policies for ecosystem services. Proc Natl Acad Sci USA 105: 9477–9482 pmid:18621700
74. National People's Congress of the People's Republic of China. (2014) China's legislature adopts revised Environmental Protection Law. http://www.npc.gov.cn/englishnpc/news/Legislation/2014-04/25/content_1861275.htm via the Internet. Accessed 14 March 2015.
75. Sun B, Zhang L, Yang L, Zhang F, Norse D, et al. (2012) Agricultural non-point source pollution in china: causes and mitigation measures. Ambio 41: 370–379. pmid:22311715
76. Zhang W, Dou Z, He P, Ju X-T, Powlson D, et al. (2013) New technologies reduce greenhouse gas emissions from nitrogenous fertilizer in China. Proc Natl Acad Sci USA 110: 8375–8380. pmid:23671096
77. Ju X-T, Xing G-X, Chen X-P, Zhang S-L, Zhang L-J et al. (2009) Reducing environmental risk by improving N management in intensive Chinese agricultural systems. Proc Natl Acad Sci USA 106: 3041–3046. pmid:19223587
78. OECD (2015) Producer and consumer support estimates database. Available http://www.oecd.org via the Internet. Accessed 14 March 2015.
79. Hodgson JA, Kunin WE, Thomas CD, Benton TG, Gabriel D (2010) Comparing organic farming and land sparing: optimizing yield and butterfly populations at a landscape scale. Ecol Lett 13:1358–1367. pmid:20825453
80. Phalan B, Onial M, Balmford A, Green RE (2011) Reconciling food production and biodiversity conservation: land sharing and land sparing compared. Science 333: 1289–1291. pmid:21885781
81. Godfray HCJ, Garnett T (2014) Food security and sustainable intensification. Phil Trans R Soc Lond 369: 20120273.
82. 82.82.Fischer J, Abson DJ, Butsic V, Chappell MJ, Ekroos J, Hanspach J, et al. (2014) Land sparing versus land sharing: moving forward. Conserv Lett 7: 149–157.
83. Frishkoff LO, Karp DS, M'Gonigle LK, Mendenhall CD, Zook J, Kremen C, et al. (2014) Loss of avian phylogenetic diversity in neotropical agricultural systems. Science 345: 1343–1346. pmid:25214627
84. Mueller ND, Gerber JS, Johnston M, Ray DK, Ramankutty N, et al. (2012) Closing yield gaps through nutrient and water management. Nature 490: 254–257. pmid:22932270
85. Alix-Garcia JM, Shapiro EN, Sims KRE (2012) Forest conservation and slippage: evidence from Mexico's national payments for ecosystem services program. Land Econ 88: 613–638.
86. Stavins RN, Whitehead BW (1992) Pollution charges for environmental protection: a policy link between energy and environment. Annual Review of Energy and the Environment 17: 187–210.
87. Newson SE, Johnston A, Renwick AR, Baillie SR, Fuller RJ (2012) Modelling large-scale relationships between changes in woodland deer and bird populations. J Appl Ecol 49: 278–286.
88. Chamberlain D E, Fuller RJ, Bunce RGH, Duckworth JC, Shrubb M (2000) Changes in the abundance of farmland birds in relation to the timing of agricultural intensification in England and Wales. J Appl Ecol 37: 771–788.
89. Bajželj B, Richards KS, Allwood JM, Smith P, Dennis JS, et al. (2014) Importance of food-demand management for climate mitigation. Nature Clim Change 4: 924–929.
90. Tilman D, Clark M (2014) Global diets link environmental sustainability and human health. Nature 515: 518–522. pmid:25383533
91. Pandya R (2014) Milk without the moo. New Sci 222: 28–29.
92. Gilroy JJ, Edwards FA, Uribe CAM, Haugaasen T, Edwards DP (2014) Surrounding habitats mediate the trade-off between land-sharing and land-sparing agriculture in the tropics. J Appl Ecol 51: 1337–1346.
93. Hobbs PR, Sayre K, Gupta R (2008) The role of conservation agriculture in sustainable agriculture. Phil Trans R Soc Lond B 363: 543–555.

94. Chen X-P, Cui Z-L, Vitousek PM, Cassman KG, Matson PA, et al. (2011) Integrated soil–crop system management for food security. Proc Natl Acad Sci USA 108: 6399–6404. pmid:21444818
95. Tester M, Langridge P (2010) Breeding technologies to increase crop production in a changing world. Science 327: 818–822. pmid:20150489
96. Bateman IJ, Harwood AR, Mace GM, Watson RT, Abson DJ, et al. (2013) Bringing ecosystem services into economic decision-making: land use in the United Kingdom. Science 341: 45–50. pmid:23828934
97. Anderson K, Nelgen S (2013) Updated national and global estimates of distortions to agricultural incentives, 1955 to 2011. Available: http://www.worldbank.org/agdistortions via the Internet. Accessed 2015 June 1.
98. Anderson K, Kurzweil M, Martin W, Sandri D, Valenzuela E (2008) Measuring distortions to agricultural incentives, revisited. World Trade Rev 7: 1–30.

CHAPTER 19

Sustainable Water Management for Urban Agriculture, Gardens and Public Open Space Irrigation: A Case Study in Perth

Raju Sharma Dhakal[1*], Geoff Syme[1], Edward Andre[1], Charles Sabato[2]

[1]*Centre for Planning, Edith Cowan University, Perth, Australia*

[2]*Water Corporation, Perth, Australia*

ABSTRACT

Urban agriculture has been increasingly popular as a form of modern agriculture in urban settings. It includes community gardens, fruit orchards, home gardens, veggie patches, public open spaces, reserves, urban forest, and recreational landscaping. However, irrigation using urban water supply has been identified as a major constraints for the development of urban agriculture. This study presents a sustainable water management trial at Butler, a northern sub-urban development in Perth, Western Australia, for urban irrigation. The trial system consists of a number of water saving features including untreated fit-for-purpose groundwater supplied via a third pipe network, drip irrigation, local weather station, soil moisture sensors connected with a local weather station, night time irrigation, soil enhancement with conditioning and mulching, and use of native plants and vegetation. The trial outcome was compared against controlled areas in terms of irrigation efficiency and sustainable water management for urban agriculture. The study demonstrated that a fit-for-purpose irrigation along with water sensitive land management could be a sustainable alternative for urban agriculture that would achieve a significant water saving and irrigation efficiency at urban settings. However, quality of untreated groundwater can be an issue while utilizing it for irrigation, but the research has shown that it can be managed with innovative irrigation techniques. This indicates that the fit-for-purpose irrigation system with water sensitive land management practices would be highly supportive in sustainable development of urban agriculture, vegetation and recreational landscaping.

Keywords: Urban Agriculture, Sustainable Irrigation System, Integrated Land and Water Management

1. INTRODUCTION

Urban agriculture has become increasingly popular across the world that includes a variety of activities: community gardens and fruit orchard, home gardens and veggie patches, urban forest, public open spaces, reserves, urban forest and recreational landscaping [1] -[5] . Urban agriculture differs from traditional agriculture as it is integrated into densely populated areas with limited land for food production and recreation space [4] . Urban agriculture can bring diverse vegetative structures back into urban system, support local bio-diversity, and provide ecosystem services across fragmented habitats and spatial levels [6] that in turn can reduce the impact of climatic variability [7] . In addition to food production, urban agriculture can cater for a wide range of urban community needs, including cultivation of vegetables, medicinal plants, spices, mushroom, fruit trees, ornamental plants, and other productive plants [5] [8] . The production of crop and agricultural goods, within and around cities, with a motivation of personal consumption or income generation [9] [10] thus integrates the local, urban and suburban economic and ecological system [1] .

Urban agriculture has a vital role in enhancing food production and bio-diversity, but that comes with significant costs and constraints [11] [12] . Albeit not required, urban agriculture utilizes significant amount of scheme water for growing foods and crops, which is usually expensive than agricultural water supplies [4] [7] . On the other hand, reducing outdoor use of scheme water for agriculture activities is a major component of many strategies to reduce urban water use and ensure reliable indoor water supply [4] . For instance, almost 40% of the scheme water is consumed for agricultural irrigation activities in urban and regional councils of Perth metropolitan. In drought period, the agricultural irrigation is normally restricted with a variety of bans (daytime and winter sprinkler ban), and demand management practices [13] -[16] . As a result, urban agricultural activities are not always supplied with sufficient irrigation, especially when they require most, hence affecting the continuity of food production and other agricultural activities. This condition is expected to be continued in WA, since a gap between water demand and supply is predicted to be 120 GL by 2030 [15] . In this context, an alternative irrigation system utilizing locally available fit-for-purpose water sources (e.g. groundwater) could have multiple benefits from helping water authorities in augmenting water supply and ensuring cheap and reliable irrigation for urban agricultural activities.

Previous studies in Australia and other countries have shown that alternative irrigation systems using groundwater and recycled wastewater are widely accepted for urban and regional agriculture, forestry and recreational landscaping [7] [17] -[22] . Wong [19] suggests that agriculture, garden and park irrigation are important avenues to utilize recycled water and nutrients within the build environment. This supports the integrated land and water management practices at respective levels by reducing water import into and net discharge out of the environment. Further, well-established urban agriculture activities would have potential to mitigate and adapt the climate change by reducing the urban heat island effects, managing micro-climate, and enhancing biodiversity and ecology [7] .

Similar studies have shown a growing acceptance among Australian communities towards alternative irrigation systems [17] [18] [22] -[24] . Hurlimann [17] observed that the majority of Mawson Lake community in South Australia accepted the irrigation use of recycled water as a better option compared to drinking water for their home gardens, veggie patches and other urban agriculture activities. Similarly, Australia-wide case studies conducted by Davis and Farrelly [22] -[24] have shown that a number of alternative water systems can be best utilized for efficient irrigation system. In context of WA, a number of studies consider groundwater as a reliable and sustainable source for outdoor irrigation activities [16] [18] [25] -[27] . These studies indicate that using groundwater as an alternative source for urban agriculture is technically feasible and receiving growing community acceptance all over Australia.

In this context, a suburban area in Perth ("The Green") was selected for this study that has alternative irrigation system via third pipe network. The system utilized local groundwater from shallow aquifer (20 - 60 m below the ground surface). 5 communal bores abstract the groundwater and supply to the irrigation ring without any treatment for irrigating the private gardens and the public open space in every alternate day [28] . This innovative irrigation system was investigated in terms of irrigation efficiency and its utility for urban agriculture. For that, total household drinking water and irrigation consumption, and total public open spaces usage in "The Green" were analyzed and compared with that of the surrounding suburban areas and Perth metropolitan. Further, public acceptance towards the groundwater as an alternative irrigation source was analyzed.

2. DESCRIPTION OF THE STUDIED SYSTEM

This study was carried out in "The Green" suburban community at Butler, Western Australia (Figure 1), where urban agricultural practices and recreational landscaping were being irrigated with an efficient irrigation system. Butler is a Greenfield suburban area of City of Wanneroo, which is 35 KM north of Perth CBD. The development of "The Green" began in 2005 and still to finish. "The Green" was compared as an experimental area with its neighbouring suburban development, mainly Ridgewood as control and Perth metropolitan city. Ridgewood is an established metropolitan suburban area having similar geography, climate, and socio-demography; but lacking the groundwater irrigation system and having a separate activity locus.

Figure 1. Location of the research area (Source: Contrywide Publication, Google Maps).

The irrigation system utilised the local groundwater without treatment, hence only used for irrigating home gardens, veggie patches, fruit orchards, community gardens, verges, public parks and ovals, urban forest, plantations, reserves, and bushlands [22] [29] . The irrigation system incorporated water smart technologies and water sensitive land management practices for the efficient consumption and optimisation of the groundwater supplied for outdoor agricultural activities, which include:

1) Auto-operated irrigation system

a) Weather station;

b) Soil moisture sensors.

2) Irrigation techniques for achieving efficiency and reducing human contacts

a) Night time irrigation;

b) Drips and sprinklers irrigation.

3) Soil conditioning and mulching

4) Water efficient agricultural activities at home

5) Water efficient agriculture, forestry, and recreational landscaping at community

2.1. Automatic Irrigation System

All agricultural and outdoor activities in "The Green" were automatically supplied with the local groundwater on the basis of local weather condition. Real time weather information was obtained from a local meteorological station (Figure 2(a)); such as temperature, rainfall, relative humidity, wind velocity, etc. This information was coupled with soil moisture information received from in-situ soil moisture sensors (Figure 2(b)) across "The Green" in controlling and operating the irrigation system. Thus established automated irrigation system not only provided irrigation virtually always when required, but also optimized the irrigation supply by avoiding overwatering, and irrigation after rain.

2.2. Efficient Irrigation Techniques

The gardens, parks, and urban agricultural activities in "The Green" were provided night time irrigation (10 p.m.-6 a.m.) that avoided evaporation loss, and also reduced possible health hazards to humans a direct contact with the groundwater. A number of efficient irrigation techniques were utilised, such as: unique pop-up sprinklers, and subsurface drip reticulation.

The sprinklers were designed for spraying bigger droplets that would avoid wind-drifting and supplying enough water even in shorter operation windows. Similarly, subsurface drip- pipes (Figure 3(a)) supplied irrigation near the root zone of crops that not only improved irrigation efficiency but also reduce evaporation loss.

2.3. Soil Conditioning and Mulching

When preparing the garden beds and lands for other agricultural activities, the soil condition was improved with sufficient soil mixture, soil conditioner, and mulch as shown in Figure 3(a) and Figure 3(b). This would not only enhance the soil quality, fertility, as well as the water holding capacity but also reduce the evaporation loss from the surface.

2.4. Water Efficient Agricultural Activities at Home

Water efficient food production and gardening activities were encouraged at each household to reduce irrigation use. As shown in Figures 4(a)-(c), vegetable gardens and front gardens were established with less water demanding crops, vegetable, hardy turf, and bush. Mostly native plants were used, lawn areas were reduced to improve irrigation efficiency.

(a) (b)

Figure 2. A local meteorological station (a) and soil moisture sensor (b) associated with the third-pipe irrigation system in "The Green".

(a) (b)

Figure 3. Photographs of garden bed preparation (a) and mulching (b) in "The Green".

2.5. Water Efficient Agriculture, Forestry, and Recreational Landscaping at Community

In "The Green", community gardens, public parks, and recreational landscaping were designed in water efficient way. Usual sprinklers were replaced with subsurface drip irrigation practices (Figure 3(a) and Figure 5(c)) and the lawn areas were dropped and replaced with less water demanding natives and bushes (Figure 5(a)). Trees and bushland (Figure 5(b)) were retained in the parks, so that these parks could be utilized for multiple purposes, such as: playground and recreational activities, community gardening, improving bio-diversity, urban cooling, and community appearance.

Furthermore, a higher density development was adopted in "The Green" that reduced residential garden areas and eventually the residential irrigation demand. The reduced residential gardens were, however, compensated with increased community gardens, public open spaces and reserves [28] [30] . In addition, wide range of water sensitive land management practices were adopted, namely: porous pavement, bio-retention trenches and basins, grassed swales, terraced gardens, and artificial water bodies [22] . These practices would help reducing irrigation demand, improving irrigation efficiency, facilitating aquifer recharge, and recycling urban water and nutrients within the build environment. These innovative designs facilitated integrated and sustainable water management at local and urban level. The innovative irrigation and water sensitive land management practices were expected to be an important avenue for ensuring efficient and integrated land and water management [14] , and improving neighbourhood appearance and liveability [31] at various spatial level.

Figure 4. Photographs of veggie patches (a) and (b), and home garden (c) in "The Green".

Figure 5. Public parks with (a) native plants, (b) trees and bushland, and (c) recreational landscaping in "The Green".

3. RESEARCH METHOD

A case study with mixed methodology was adopted to evaluate the agricultural irrigation efficiency and water conservation as a result of the fit-for-purpose irrigation system and water efficient urban development. This paper utilizes the secondary data on drinking water consumption, irrigation usage in gardens and parks and land use pattern to analyze the scheme water saving and irrigation efficiency at "The Green". The drinking water consumption and irrigation usage data were obtained from the Western Australian Water Corporation and land use data were obtained from the City of Wanneroo. At the same time, the qualitative information from survey and stakeholders interviews were analyzed to understand community acceptance and socio-economic challenges associated with the alternative irrigation system. The data were sorted, managed and analyzed using computer software: Microsoft Excel 2007, IBM SPSS 22, IBM SPSS Amos 22 and NVivo 9.

The drinking water supply was metered and measured in KL/year/household, whereas the groundwater irrigation supply was not metered. Hence a bulk amount of residential groundwater consumption was divided by the number of households connected to the irrigation system to get the average groundwater usage (in KL) per household per year. Similarly, the annual public open space irrigation usage was measured with the meter installed in each public parks and verges and given in ML/year.

The total household water usage was calculated as a sum of average drinking water consumption and irrigation usage. The comparative analysis of drinking water consumption data across the studied areas explained the drinking water conservation, whereas the analysis of total domestic water (Drinking and groundwater) usage data provided the net water efficiency situation at "The Green". The analysis of drinking water conservation was not restricted only for the survey participants. It utilized the data for the whole studied suburban areas to explore a complete picture of water conservation and irrigation efficiency scenarios.

4. RESULTS AND DISCUSSIONS

4.1. Water Conservation Scenario

In "The Green", both drinking water and groundwater irrigation were supplied; whereas in Ridgewood and Perth metro region, only drinking water was supplied. "The Green" was started since 2007 but the number of households connected to the groundwater irrigation system was very low until 2010, hence only the data from 2011 were used for the analysis. In Table 1, the total household water consumption in "The Green" (drinking water and groundwater irrigation) is compared with that of Ridgewood and Perth metropolitan (only drinking water).

The average drinking water consumption in "The Green" was significantly lower compared to Ridgewood and Perth metropolitan for all times. A typical household in "The Green" consumed only half of the drinking water used by an average Perth Metropolitan household. Such a significant water savings in "The Green" was mainly because of groundwater irrigation system that replaced the drinking water usage in garden irrigation and other outdoor agricultural activities.

As shown in Table 1, the total household water consumption in "The Green" in 2012 appeared to be significantly higher than neighbouring suburban areas and metropolitan average. It could be attributed to increased usage of groundwater (188 KL) for new garden establishment, and other once-off usage such as construction works. It became further evident from continuous reduction in total water consumption in next two consecutive years (Table 1).

It should be noted that a newly established garden or lawn in Perth metropolitan could get irrigation exemption for up to 42 days in summer and 35 days in winter. During exemption period, a significant amount of water (10 mm/day in summer, and 7.5 mm/day in winter) would be consumed [34] . This was supposed to be the main reason for the higher amount of groundwater consumption in "The Green" given the number of newly established garden and lawns. This figure could get higher, but the water efficient irrigation system and land management practices help to restrain it.

As shown in Figure 6, the household irrigation consumption in "The Green" appeared to be significantly higher than allocated, whereas the public parks irrigation consumption figures were far below than the allocation limit. This could be linked again to the large number of irrigation exemptions for newly established gardens across "The Green" suburb in 2013, which was supported with the reduced consumption in successive years (Table 1). Whereas the groundwater consumption for public open space irrigation in "The Green" was significantly lower than the allocation limit as well as the average public open space irrigation in surrounding suburban areas and metropolitan figures. This indicates that the water efficient irrigation technology and the water sensitive land management practices could significantly reduce the irrigation demand of public parks, verges, and other green infrastructures at the urban setting. The findings supports the empirical ideas on irrigation efficiency and urban water conservation put forth by Wong [19] , Water Corporation [15] , and Barron, Wendling, Tucker, Green, Devkota and Donn [18] .

Table 1. Water consumption (KL/household) in "The Green", Ridgewood, Butler, and Perth metropolitan region [32] [33] .

Areas\Years	2005	2006	2007	2008	2009	2010	2011	2012	2013	2014
Perth Metro*	375.0	387.5	362.5	362.5	352.5	347.5	344.3	331.5	336.6	347.5
Ridgewood	265.9	178.7	165.0	253.8	300.7	302.6	285.2	285.5	292.6	282.9
Butler (except "The Green")	258.7	282.6	371.1	391.0	364.2	346.2	333.6	290.9	269.5	278.3
The Green-Drinking Water				6.7	37.9	91.3	116.9	154.4	167.6	172.3
-Groundwater							54.1	188.8	167.5	115$^{\Psi}$
-Total				6.7	37.9	91.3	171.0	343.3	335.1	287.3$^{\Psi}$

*Household water consumption for Perth Metropolitan is calculated as a multiplication of average household size (2.6) with the average per-capita water consumption for each year. $^{\Psi}$The allocated groundwater for residential irrigation in "The Green" is approximately 115 KL/household for 2013/14.

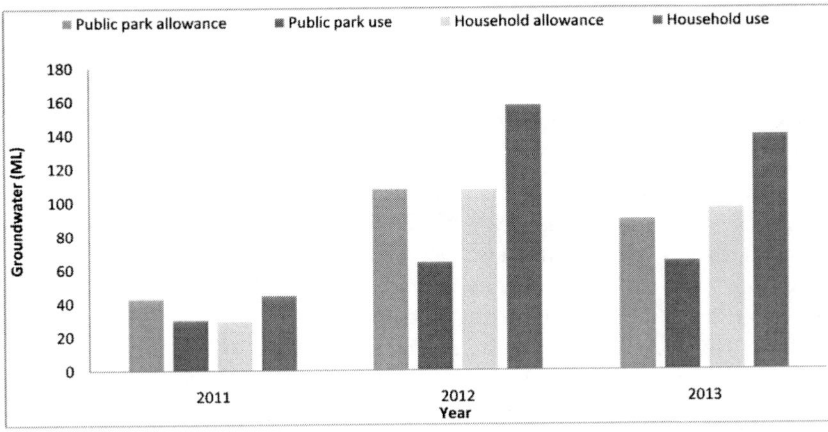

Figure 6. Total groundwater allocation and consumption in "The Green".

It is equally important to note that the irrigation in "The Green" is sourced from superficial aquifer, which is directly recharged by rainfall. The water sensitive land management practices implemented in "The Green" were reported to increase the aquifer recharge by 383 ML per year than that of pre-development conditions. It was also noticed that after abstracting 209 ML/year for groundwater irrigation network, still there would be an increase in net aquifer recharge by 174 ML/year [35] . Such an increase in aquifer recharge indicates that the groundwater would be a sustainable source for irrigating urban agriculture, vegetation, and recreational land-scaping as shown by previous studies of Western Australian Planning Commission [14] , and Nolasco [4] .

The irrigation trial has been operated for seven years but still approximately 20% of blocks are to be developed. Though new blocks are yet to be connected, the trend in water saving and irrigation efficiency is quite promising. Almost 50% less drinking water demand at residential settings and significant reduction in irrigation for agricultural activities in "The Green" than surrounding suburban areas and metropolitan average clearly indicate that the trialled system will achieve its target even at the full residential development of the suburb. However, an overconsumption of groundwater that was evident in "The Green" could be linked mainly to irrigation exemptions and other once-off usage. These transitional irrigation inconsistencies, as observed in Mawson Lakes Australia [17] , would gradually settle down after the suburban areas completely developed and irrigation system fully connected. Having said that, the complete analysis for the factors and parameters associated with the success of the irrigation trial requires completion of the residential development and longitudinal data on usage and behaviour towards water and irrigation systems. Joint effort of Urban Agriculturist, water providers and urban planners.

4.2. Inherent Socio-Economic Challenges

The community perceptions were found to be positive towards the efficient irrigation system and a majority of the consumers were reported to be satisfied with the system in "The Green". The higher level of satisfaction was mainly associated with the automatic irrigation operation, water savings, environmental benefits, and the trust on irrigation service providers. However, there were some social and economic challenges, mainly lack of clarity in institutional roles, and management of consumer expectations that need to be addressed for better management of potential alternative irrigation systems. This study shows that the fit-for-purpose irrigation with integrated land and water management brings the community members, government authorities, and service providers together to plan, implement and develop sustainable urban agricultural practices, and share mutual benefits.

4.3. Challenges in System Replication

The qualitative data analysis clearly shows that the majority of residents participated positively towards the efficient irrigation system for food production and agricultural activities at their home and community. Further, it shows an increasing awareness towards the efficient gardens and irrigation system in "The green" community. This is an important finding for the future development of urban agricultural activities with reliable efficient irrigation systems in similar location and climate across Australia.

Based on the findings of this study, alternative irrigation system sourced from rainwater has been successfully implemented in a neighbouring suburban area (Evermore Height, Alkimos). Successive new alternative irrigation projects in the Perth metropolitan and regional areas have also being formulated and developed taking Butler trial as a successful urban agriculture irrigation model [30] . These findings and successful replications clearly demonstrate that the fit-for-purpose irrigation system is instrumental in sustainable development of urban agriculture across Australia.

5. CONCLUSION

An alternative fit-for-purpose irrigation system was trialled in "The Green" community at Butler, Perth to understand sustainable water management for urban agriculture. The quantitative and qualitative data were collected and analyzed. The study suggested that a fit-for-purpose irrigation was a sustainable alternative for urban agriculture. Additionally, it offers significant irrigation efficiency in urban agriculture, and consequent reduction in water demand at urban residential settings. The water quality issues associated with fit-for-purpose irrigation appear to be nominal and do not affect the community acceptance. This indicates that the fit-for-purpose irrigation system with water efficient land management practices is highly encouraging for the sustainable development of urban agriculture and recreational landscaping across spatial levels.

ACKNOWLEDGEMENTS

Authors greatly acknowledge the key contribution of WA Water Corporation who provided relevant data and the fund to undertake this study. We thank Satterley Property Group for their financial contribution; Centre for Planning at Edith Cowan University for the operational management of the study. We are also grateful for Mr. Tim Perkins (Centre for Planning, ECU) and Dr. Bishnu Devkota (Infra Tech Pacific/WorleyParson, past-WA Water Corporation) for their generous comments on this paper.

CITE THIS PAPER

Raju Sharma Dhakal, Geoff Syme, Edward Andre, Charles Sabato (2015) Sustainable Water Management for Urban Agriculture, Gardens and Public Open Space Irrigation: A Case Study in Perth. *Agricultural Sciences*,**06**,676-685. doi: 10.4236/as.2015.67065

REFERENCES

1. Mougeot, L.J.A. (2010) Agropolis: The Social, Political and Environmental Dimensions of Urban Agriculture. IDRC, London.

2. Zezza, A. and Tasciotti, L. (2010) Urban Agriculture, Poverty, and Food Security: Empirical Evidence from a Sample of Developing Countries. Food Policy, 35, 265-273.
3. Urban Agriculture Australia (2013) What Is Urban Agriculture and Why It Is Important? Urban Agriculture Australia, Canberra.
4. Nolasco, J. (2013) Sustainable Water Management for Urban Agriculture: Planting Justice. Pacific Institute, Oakland.
5. Lin, B.B., Philpott, S.M. and Jha, S. (2015) The Future of Urban Agriculture and Biodiversity-Ecosystem Services: Challenges and Next Steps. Basic and Applied Ecology, 16, 189-201.
6. Lin, B.B. and Fuller, R.A. (2013) FORUM: Sharing or Sparing? How Should We Grow the World's Cities? Journal of Applied Ecology, 50, 1161-1168.
7. Lwasa, S., Mugagga, F., Wahab, B., Simon, D., Connors, J.P. and Griffith, C. (2015) A Meta-Analysis of Urban and Peri-Urban Agriculture and Forestry in Mediating Climate Change. Current Opinion in Environmental Sustainability, 13, 68-73.
8. Lovell, S.T. (2010) Multifunctional Urban Agriculture for Sustainable Land Use Planning in the United States. Sustainability, 2, 2499-2522.
9. Guendel, S. and Richards, W. (2002) Peri-Urban and Urban Livestock Keeping in East Africa—A Coping Strategy for the Poor, Scoping Study Commissioned by DFID. Natural Resources Institute, Aylesford.
10. Nugent, R. (2000) The Impact of Urban Agriculture on the Household and Local Economies. In: Bakker, N., Dubbeling, M., Gündel, S., Sabel-Koshella, U. and de Zeeuw, H., Eds., Growing Cities, Growing Food. Urban Agriculture on the Policy Agenda, Zentralstelle für Ern&aumlhrung und Landwirtschaft (ZEL), Feldafing, 67-95.
11. Mok, H.-F., Williamson, V., Grove, J., Burry, K., Barker, S.F. and Hamilton, A.J. (2014) Strawberry Fields Forever? Urban Agriculture in Developed Countries: A Review. Agronomy for Sustainable Development, 34, 21-43.
12. Cohen, N. and Reynolds, K. (2015) Resource Needs for a Socially Just and Sustainable Urban Agriculture System: Lessons from New York City. Renewable Agriculture and Food Systems, 30, 103-114.
13. Water Corporation (2010) Perth Residential Water Use Study 2008/2009. Water Corporation, Perth.
14. Western Australian Planning Commission (2008) Better Urban Water Management. Western Australian Planning Commission, Perth.
15. Water Corporation (2009) Water Forever: Towards Climate Resilience. Water Corporation, Perth.
16. Smith, A., Pollock, D. and McFarlane, D. (2005) Opportunity for Additional Self Supply of Groundwater from the Superficial Aquifer beneath Metropolitan Perth. CSIRO: Water for a Healthy Country National Research Flagship, Canberra.
17. Hurlimann, A. (2008) Community Attitudes to Recycled Water Use: An Urban Australian Case Study Part 2. CRC for Water Quality and Treatment, Salisbury.
18. Barron, O., Wendling, L., Tucker, D., Green, M., Devkota, B. and Donn, M. (2010) Investigation of Techniques to Better Manage Western Australia's Non-Potable Water Resources. CSIRO: Water for a Healthy Country National Research Flagship, Canberra.
19. Wong, T.H. (2007) Water Sensitive Urban Design—The Journey Thus Far. Australian Journal of Water Resources, 10, 213-222.
20. Po, M., Nancarrow, B.E., Leviston, Z., Porter, N.B., Syme, G.J. and Kaercher, J.D. (2005) Predicting Community Behaviour in Relation to Wastewater Reuse: What Drives Decisions to Accept or Reject? CSIRO: Water for a Healthy Country National Research Flagship, Perth.
21. Nancarrow, B.E., Leviston, Z., Po, M., Porter, N.B. and Tucker, D.I. (2008) What Drives Communities' Decisions and Behaviours in the Reuse of Wastewater. Water Science and Technology, 57, 485-491.

22. Davis, C. and Farrelly, M. (2009) Demonstration Projects: Case Studies form Perth, Australia. National Urban Water Governance Program, Monash University, Melbourne.
23. Davis, C. and Farrelly, M. (2009) Demonstration Projects: Case Studies form Melbourne, Australia. National Urban Water Governance Program, Monash University, Melbourne.
24. Davis, C. and Farrelly, M. (2009) Demonstration Projects: Case Studies form South East Queensland, Australia. National Urban Water Governance Program, Monash University, Melbourne.
25. Department of Water (2010) Predicting the Future Demand for Water Resources in Western Australia. Department of Water, Perth.
26. Dzidic, P. and Green, M. (2012) Outdoing the Joneses: Understanding Community Acceptance of an Alternative Water Supply Scheme and Sustainable Urban Design. Landscape and Urban Planning, 105, 266-273.
27. Water Corporation (2005) Integrated Water Supply Scheme Source Development Plan 2005: Planning Horizon 2005-2050. Water Corporation, Perth.
28. Water Corporation (2007) Memorandum of Understanding: "The Green" at Brighton (Brighton Village 4) Superficial Groundwater Demonstration Project. Water Corporation, Perth.
29. Water Corporation (2007) Brighton Superficial Groundwater Supply Demonstration Project: Health Risk Management Plan. Water Corporation, Perth.
30. Satterley Property Group (2010) The End to the Waste of Precious Water Starts Here: Smartwater Initiative, The Green. Satterley Property Group, Perth.
31. Western Australian Planning Commission (2007) Liveable Neighbourhood: A Western Australian Government Sustainable Cities Initiatives. Western Australian Planning Commission, Perth.
32. Australian Bureau of Statistics (2013) 2011 Census Population and Housing: Greater Perth—Basic Community Profile (Catalogue Number 2001.0). Australian Bureau of Statistics, Canberra.
33. Water Corporation (2015) Reducing Water Use. Water Corporation, Perth.
34. Water Corporation (2015) Exemptions. Water Corporation, Perth.
35. GHD (2006) Satterley Property Group: Report on Brighton Stage 4. Hydrological Assessment. GHD, Perth.

Index